Illustrated Plants of Florida and the Coastal Plain

Maupin House Publishing
Gainesville

Illustrated Plants
of Florida and
the Coastal Plain

by Dr. David W. Hall

with illustrations by
Edward H. Stehman

based on the collections of
Leland and Lucy Baltzell

ILLUSTRATED PLANTS OF FLORIDA AND THE COASTAL PLAIN
A Maupin House Book/July 1993

Composition by Mac's Output, Inc., Gainesville, under the
supervision of Kenneth R. Skillman.
Special thanks to Kellie McMaster and Mike Podolsky for
all of their hard work and dedication.

Library of Congress Cataloging-in-Publication Data

Hall, David W. (David Walter), 1940–
    Illustrated plants of Florida and the coastal plain / by David W.
Hall ; with illustrations by Edward H. Stehman.
       p.   cm.
    "Based on the collections of Leland and Lucy Baltzell."
    Includes bibliographical references (p.   ) and indexes.
    ISBN 0-929895-40-1 ; $19.95
    1. Botany--Florida--Pictorial works. 2. Coastal plants--South
Atlantic States--Pictorial works. 3. Coastal plants--South Atlantic
States--Identification. I. Stehman, Edward H. II. Title.
QK154.H26 1993
581.9759--dc20   93-11817
                                             CIP

ISBN 0-929895-40-1

Maupin House
P.O. Box 90148
Gainesville, Florida 32607-0148

PRINTED IN THE UNITED STATES OF AMERICA

This book is lovingly dedicated to my mother and father,
Lucy Frances Baltzell and Leland McMaster Baltzell.

— Martha Baltzell Stone

This book describes and illustrates the wildflowers of Florida and the Southeastern Coastal Plain from Lake Okeechobee in peninsula Florida northward. Most common shrubs and herbaceous plants are included but grasses, rushes and trees are excluded.

Plants are arranged alphabetically by family. The intent is for users to simply look through the illustrations for a match. The description of the plant can then be consulted for confirmation. As users become more familiar with plant families, identification will become faster. Two indexes are included for reference.

Leland and Lucy Baltzell collected the specimens on which this book is based. A map of each county was used to locate collection sites. Care was taken not to collect rare and endangered species. Instead these protected plants would be sketched in the field and their locations carefully noted.

Edward H. Stehman, an engineer who painted oils and sketched, created the illustrations of the specimens which were collected from 1966 through 1989. Since photographs frequently do not show accurate details, the collectors chose to have the specimens illustrated.

Dr. Daniel B. Ward, then director of the Herbarium of the University of Florida, was enthusiastic and encouraging about the work, and Dr. David W. Hall subsequently supervised the collections which totaled over 11,800 specimens during 23 years. Dr. Hall supervised Maryetta Cook, Brenda Herring, Trudy Lindler, and Lisa Modola, who, along with him, spent hundreds of hours working on this project. Their professional skills and guidance were invaluable.

Others who contributed to the project were Kent D. Perkins, Walter S. Judd, J. Dan Skean, and Dr. Winston C. Baltzell.

David W. Hall, Ph.D., was born in New Orleans, Louisiana, and was raised in Augusta, Georgia. He is currently a senior scientist with KBN Engineering and Applied Sciences in Gainesville. He joined KBN in January 1991 from the University of Florida where he was Director of the Plant Identification and Information Services.

He holds bachelor's and master's degrees from Georgia Southern University and the Doctor of Philosophy degree from the University of Florida. He is a recognized expert in the field of plant identification and has published 6 books and over 100 articles.

In his 19 years with the University he specialized in the identification and biology of weeds and grasses, ornamental uses of native plants, and forensic botany. At KBN he is responsible for plant identification, endangered species, wetland jurisdictions, general ecology, and forensic botany. He has given talks and workshops throughout Florida and the United States.

Leland McMaster Baltzell was born on February 13, 1901, in LaBelle, Missouri. He lived and worked for 38 years in Reno, Pennsylvania, where he was employed as a chemical engineer by Penzoil. He was educated at Westminster College at Fulton, Missouri, with additional work at Missouri University and Texas A & M University.

Lucy Frances Baltzell was born on December 13, 1907, in Henderson, Kentucky. During their years in Reno, Lucy worked part-time for Dr. Herbert Wahl, a botanist, at Pennsylvania State University, although she considered herself primarily a house-wife.

The Baltzells lived in Florida seasonally from 1963 through 1965. Upon Mr. Baltzell's retirement from Penzoil, they moved to Leesburg, Florida where they still live. Having done some col-lecting of wildflowers for Dr. Wahl in Pennsylvania, they soon became interested in the wildflowers of Florida. They spent many satisfying years collecting Florida plants.

While collecting specimens, they wore safari hats, snake proof boots, and leather aprons which held magnifying glasses, knife, scissors, and a pistol. (Lucy was a crack shot but they never had to use the pistol.) Because they often found themselves off the beaten track, they had interesting experiences. Occasionally they encountered suspicious groups of men deep in the woods — some perhaps paramilitary and others perhaps engaged in legally questionable activities. For their part, they were frequently ques-tioned by bemused law enforcement officers.

The Baltzell's abiding interest in plants has led to this book. This intense pursuit provided exciting plant discoveries in secluded woods and along open roadsides, filling their retirement years with joy and adventure.

WEST

NORTH

CENTRAL

SOUTH

Contents

NOTE: Asterisks (*) in plant descriptions refer to arrows in illustrations.

> **ACANTHUS FAMILY**   *ACANTHACEAE*
>
> Trees, shrubs or herbs; leaves opposite; flowers bisexual;
> sepals 4 or 5; petals 4 or 5; ovary superior; fruit a capsule.

PHILIPPINE VIOLET
Barleria cristata L.

Perennial, to 1 m tall; stems hairy; leaves 5-10 cm
long; petals to 5 cm long, blue to white. Native
to India. Escapes from cultivation. Disturbed
sites. Rare SF, CF, NF. Spr-sum.

CRIMSON DICLIPTERA
Dicliptera assurgens (L.) Juss.

Perennial, 0.3-1 m tall; leaf blades 2-10 cm long;   (not shown)
petals 2-2.5 cm long, crimson, red, curving; tube
longer than lips; capsules 7-8 mm long.
Hammocks. Infreq. SF, CF. All yr.

LAVENDER DICLIPTERA
Dicliptera brachiata (Pursh) Spreng.

Perennial, 0.3-1 m tall; stems smooth to hairy;
leaf blades 2-14 cm long; petals 15-20 mm long,
pink, lavender or purple; tube and lips equal in
length; capsules to 5 mm long. Low, swampy
sites. Infreq. CF, NF, WF. W to Tex., N to Va.
Spr-fall.

SWAMP TWIN FLOWER
Dyschoriste humistrata (Michx.) Kuntze

Perennial, 10-30 cm tall; stems erect to ascending; leaves to 5 cm long; flowers mostly sessile;
petals 1-1.5 cm long, violet or white; capsules to
1 cm long. Moist to wet sites. Freq. CF, NF, WF.
N to S.C. Spr-sum.

(not shown)

TWIN FLOWER
Dyschoriste oblongifolia (Michx.) Kuntze

Perennial, 10-35 cm tall; leaves 2.5-4.5 cm long;
flowers mostly sessile; petals 2-3 cm long, blue or
purple; capsules to 1.5 cm long. Dry to moist
sites. Common all Fla. N to S.C. Spr-fall.

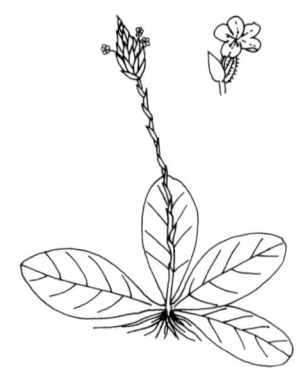

SCALE-STEM
Elytraria caroliniensis (J. F. Gmel.) Pers.

Perennial; basal leaves 5-20 cm long; upper leaves
greatly reduced, almost scale-like; flower stems 10-
60 cm tall; spikes 2-6 cm long, with white to blue
flowers; stamens 2; capsules 4-5 mm long. Low,
woody sites. Infreq. all Fla. N to S.C. Spr-fall.

WATER-WILLOW
Justicia americana (L.) Vahl

Perennial, 0.3-1 m long or tall; leaves 5-15 cm
long; *petals to 1.5 cm long, purple to white;
*capsules to 2 cm long. Wet mucky sites. Infreq.
WF. W to Tex., N to Canada. Spr-sum.

NARROWLEAF WATER-WILLOW
Justicia angusta (Chapm.) Small

Perennial, 20-50 cm tall; leaves 2-7 cm long, to
0.5 cm wide; flowers in spikes to 20 cm long;
petals to 3 cm long, purple and white; capsules to
1.2 cm long. Low wet sites. Common all Fla. Spr.

COOLEY'S JUSTICIA
Justicia cooleyi Monachino & Leonard

(not shown)

Perennial, to 50 cm tall; stems covered with dense
white (partly glandular) hairs; leaves to 7 cm long,
to 3.5 cm wide; flowers in panicles to 7 cm long;
petals to 1.2 cm long, red-purple; capsules 5-10
mm long. Hammocks. Rare CF. Sum-fall.

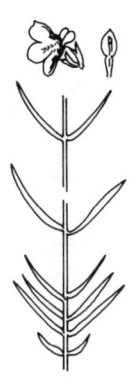

LARGE-FLOWERED WATER-WILLOW
Justicia crassifolia (Chapm.) Small

Perennial, 20-40 cm tall; leaves 3-15 cm long;
petals 2-3 cm long, purple; capsules to 2.5 cm
long. Wet sites. Infreq. WF. Spr.

BOTTOMLAND WATER-WILLOW
Justicia ovata (Walt.) Lindau

Perennial, 10-30 cm tall; leaves 3-10 cm long, to
3 cm wide, elliptic to ovate; flowers in pairs, in
spikes to 10 cm long; petals 1-2 cm long, white
to pale purple; capsules to 1.5 cm long. Low wet
sites. Infreq. CF, NF, WF. W to Tex., N to Va.
Spr-fall.

ONE-FLOWER WATER-WILLOW
Justicia ovata (Walt.) Lindau
var. *lanceolata* (Chapm.) R. W. Long

Similar to Justicia ovata except: leaves to 2 cm
wide, linear or lanceolate; flowers solitary in
spikes. Low wet sites. Infreq. WF. W to Tex.,
E to Ga. Spr-sum.

MEXICAN RUELLIA
Ruellia brittoniana Leonard ex Fern.

Perennial, to 1 m tall; stems erect; leaves to 30 cm
long, less than 2 cm wide, in long-stalked cymes;
petals 3-3.5 cm long, violet to purple; capsules to
2.5 cm long. Native to Mexico. Moist to wet
sites. Infreq. all Fla. W to Tex., N to S.C. Spr-fall.

WILD PETUNIA
Ruellia caroliniensis (J. F. Gmel.) Steud.

Perennial; 20-90 cm tall; stems lying flat or erect,
hairy; leaves 2-10 cm long, hairy; flowers sessile;
petals 4-7 cm long, blue, lavender or white; capsules 1-1.8 cm long. Dry sites. Common all Fla.
W to Tex., N to Canada. Spr-fall.

(not shown)

Ruellia malacosperma Greenm.

Perennial, to 1 m tall; leaves to 12 cm long, over
2 cm wide; flowers in long-stalked cymes; petals
4-5 cm long, purple or blue; capsules to 2-2.5 cm
long. Escapes from cultivation. Disturbed sites
Rare SF. Spr-fall.

BLACK-EYED CLOCKVINE
Thunbergia alata Bojer

Perennial, climbing to 1 m long; leaves 3-11 cm long; petals to 2 cm long, to 3.5 cm wide, yellow, orange or cream; *capsules to 3 cm long. Native to tropical Africa. Disturbed sites. Infreq. SF, CF, NF. Sum, all yr S.

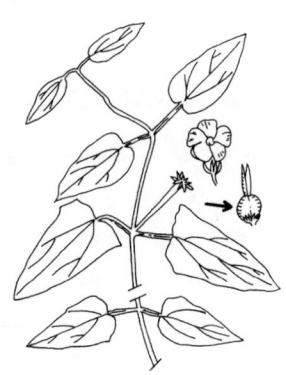

WHITE THUNBERGIA
Thunbergia fragrans Roxb.

Perennial climbing vine, to 2 m long; leaves 4-12 cm long; petioles not winged; flowers 3-5 cm wide, white; capsules to 2 cm long. Native to India. Cultivated and escapes. Woody sites. Infreq. SF, CF. All yr.

SKY-VINE
Thunbergia grandiflora Roxb.

Perennial climbing vine, to over 9 m long; leaves to 20 cm long; flowers 7-9 cm long, 6-8 cm wide, blue or white; capsules to 2 cm long. Native to India. Cultivated and escapes. Disturbed areas. Rare SF, CF. All yr.

AGAVE FAMILY *AGAVACEAE*

Shrubs, trees or herbs, with underground runners; leaves mostly basal; flowers bisexual or unisexual (dioecious or monoecious); floral parts (tepals) in 2 whorls of 3; ovary superior or inferior; fruit a capsule or berry.

RATTLESNAKE MASTER
Polianthes virginica (L.) Shinners
[*Manfreda virginica* (L.) Salisb.]

Perennial, 0.8-1.8 m tall; basal leaves 10-30 cm long, fleshy; *petals to 2.5 cm long, green, very fragrant at night; *capsules to 2 cm in diam. Dry, rocky woods. Infreq. NF, WF. W to Tex., N to N.C. Sum.

MOTHER-IN-LAW'S TONGUE
Sansevieria hyacinthoides (L.) Druce
[*S. thyrsiflora* Thunb.]

Perennial; *leaves 0.4-1 m tall, tufted on rhizome; *flowers to 3.5 cm long, white or green; *berries to 8 mm in diam., red. Native to South Africa. Remaining from cultivation and escaping into disturbed areas. Infreq. SF, CF, NF. Fall.

SPANISH BAYONET
Yucca aloifolia L.

Perennial, 1-2.5 m tall; stems branching; leaves 40-60 cm long, sharp-pointed; *petals to 6 cm long, white; *berries to 9 cm long. Native along Fla. coast. Cultivated and escapes from cultivation. Dunes, sandhills. Infreq. all Fla. W to La., N to N.C. Spr-sum.

BEAR-GRASS
Yucca filamentosa L.

Perennial; leaves 30-60 cm long; leaf margins
with fine long threads; flower stalks 1.5-4 m tall;
petals to 4.5 cm long, white; capsules to 6 cm
long. Dry sites. Freq. CF, NF, WF. W to Miss.,
N to N.C. Spr-sum.

BEAR-GRASS
Yucca flaccida Haw.
[*Y. filamentosa* L.]

Perennial, 1.5-4 m tall; leaves 30-60 cm long;
leaf margins often with long threads; tepals 2-6
cm long, white or greenish; *flowers in terminal
panicle; capsules 5-6 cm long. Dry sites. Freq.
all Fla. N to N.C. and N.J., W to Miss. Spr-sum.

CARPET WEED FAMILY   *AIZOACEAE*

Succulent herbs or shrubs; leaves opposite, alternate or
whorled; flowers bisexual; sepals 4-8; petals mostly 0; ovary
superior or inferior; fruit a capsule, nut or berry.

CARPET WEED
Mollugo verticillata L.

Annual, 4-30 cm long; stems prostrate, branching; leaves 1-3 cm long, whorled; sepals green
with white margins; *capsules to 3 mm long;
seeds many. Disturbed sites. Freq. all Fla. W to
Tex., N to Canada. Spr-fall.

SEA PURSLANE
Sesuvium portulacastrum L.

Perennial, to 2 m tall, succulent to partially
woody; stems prostrate to ascending; leaves 1.5-4
cm long, opposite; flowers stalked; sepals 5,
petal-like, green on outside, pink on inside; capsules 8-10 mm long, opening circularly (circumscissile), stalked. Seashores. Freq. all Fla. N to
N.C. Spr-fall, all yr S.

WATER-PLANTAIN FAMILY   *ALISMATACEAE*

Aquatic herbs; leaves strap-shaped (phyllodial) or with stalks
and blades; flowers bisexual or unisexual (dioecious); sepals
3; petals 3; ovary superior; fruit a head of achenes.

ENGELMANN'S ARROWHEAD
Sagittaria engelmanniana J. G. Smith

Perennial; young leaves phyllodial; intermediate
leaves long-petioled and linear; later leaves sagittate,
to 25 cm long; flower stems to 80 cm tall; flowers in
3-12 whorls; petals to 1.5 cm long; fruiting heads to
1.5 cm wide; bracts of pedicels acute, acuminate,
nearly as long as pedicels; achenes to 4.5 mm long.
Marshes, swamps, ditches. Very infreq. WF. W to
Miss., N to Ind., E to Mass. Sum-fall.

(not shown)

SLENDER ARROWHEAD
Sagittaria graminea Michx.

Perennial; young leaves phyllodial; older leaves
with petioles and blades to 50 cm long; flower
stems to 60 cm tall; flowers in 2-12 whorls; petals
0.6-2 cm long; *fruiting heads to 1 cm wide;
*achenes to 0.4 mm long. Watery sites. Common
all Fla. W to Tex., N to Canada. Spr-fall.

NARROW LEAVED ARROWHEAD
Sagittaria isoetiformis J. G. Smith

Perennial; leaves 5-50 cm long, phyllodial;
flower stems to 80 cm tall; flowers in 1-4 whorls;
petals to 5 mm long; fruiting heads to 1 cm wide;
achenes to 1.5 mm long. Low, wet sites.
Common all Fla. N to Ala., E to N.C. Spr-fall.

SPRING-TAPE
Sagittaria kurziana Glueck

(not shown)

Perennial; leaves to 7 m long, strap-shaped, 3- to
5-ribbed; flower stalks as long as leaves; petals 9-
14 mm long, white; achenes to 2 mm long.
Springs, spring runs. Infreq. CF, NF. Spr-sum.

LANCE LEAF ARROWHEAD
Sagittaria lancifolia L.

Perennial; young leaves phyllodial; older leaves
with petioles and blades to 80 cm long; flower
stems to 1.5 m tall; flowers in 5-12 whorls; petals
1-2 cm long; *fruiting heads to 1.5 cm wide; achenes to 2.2 mm long. Low, wet sites. Common
all Fla. W to Tex., N to Del. Spr-fall.

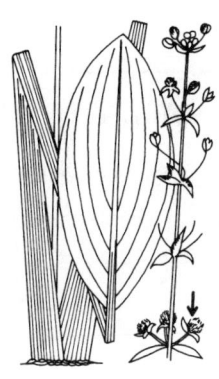

COMMON ARROWHEAD
Sagittaria latifolia Willd.

Perennial; young leaves phyllodial; older leaves
to 1 m long, hastate or sagittate; flower stalks to
80 cm tall; flower whorls 2-10; petals to 2.5 cm
long; fruiting heads 1-3 cm in diam.; achenes 2-4
mm long. Wet sites. Freq. all Fla. W to Calif., N
to Maine. Sum-fall.

FLOATING LEAF ARROWHEAD
Sagittaria stagnorum Small

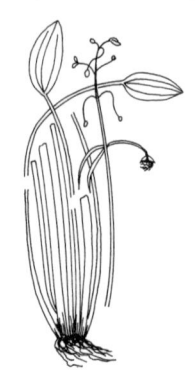

Perennial; leaves phyllodial in deep water; leaves
with petioles and blades to 95 cm long, floating
in shallow water; flower stems to 30 cm tall;
flowers in 2-6 whorls; petals to 6 mm long; fruiting heads to 8 mm in diam.; achenes to 1.5 mm
long. Fresh water sites. Freq. all Fla. W to Ala.,
N to Mass. Spr-fall.

WATER ARROWHEAD
Sagittaria subulata (L.) Buch.
[*S. lorata* (Chapm.) Small]

Perennial; leaves 2-30 cm long, strap-like; flower
stalks 5-40 cm tall; petals 4-8 mm long, white;
achenes to 2 mm long. Ponds, spring runs, tidal
waters. Freq. all Fla. W to Ala., E to Mass. Sum-
fall, all yr S.

AMARANTHUS FAMILY   *AMARANTHACEAE*

Herbs or shrubs; leaves alternate or opposite; flowers bisexual
or unisexual (monoecious or dioecious); sepals 2-5; petals 0;
ovary superior; fruit a capsule, utricle or berry.

ALLIGATOR-WEED
Alternanthera philoxeroides (Mart.) Griseb.

Perennial, 0.3-1.5 m long; stems lying flat, creeping or trailing, smooth; leaves 5-12 cm long,
slightly fleshy, rounded or obtuse; globose spikes
axillary or terminal; stamens 5; sepals 5, to 6 mm
long, white or green. Native to South America.
Low, wet sites; also weedy in some crops and
gardens. Freq. all Fla. W to Tex., E to Va. All yr.

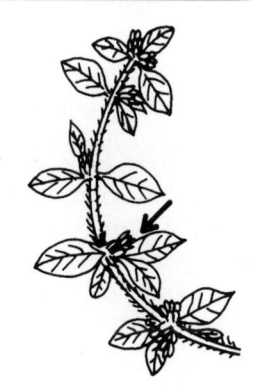

SMOOTH CHAFF-FLOWER
Alteranthera polygonoides (L.) R. Brown

Perennial, 10-50 cm long; stems prostrate; leaves
5-20 mm long; *spikes sessile, axillary; sepals 5,
to 4 mm long, white; utricles to 2 mm long, glob-
ular-obovoid. Disturbed areas. Infreq. all Fla.
W to La., E to N.C. Spr-fall, all yr S.

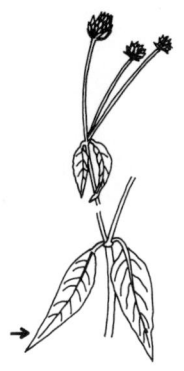

POINTED LEAF CHAFF-FLOWER
Alternanthera ramosissima (Mart.) Chod.

Perennial, to 4 m long; stems diffusely branched;
leaves 2-9 cm long, papery, *acuminate; sepals
4-5 mm long, white or green. Coastal hammocks. Infreq. SF, CF. All yr.

STALKLESS CHAFF-FLOWER
Alternanthera sessilis (L.) R. Br.

Perennial, 30-50 cm tall; stems creeping to erect;
leaves to 5 cm long, linear-lanceolate, rounded
or acute; spikes sessile; sepals 4-5 mm long,
white or green. Wet sites. Rare CF, WF. Sum-
fall, all yr S.

(not shown)

GIANT AMARANTH
Amaranthus australis (Gray) Sauer

Annual, 1.5-9 m tall; stems smooth; leaves to 30
cm long; male and female flowers on separate
plants (dioecious), each to 3 mm long, yellow to
green; utricles to 2 mm long. Near wet sites.
Freq. all Fla. W to Tex. Sum-fall, all yr S.

GULF COAST AMARANTH
Amaranthus crassipes Schlecht.

Annual, 10-60 cm long; stems smooth; leaves 1-
10 cm long; flowers monoecious, to 2 mm long, green to white; utricles to 2 mm long. Native to tropical America. Disturbed sites. Rare SF (Key West). S.C. and Gulf Coast Ala. to Tex. All yr. Possibly extirpated, not collected in recent yrs.

(not shown)

SMOOTH PIGWEED
Amaranthus hybridus L.

Annual, 0.6-3 m tall; stems smooth; leaves 3-15 cm long; male and female flowers on same plant (monoecious), each to 2 mm long, red or green; utricles to 2.5 mm long. Disturbed and cultivated sites. Freq. all Fla. All eastern U.S. Sum-fall.

SPINY AMARANTH
Amaranthus spinosus L.

Annual, 0.2-1.2 m tall; *stems spiny; leaves 1.5-
10 cm long; flowers monoecious, to 3 mm long, green to white; utricles to 2 mm long. Disturbed sites. Freq. all Fla. W to Tex., N to Canada. Sum-fall.

SAMPHIRE
Blutaparon vermiculare (L.) Mears
[*Philoxerus vermicularis* (L.) R. Br.]

Perennial, to 2 m long; stems succulent, lying flat to ascending; leaves 1-6 cm long, opposite, sessile, semi-terete; spikes 1-3 cm long; sepals 5, 3-5 mm long, silvery white; stamens 5; utricles to 1 mm long. Coastal sites. Freq. SF, CF, NF. W to Tex. All yr.

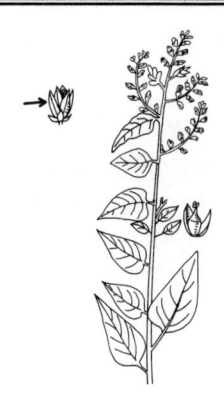

SLENDER COCK'S-COMB
Celosia nitida Vahl

Perennial, 0.3-1.1 m long, straggling; leaves 1.5-
7 cm long; sepals 4-5 mm long, yellow to white;
*utricles enclosed by sepals. Coastal hammocks.
Rare SF, CF. W to Tex. Sum, all yr S.

AFRICAN COCK'S-COMB
Celosia trigyna L.

Annual, 0.3-1.2 m tall, spreading, vine-like;
leaves to 8 cm long; sepals 5, to 3 mm long, 1-
nerved, greenish white; stamens 5; style shorter
than stigmas; utricles circumscissile. Native to
Africa. Disturbed areas. Infreq. CF, NF. Sum-
fall.

COTTON WEED
Froelichia floridana (Nutt.) Moq.

Annual, 0.4-1.8 m tall; stems hairy; leaves 3-12
cm long, with soft hairs; flowers in spikes to 6 cm
long; sepals green to white; utricles to 5 mm
long. Dry sites. Common all Fla. W to Tex., N
to Del. Sum-fall.

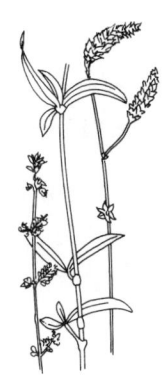

GLOBE AMARANTH
Gomphrena serrata L.
[*G. dispersa* Standl.]

Perennial or annual; stems diffuse, lying flat,
hairy; leaves 2-5 cm long; flowers perfect, white
or purple-red; sepals 5, hairy; bractlets 5-6 mm
long, white or purplish red; utricles 1-2 mm long.
Disturbed sites. Common all Fla. W to Tex.
Sum-fall, all yr S.

BLOODLEAF
Iresine diffusa Humb. & Bonpl. ex Willd.
[*I. celosia* L.]

Annual, 1-3 m tall; stems clambering to erect;
leaves 3-14 cm long; flowers unisexual; spikes
paniculate, white, pink or green; sepals 5, 1-1.5
mm long, 3-nerved; female flowers woolly at
base; utricles shorter than sepals. Disturbed sites.
Freq. SF, CF, WF. W to La., E to N.C. Sum-fall,
all yr S.

PERENNIAL BLOODLEAF
Iresine rhizomatosa Standl.

(not shown)

Perennial, 0.5-1.5 m tall; stems erect or reclining;
leaves 6-15 cm long; flowers unisexual; spikes
paniculate, white, pink or green; sepals 5, 1-1.3
mm long, 3-nerved; utricles as long as sepals.
Low, wet sites. Infreq. WF. W to Tex., N to
Kans., E to Md. Sum-fall.

AMARYLLIS FAMILY   *AMARYLLIDACEAE*

Herbs arising from swollen underground stems, bulbs or
rhizomes; leaves linear; flowers bisexual; floral parts in
whorls of 3; ovary inferior; fruit a capsule or berry.

WILD ONION
Allium canadense L.

Perennial, 20-60 cm tall; leaves 10-30 cm long;
flower clusters with flowers which form small
*bulbs; petals to 7 mm long, pink or white; cap-
sules to 3 mm long. Moist thickets, fields. Infreq.
CF, NF, WF. W to Tex., N to Canada. Spr-fall.

STRING-LILY
Crinum americanum L.

Perennial; basal leaves 0.6-1.5 m long; flower
stem leafless, 30-80 cm tall; flowers 2-6 in clusters, fragrant; petals 5-14 cm long, white; capsules 4-6 cm thick, 3-lobed. Low, wet sites.
Freq. all Fla. N to Ga., W to Tex. Spr-fall.

SPIDER-LILY
Hymenocallis crassifolia Herb.

Perennial; leaves to 50 cm long, to 3 cm wide;
flower stems 40-60 cm tall; flowers 2 or 3; crown
to 4.5 cm in diam., white. Lake and stream sides.
Infreq. CF, NF. N to N.C. Spr.

MANY-FLOWERED SPIDER-LILY
Hymenocallis latifolia (Mill.) Roem.

Perennial; leaves to 80 cm long, to 7 cm wide;
flower stems to 80 cm tall; flowers 2-16; crown
to 3.5 cm in diam., white. Wet sites. Infreq. SF,
CF, NF. Spr-sum.

(not shown)

ALLIGATOR-LILY
Hymenocallis palmeri S. Wats.

Perennial; leaves to 40 cm long, to 1 cm wide;
flower stems to 40 cm tall; flowers solitary;
crown to 5 cm wide, white. Wet sites. Infreq.
SF, CF, NF. Spr-sum.

TWO-FLOWERED SPIDER-LILY
Hymenocallis rotatum Le Conte

Perennial; leaves to 50 cm long; flower stems to
31 cm tall; flowers 2, terminal; crown to 6.5 cm
in diam., white. Flatwoods. NF. Spr. Not
reported recently.

COMMON STAR-GRASS
Hypoxis juncea J. E. Smith

Perennial; leaves to 30 cm long, to 0.8 mm wide,
thread-like, with margins rolled under; flower
stems to 21 cm tall, hairy; petals to 1.2 cm long,
yellow; capsules to 8 mm long; seeds black, covered with round projections. Moist to wet sites.
Common all Fla. N to N.C. Spr-sum.

SWAMP STAR-GRASS
Hypoxis leptocarpa (Engelm. & Gray) Small

Perennial; leaves 10-55 cm long, to 12 mm wide;
petals 6.5-8 mm long, yellow; capsules 6-11 mm
long; seeds with dull projections. Damp to wet
sites. Common all Fla. W to Tex., N to N.C.
Spr-sum.

SMALL-FLOWERED STAR-GRASS
Hypoxis micrantha Pollard

Perennial; leaves linear, 8-20 cm long, to 2.5 mm
wide; flower stems 5-10 cm tall, hairy; petals 0.4-
1 cm long, yellow; capsules 4-8 mm long; seeds
dark brown, covered with sharp projections.
Moist sites. Common all Fla. W to Tex., N to
Va. Spr.

ATAMASCO-LILY
Zephyranthes atamasco (L.) Herb.

Perennial; leaves 20-40 cm long, to 4 mm wide,
basal, concave, with sharp margins; flower stems
to 35 cm tall; petals to 8 cm long, white or pinkish within; *styles longer than stamens. Low
sites. Freq. CF. W to Miss., E to Va. Spr.

RAIN-LILY
Zephyranthes simpsonii Chapm.

(not shown)

Perennial; leaves to 30 cm long, to 2 mm wide,
basal, with smooth margins; flower stems to 25
cm tall; petals to 6 cm long, white with pink or
purple stripes; styles shorter than stamens. Low
sites. Freq. SF, CF, NF. Wint-spr.

ZEPHYR-LILY
Zephyranthes treatiae S. Wats.

Perennial; leaves 10-30 cm long, to 3 mm wide,
nearly round, with rounded margins; flower
stems to 20 cm tall; petals to 8 cm long, white to
pink or purple. Low sites, swamps. Freq. CF,
NF, WF. E to Ga. Spr.

(not shown)

> ## SUMAC FAMILY   *ANACARDIACEAE*
>
> Vines, shrubs or trees, with resinous sap; leaves mostly alternate; flowers bisexual or unisexual; sepals 4 or 5; petals 4 or 5; ovary superior; fruit a drupe.

WINGED SUMAC
Rhus copallina L.

Perennial shrub, to 10 m tall; stems hairy; leaf rachis winged; leaflets 9-23, 2-10 cm long; petals mostly 5, to 3 mm wide, green to white; drupes to 4 mm in diam., often red-pink, hairy. Dry, woody, often disturbed sites. Common all Fla. W to Tex., N to Minn. and Maine. Spr-sum.

SMOOTH SUMAC
Rhus glabra L.

Perennial shrub, to 6 m tall; stems smooth; leaves wingless; leaflets 13-31, to 14 cm long; petals to 3 mm long, green or white; drupes to 4 mm in diam., red, velvety. Dry rich sites. Infreq. WF. N to Canada. Spr-sum.

POISON IVY
Rhus radicans L.
[*Toxicodendron radicans* (L.) Kuntze]

Perennial vine, climbing by aerial roots; leaflets 3, to 6 cm long; sepals 5; petals 5, green to white; drupes to 5 mm in diam. Thickets, woods, swamps, sandhills. Freq. all Fla. W to Ariz., N to Canada. Spr-sum.

POISON OAK
Rhus toxicodendron L.

Perennial shrub, to 1 m tall; twigs hairy; leaflets 3, 4-12 cm long; petals to 2 mm long, white; *drupes to 7 mm in diam., hairy to smooth. Dry woods. Infreq. NF, WF. W to Tex., N to N.J. Spr-sum.

POISON SUMAC
Rhus vernix L.
[*Toxicodendron vernix* (L.) Kuntze]

Perennial shrub, to 7 m tall, very poisonous; stems smooth; leaf rachis not winged; leaflets 7-15, to 12 cm long; petals to 2 mm wide, yellow; drupes to 7 mm in diam., smooth, white. Swampy sites. Common CF, NF, WF. W to Tex., N to Minn. and Maine. Spr-sum.

CUSTARD-APPLE FAMILY ANNONACEAE

Shrubs or trees; leaves alternate; flowers bisexual; sepals 3 or 4; petals 6 or 8; ovary superior; fruit a berry.

NARROW-LEAF PAWPAW
Asimina angustifolia Raf.

(not shown)

Perennial small shrub, to 1.5 m tall; leaves 6-10 cm long, mostly linear, smooth; flowers emerging after new leaves appear; outer petals 4-6 cm long, white; inner petals 4-5 cm long, white; berries 4-8 cm long. Flatwoods. Infreq. CF, NF, WF. N to Ala. and Ga. Spr-sum.

POLECAT BUSH
Asimina incarna (Bartr.) Exell

Perennial shrub, to 1.5 m tall; leaves 5-8 cm
long, elliptic to obovate, woolly when young;
flowers emerging with or before leaves; outer
petals 4-6 cm long, white; inner petals to 2 cm
long, white with yellowish base; berries 4-8 cm
long, green-yellow. Pinelands. Freq. CF, NF,
WF. N to Ga. Spr.

FLAG PAWPAW
Asimina obovata (Willd.) Nash

Perennial, to 3 m tall; leaves 4-14 cm long; sepals 3, deciduous; petals 6; outer petals 3, 6-11 cm
long; inner petals 3, 3-6 cm long; all petals white
or green-yellow, with maroon spots at base; flowers emerging after new leaves appear, terminal or
on branches axillary to new leaves; berries 1.5-6
cm long. Dry pinelands, scrub. Freq. CF, NF.
Endemic Fla. Spr.

SMALL-FRUITED PAWPAW
Asimina parviflora (Michx.) Dunal

Perennial shrub, to 4 m tall; leaves to 18 cm
long, obovate, pointed; outer petals 7-10 mm
long; petals green-purple; berries 2-6 cm long.
Woody sites. Freq. CF, NF, WF. W to Miss., E to
N.C. Spr.

(not shown)

DWARF PAWPAW
Asimina pygmaea (Bartr.) Dunal

Perennial shrub, 20-60 cm tall; stems ascending
or arching; leaves 4-15 cm long, with netted
venation; leaf margins revolute; outer petals to 3
cm long, white to pink; inner petals to 2 cm long,
maroon to purple; berries 3-5 cm long, curved,
yellow-green. Woods, disturbed sites. Freq. CF,
NF, WF. N to Ga. Spr.

FLATWOODS PAWPAW
Asimina reticulata Chapm.

Perennial shrub, to 1 m tall; leaves 2.5-9 cm
long, elliptic, sparsely covered with orange hairs
above, densely hairy below when young; flowers
emerging with or before leaves; *outer petals 3-
4 cm long, yellow-white; inner petals 2-2.5 cm
long, yellow-white with purple base; *berries 2.5-
9 cm long, yellow-green. Scrub, pinelands.
Common SF, CF, NF. Spr.

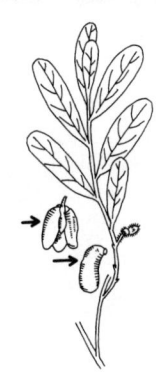

DOGBANE FAMILY   *APOCYNACEAE*

Herbs, vines, shrubs or trees, with milky latex; leaves opposite,
alternate or whorled; flowers bisexual; sepals 5; petals 5; ovary
superior; fruit a pair of follicles.

COMMON ALLAMANDA
Allamanda cathartica L.

Perennial, 1-3 m tall; twigs hairy; leaves 5-15 cm
long, opposite or whorled; petals 7-10 cm long,
yellow; capsules 4-6 cm in diam., prickly, poisonous. Native to tropical America. Escapes from
cultivation. Disturbed sites. Infreq. SF, CF, NF.
Spr-fall.

BLUE DOGBANE
Amsonia ciliata Walt.

Perennial, 0.6-1.3 m tall; stems branching, hairy;
leaves 4-7 cm long; petals to 8 mm long, blue,
purple or white; follicles 8-16 cm long. Dry sites.
Freq. CF, NF, WF. W to Tex., N to N.C. Spr-
sum.

BLUE STAR
Amsonia rigida Shuttlw. ex Small

Similar to *Amsonia tabernaemontana* except:
leaves to 8 cm long; *flowers smooth outside.
Wet woods, marshes. Infreq. NF, WF. W to La.,
E to Ga. Spr-sum.

TEXAS STAR
Amsonia tabernaemontana Walt.

Perennial base with several stems 0.3-1.3 m tall;
stems smooth; leaves to 16 cm long, oval; *flowers purple-blue, hairy outside; follicles 8-12 cm
long. Low, woody sites. Rare WF. N to Va.,
W to Tex., inland to Ill. Spr.

MADAGASCAR PERIWINKLE
Catharanthus roseus (L.) G. Don

Perennial, 20-70 cm tall; leaves 4-8 cm long;
*petals to 3 cm long, white, pink or purple; folli-
cles 2-3 cm long. Native to Madagascar. Open
sites. Freq. SF, CF, NF. Spr-sum.

OLEANDER
Nerium oleander L.

Perennial tree, to 10 m tall; sap poisonous; leaves
6-12 cm long; flowers 3-4.5 cm wide, white to
red; follicles 10-20 cm long. Native to
Mediterranean region. Cultivated, persists, rarely
escapes. Disturbed sites. Infreq. SF, CF, NF.
W to La. Spr-fall.

HOLLY FAMILY   *AQUIFOLIACEAE*

Trees or shrubs; leaves alternate; stipules present; flowers mostly unisexual (dioecious); sepals 4-9; petals 4-9; ovary superior; fruit a berry.

SAND HOLLY or CAROLINA HOLLY
Ilex ambigua (Michx.) Torr.
[*I. bushwellii* Small]

Perennial shrub or tree, to 6 m tall; twigs purple, smooth; leaves 2-4 cm long, oval, deciduous; leaf margins with fine teeth at tip; sepals 4-5; petals 4-5, white; drupes to 10 mm in diam., red; seeds smooth. Hammocks. Freq. CF, NF, WF. W to Tex., N to N.C.. Spr.

DAHOON HOLLY
Ilex cassine L.

Perennial shrub or small tree, to 12 m tall; bark gray; leaves to 10 cm long, leathery; leaf margins revolute; sepals 4; petals 4, to 4.5 mm in diam., white; drupes to 7 mm in diam., red or rarely yellow. Swamps, hammocks. Freq. all Fla. W to La., E to Va. Spr.

LARGE GALLBERRY
Ilex coriacea (Pursh) Chapm.

(not shown)

Perennial shrub, 1-4 m tall; leaves to 8 cm long; leaf margins with few teeth; sepals 5-9; petals 5-9, to 9 mm wide, white; drupes to 7 mm wide. Swamps. Infreq. CF, NF, WF. W to Tex, E to Va. Spr.

GALLBERRY
Ilex glabra (L.) A. Gray

Perennial shrub, 0.5-2 m tall; twigs hairy; leaves
1-5 cm long; leaf margins with few teeth at tip;
sepals 5-8; petals 5-8, to 7 mm wide, white; drupes to 7 mm in diam., black. Flatwoods. Freq.
all Fla. W to Tex., N to Canada. Spr-sum.

SMOOTH WINTERBERRY
Ilex laevigata (Pursh) Gray

(not shown)

Similar to *Ilex verticillata* except: sepal margins
entire; drupes orange-red. Wet bogs, swamps.
Ga., N to Maine. Spr-sum.

SCRUB HOLLY
Ilex opaca Ait.
var. *arenicola* (Ashe) Ashe
[*I. arenicola* Ashe]

Perennial shrub or small tree, to 5 m tall; leaves
2.5-6 cm long; leaf margins revolute, with spine-
like teeth; drupes over 7 mm in diam., red.
Scrub. Infreq. CF, NF. Spr.

WINTERBERRY
Ilex verticillata (L.) Gray

(not shown)

Perennial shrub, to 8 m tall; leaves 2-10 cm long,
deciduous, elliptic-obovate; petioles 0.5-2 cm
long; sepals 5-8, with ciliate margins; petals 5-8,
white; drupes to 8 mm in diam., red; seeds
ribbed. Swampy sites. Infreq. WF. W to Tex.,
N to Canada. Sum.

YAUPON
Ilex vomitoria Ait.

Perennial shrub or tree, to 8 m tall; stems hairy;
leaves 1-3 cm long; leaf margins scalloped with
rounded teeth; sepals 4; petals 4, to 5 mm wide,
white or yellow; drupes to 8 mm in diam., red,
occasionally yellow. Well-drained sites. Freq.
CF, NF, WF. W to Tex., E to Va. Spr.

ARUM FAMILY 　 *ARACEAE*

Rhizomatous or tuberous herbs, climbing vines or woody
plants; leaves basal or growing on stems; flowers bisexual or
unisexual (monoecious); flowers borne on spadix subtended
by spathe; floral parts 4-6 or absent; ovary superior; fruit a
berry or utricle.

GREEN DRAGON
Arisaema dracontium (L.) Schott

Perennial, 0.5-1 m tall; leaf 1, divided into 5-15
leaflets 10-30 cm long, linear to obovate; spathe   (not shown)
to 12 cm long, green; spadix extending beyond
spathe; berries in clusters 6-8 cm long, red-
orange. Moist to wet, woody sites. Infreq. CF,
NF, WF. W to Tex., N to Canada. Spr-sum.

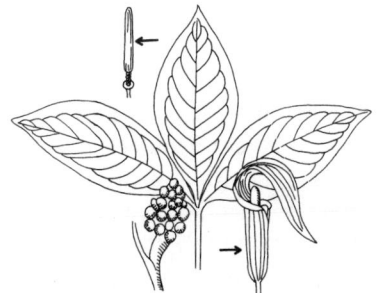

JACK-IN-THE-PULPIT
Arisaema triphyllum (L.) Schott

Perennial, 20-80 cm tall; leaves 1 or 2, divided
into 3 leaflets 8-30 cm long, obovate; *spathes 4-
12 cm long, green or green with purple-brown
stripes; *spadix not extending beyond spathe;
berries in clusters 2.5-15 cm long, red. Moist to
wet, woody sites. Freq. all Fla. W to Tex., N to
Canada. Spr-sum.

WILD TARO
Colocasia esculenta (L.) Schott

Perennial; leaves to 50 cm long, oval to arrow-shaped; petiole attached to center of leaf under-side (peltate); *spadix to 50 cm tall, with female flowers below and male flowers above. Native to Pacific Islands. Cultivated. Aquatic sites. Freq. all Fla. N into southeastern states. Sum-fall.

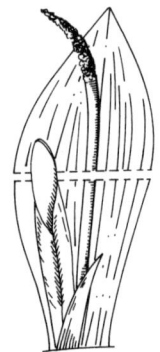

GOLDEN CLUB
Orontium aquaticum L.

Perennial, 15-60 cm tall; leaves 10-40 cm long, to 10 cm wide; spadix to 30 cm long; flowering portion of spadix to 7 cm long, yellow; spathe to 30 cm long; flowers perfect (male and female flowers present), on bottom of spadix; sepals and petals 6; berries blue-green; seeds 6-10 mm in diam. Aquatic sites. Infreq. CF, NF, WF. W to La., N to Mass. Wint-spr.

SPOON FLOWER
Peltandra sagittifolia (Michx.) Morong

Perennial; leaves 15-25 cm long, arrow-shaped; *spathe 6-12 cm long, white; *spadix to 5 cm long; berries to 1 cm in diam., red. Swamps, bogs. Infreq. CF, NF, WF. W to Miss., N to N.C. Spr.

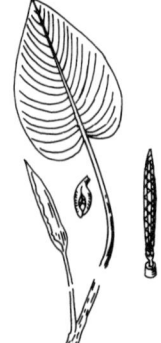

GREEN ARUM
Peltandra virginica (L.) Schott & Endl.

Perennial; leaves 10-50 cm long; male and female flowers separate but on same plant (monoecious); spathes 8-28 cm long, green with white edges; spadix to 20 cm long, white, yellow or orange, with female flowers below and male flowers above; berries 6-15 mm in diam., green or brown. Swamps, wet sites. Freq. all Fla. W to Tex., N to Canada. Spr-sum, spr-fall S.

WATER LETTUCE
Pistia stratiotes L.

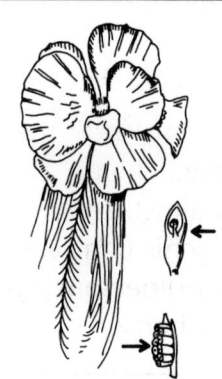

Perennial, floating; leaves 3-15 cm long, in basal
rosettes, hairy; flowers unisexual, on spadix, with
female flowers below and male flowers above;
*spathes to 1.5 cm long, hairy, green-white;
*seeds cylindrical. Ponds, lakes, streams.
Common all Fla. W to Tex. All yr.

GINSENG FAMILY   *ARALIACEAE*

Trees, shrubs, vines or herbs; leaves mostly alternate; stipules
present; flowers bisexual or unisexual (monoecious or dioecious); sepals mostly 4 or 5; petals mostly 3 or 5; ovary inferi-
or; fruit a drupe or berry.

DEVIL'S WALKINGSTICK
Aralia spinosa L.

Perennial shrub, to 8 m tall; stems prickly; leaves
to 1.2 m long, pinnately or ternately compound;
*petals 5, 2-3 mm long, white; *drupes to 5 mm
in diam., black, oval to round. Woods, low sites.
Freq. CF, NF, WF. W to Tex., N to N.J. Sum.

PALM FAMILY   *ARECACEAE*

Nonwoody trees or shrubs; leaf blades folded like fan with each blade attached on each side of central rachis (feather palms), linear, shaped like fish tail, or clustered around short rachis (fan palms); flowers bisexual or unisexual (monoecious or dioecious); sepals 3; petals 3; ovary superior; fruit a drupe or berry.

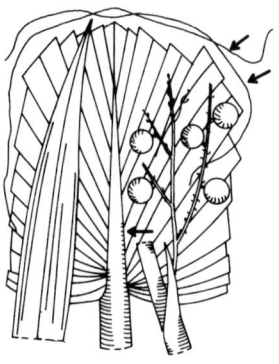

SCRUB PALMETTO
Sabal etonia Swingle ex Nash

Perennial, to 2 m tall; stem underground; leaves to 1 m long, fanlike with *thread-like extensions; leaf *midrib extending nearly through blade; spadix 50-80 cm long; drupes to 17 mm in diam., bluish black. Dry sites. Infreq. SF, CF, NF. Endemic. Spr-sum.

DWARF PALMETTO
Sabal minor (Jacq.) Pers.

Perennial, to 8 m tall; stems underground; leaves fanlike; leaf midrib short, to 3 cm long; spadix 1-2 m long; drupes to 7 mm in diam., black. Moist to wet woods. Freq. all Fla. W to La., E to N.C. Spr-fall.

(not shown)

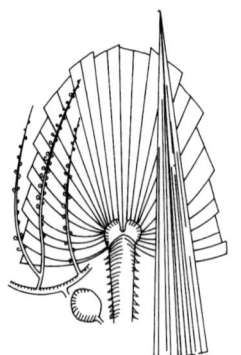

SAW PALMETTO
Serenoa repens (Bartr.) Small

Perennial, to 7 m tall; stems underground, creeping, leaning to erect, branching; leaves fanlike, with no midrib; leaf petioles spiny; petals to 4.2 mm long; drupes to 20 mm long, black. Indian food. Dry to wet woods. Freq. all Fla. W to La., E to S.C. Spr-sum.

BIRTHWORT FAMILY   *ARISTOLOCHIACEAE*

Herbaceous or woody vines; leaves alternate; flowers bisexual; sepals 1, 3 or 5; petals 0 or rarely 3; ovary mostly inferior; fruit a capsule.

CALICO FLOWER
Aristolochia littoralis Parodi

Perennial climbing vine; leaves 5.5-6 cm long; flowers 1-lobed, purple and cream, splotched; capsules to 5 cm long, 6-valved; seeds numerous. Native to tropical America. Escapes from cultivation. Disturbed wooded sites. Infreq. all Fla. Spr-fall.

MILKWEED FAMILY   *ASCLEPIADACEAE*

Herbs, vines or shrubs, with milky latex; leaves opposite, whorled or alternate; flowers bisexual; sepals 5; petals 5; filaments of stamens forming tube (gynostegium) enclosing pistil; appendage from base of each filament forming hood; hoods forming corona; most hoods having horn and reflexed petals; ovary superior; fruit 1 or 2 follicles; seeds usually with long, white, silky hairs.

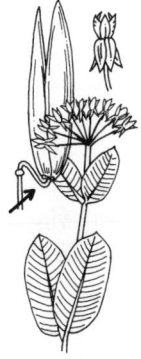

CURLY MILKWEED
Asclepias amplexicaulis J. E. Smith

Perennial, 0.4-1 m tall; leaves 6-15 cm long, 4-8 cm wide; petals to 1.1 cm long, green-red or green-purple; hoods red; *follicles 8-16 cm long. Dry sites. Infreq. CF, NF, WF. W to Tex., N to N.H. Spr-sum.

SHORT HOODED MILKWEED
Asclepias cinera Walt.

Perennial, 30-70 cm tall; stems smooth; sap
milky; leaves 3-9 cm long, linear, filiform; *petals
5-7 mm long, purple; follicles 8-10 cm long.
Wet pinelands or rarely dry ridges. Infreq. CF,
NF, WF. N to S.C. Spr-sum.

FRAGRANT MILKWEED
Asclepias connivens Baldw.

(not shown)

Perennial, to 90 cm tall; leaves to 7 cm long,
oval; petals 12-15 mm long, reflexed, green to
yellow; hoods 7-9 mm long; follicles 1.2-1.5 cm
long. Moist to wet sites. Infreq. CF, NF, WF.
W to Miss., N to Ga. Sum.

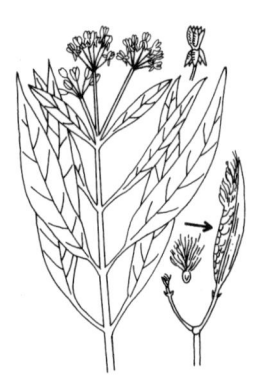

SCARLET MILKWEED
Asclepias curassavica L.

Annual, 0.3-1.5 m tall; leaves 5-12 cm long, to 1
cm wide; petals to 1 cm long, scarlet or yellow;
hoods orange; *follicles to 10 cm long. Dry to
moist or wet sites, often in disturbed areas.
Infreq. SF, CF, NF. W to Tex. All yr.

SCRUB MILKWEED
Asclepias curtissii Gray

Perennial, 50-70 cm long; stems lying flat or
ascending; leaf blades 3-6 cm long, smooth;
*petals 5.5-6 mm long, greenish white; *follicles
8-11 cm long. Scrub, dry sites. Infreq. peninsula
Fla. Endemic. Spr-fall.

SANDHILL MILKWEED
Asclepias humistrata Walt.

Perennial, 70-90 cm long or 20-70 cm tall; stems prostrate, glaucous; leaves 5-12 cm long, with pink to maroon veins; *petals to 7 mm long, gray to purple, rose or lavender; *hoods white; follicles 8-14 cm long. Sandhills, scrub. Freq. CF, NF, WF. W to La., N to N.C. Spr-sum.

SWAMP MILKWEED
Asclepias incarnata L.

Perennial, 0.6-1.5 m tall; leaves 4-17 cm long, to 4 cm wide; *petals to 4 mm long, pink to purple; hoods pink to purple; follicles to 10 cm long. Swamps, wet sites. Infreq. SF, CF, NF. W to Tex., N to Canada. Sum.

RED MILKWEED
Asclepias lanceolata Walt.

Perennial, 0.4-1.2 m tall; stems smooth; leaves 1-25 cm long, less than 1 cm wide, opposite; petals to 10 mm long, reflexed, red-orange; hoods orange to red; *follicles 7-10 cm long. Swamps, wet sites. Freq. all Fla. W to Tex., N to N.J. Sum.

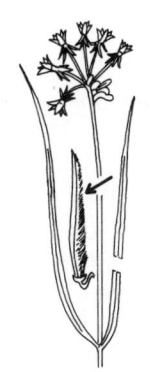

FLORIDA MILKWEED
Asclepias longifolia Michx.

Perennial, 20-80 cm tall; stems erect to ascending, purplish; leaves to 18 cm long; *petals greenish white, purplish-tipped; *hoods 2-3 mm long; *anthers longer than hoods; *follicles 9-11 cm long. Moist to wet sites. Infreq. all Fla. W to Tex., N to Del. Spr-sum.

WEAK-STEM MILKWEED
Asclepias michauxii Dcne. in DC

Perennial, 10-40 cm tall; stems lying flat, hairy; leaves 4-10 cm long; cymes solitary; corona purple, longer than stigma; petals to 5 mm long, greenish white, purple beneath; follicles 8-15 cm long, erect. Moist, woody sites. Infreq. CF, NF, WF. W to La., E to S.C. Spr-sum.

(not shown)

PEDICILLATE MILKWEED
Asclepias pedicillata Walt.

(not shown)

Perennial, 5-40 cm tall; stems hairy; uppermost leaves 2-6 cm long, linear; lower leaves bractlike; flowers greenish yellow; petals 8-12 mm long, erect, enclosing hoods and stigma; hoods 3-4 mm long; follicles 11-15 cm long. Moist to wet sites. Common all Fla. N to N.C. Spr-sum.

AQUATIC MILKWEED
Asclepias perennis Walt.

Perennial, 30-60 or 90 cm tall; stems smooth; leaves 5-12 cm long; petals 2.5-4.5 mm long, white or pink; corona 2-3 mm wide and long; horns exserted; follicles 4-8 cm long. Moist woods, swamps. Occasional CF, NF, WF. E to S.C., W to Tex., N to Mo. and Ind. Spr-sum.

LAVENDER MILKWEED
Asclepias rubra L.

Perennial, 0.3-1.2 cm tall; stems smooth; leaves 7-20 cm long, to 4 cm wide; petals 8-9 mm long, purple-red; hoods to 6 mm long, orange; *follicles 8-10 cm long. Moist to wet sites. Rare WF. W to Tex., N to N.C. Sum.

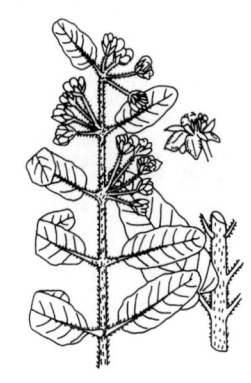

VELVET-LEAF MILKWEED
Asclepias tomentosa Ell.

Perennial, 0.3-1.2 m high; stems and leaves covered with woolly hairs; leaves 5-8 cm long; flowers yellow-green; hoods to 4 mm long; stamens
to 4 mm long; follicles 10-13 cm long. Dry sites.
Common all Fla. W to Ark., N to N.C. Spr-sum.

BUTTERFLY WEED
Asclepias tuberosa L.

Perennial, 3-60 cm tall; stems erect, ascending
to lying flat, hairy; sap clear; leaves 3-10 cm
long, alternate, linear-lanceolate, hairy on lower
surfaces; petals orange, red or yellow; *follicles
8- 15 cm long. Dry sites. Infreq. all Fla. W to
Ariz., N to Canada. Sum-fall.

FIDDLE LEAF BUTTERFLY WEED
Asclepias tuberosa L.
subsp. *rolfsii* (Britt.) Woods.

Similar to *Asclepias tuberosa* except: 10-30 cm
tall; leaves hastate or fiddle-shaped; petals
orange. Dry sites. Freq. all Fla. N to S.C.
Spr-sum, all yr S.

WHITE MILKWEED
Asclepias variegata L.

Perennial, 0.2-1 m tall; leaves 6-15 cm long;
petals 6-8 mm long, white; *hoods to 3 mm long;
*follicles 8-14 cm long. Dry woods. Infreq. NF,
WF. W to Tex., N to Ohio, E to Conn. Spr-sum.

WHORL-LEAF MILKWEED
Asclepias verticillata L.

Perennial, 30-80 cm tall; leaves 2-6 cm long,
whorled; *petals 3-4 mm long, greenish white;
hoods 1-2 mm long; horns exserted; *follicles
4.5-12 cm long. Dry sites. Occasional all Fla.
W to Tex., N to Canada. Spr-fall.

GREEN-CROWNED MILKWEED
Asclepias viridis Walt.

Perennial, 20-60 cm long; stems spreading;
leaves 6-13 cm long, to 6 cm wide; *petals 13-
15 mm long, greenish; hoods to 4.5 mm long,
purplish violet; *follicles 6-11 cm long. Dry
open sites. Infreq. all Fla. W to Tex., E to Ala.
Spr-sum.

SOUTHERN MILKWEED
Asclepias viridula Chapm.

Perennial, 30-70 cm tall; stems hairy; leaves 4-10
cm long, 1-4 mm wide; petals 4-5 mm long,
green to brown-purple; hoods to 2.8 mm long,
cream to brownish purple. Wet sites. Infreq. CF,
NF, WF. Spr-sum.

(not shown)

COASTAL CYNANCHUM
Cynanchum angustifolium Pers.

Perennial twining vine, to 1 m long; leaves 2-8
cm long, linear to linear-lanceolate; *flowers
greenish white or purplish; petals 5-7 mm wide,
only slightly united at base; follicles 4-7 cm long.
Coastal marshes, lake margins. Freq. all Fla.
coastal counties. W to Tex., N to N.C. Spr-fall.

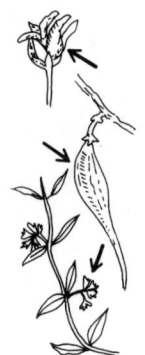

BLODGETT'S CYNANCHUM
Cynanchum blodgettii (Gray) Shinners

Perennial twining vine; leaves 2-3 cm long;
*petals to 4 mm long, green to white or purple;
*follicles to 6 cm long. Coastal hammocks.
Infreq. SF. Spr-fall.

FRAGRANT CYNANCHUM
Cynanchum northropiae (Schlect.) Alain

(not shown)

Perennial twining vine; leaves 1.5-4 cm long,
oval to elliptic; petal lobes 2-3 mm long, white,
united to middle. Hammocks, pinelands.
Infreq. SF, CF. All yr.

LEAFLESS CYNANCHUM
Cynanchum scoparium Nutt.

Perennial vine; stems diffuse; leaves 2-5 cm long,
present when young, falling off with age; *petals
to 2 mm long, greenish white; *follicles to 4.5 cm
long. Coastal hammocks. Infreq. all Fla.
N to S.C. Spr-fall.

STRANGLER VINE
Morrenia odorata (Hook. & Arn.) Lindl.

(not shown)

Perennial, climbing and running over other plants;
stems and leaves with short felty hairs; leaves to
8 cm long, arrow- to heart-shaped to oval; flowers
to 1 cm long, yellow to green; follicles to 10 cm
long, to 7 cm wide; seeds carried by wind. Native
to South America. Troublesome in orange groves.
Disturbed sites. Freq. CF, NF. Sum.

WHITE VINE
Sarcostemma clausum (Jacq.) Schult.

Perennial vine matting over shrubs and small
trees; leaves 2-8 cm long; petals to 5 mm long,
white; follicles to 7 cm long. Low sites.
Infreq. SF, CF. All yr.

BLACK MANGROVE FAMILY　　　　　*AVICENNIACEAE*

Shrubs or trees; leaves opposite; flowers bisexual; sepals 5;
petals 5; ovary superior; fruit a capsule.

BLACK MANGROVE
Avicennia germinans (L.) L.

Perennial shrub, to 25 m tall; leaves 3-12 cm
long, thick, evergreen, gray beneath; petals
white; capsules 3-5 cm long. Coastal sites.
Common south to infreq. northward in Fla.
W to Tex. All yr.

SALTWORT FAMILY   *BATACEAE*

Succulent to semi-woody shrubs; leaves opposite; flowers unisexual (dioecious); male flowers with 2 sepals and 4 petals; female flowers with no floral parts; fruit a multiple of berries.

SALTWORT
Batis maritima L. (only species)

Perennial shrub, 0.5-1.5 m long, having strong odor; stem spreading, trailing, creeping; leaves 1-3 cm long, opposite, fleshy, half round; male and female flowers separate but on same plant (dioecious), in spikes; female spikes 4- to 12-flowered; male spikes 2-lobed; fruits 1-2 cm long. Coastal sites. Freq. all Fla. W to Tex., E to S.C. Spr-sum.

BEGONIA FAMILY   *BEGONIACEAE*

Herbs or shrubs; mostly with succulent stems; leaves alternate; stipules present; flowers unisexual (monoecious); female flowers with 2-5 petals; male flowers with 2 sepals and 2 petals; ovary inferior; fruit a winged capsule.

PERPETUAL BEGONIA
Begonia semperflorens Link & Otto

Perennial, to 1 m tall; stems succulent, green or red; leaves to 13 cm long, to 7 cm wide; *male petals 4; *female petals 5; flowers pink, red or white; *capsules to 1.5 cm long, 2-winged. Native to South America. Escapes from cultivation. Low sites. Rare all Fla. All yr.

BIRCH FAMILY   *BETULACEAE*

Trees or shrubs; leaves alternate; stipules present; flowers unisexual (monoecious); male flowers in aments (catkins); female flowers in cymes; floral parts consisting of scales; ovary inferior; fruit a samara or nut.

AMERICAN HORNBEAM or BLUE BEECH
Carpinus caroliniana Walt.

Perennial small tree, to 13 m tall; trunk gray, giving appearance of flexed muscles; leaves 3-6 cm long; catkins hanging down; nuts 5-6 mm long. Low, wet woody sites. Freq. NF, CF, WF. W to Tex., N to Canada. Spr-sum.

BIGNONIA FAMILY   *BIGNONIACEAE*

Trees, shrubs or woody vines; leaves opposite, whorled or alternate; flowers bisexual; sepals 2-5; petals 5; ovary superior; fruit a capsule.

TRUMPET VINE
Campsis radicans (L.) Seem.

Perennial climbing or trailing vine; leaflets 7 or more, 2-8 cm long, opposite, deciduous; petals 5-9 cm long, red or orange; capsules 10-20 cm long; seeds winged. Woods, fields, wet sites. Freq. CF, NF, WF. W to Tex., E to N.J. Spr-fall.

CAT'S-CLAW VINE
Macfadyena unguis-cati (L.) A. Gentry
[*Doxantha unguis-cati* (L.) Rehder]

Perennial, high climbing woody vine; leaves 2-7 cm long; leaflets 2 per leaf with 3-pronged tendril (cat-claw) at tip; petals forming tube to 8 cm long, yellow; capsules to 50 cm long. Native to tropical America. Cultivated and escapes. Disturbed areas. Rare but can be locally common SF, CF, NF. Spr.

FLAME VINE
Pyrostegia venusta (Ker-Gawl.) Miers
[*P. ignea* Presl]

Perennial climbing vine; leaflets 4-8 cm long; sepals glandular; flowers in panicles that hang down; petals 5-7 cm long, red-orange; capsules to 30 cm long. Native to South Africa. Escapes from cultivation. Disturbed sites, thickets. Rare SF, CF, NF. Spr-sum.

BORAGE FAMILY · *BORAGINACEAE*

Herbs, shrubs, vines or trees; leaves alternate; flowers mostly bisexual, arranged in coiled-up cymes (scorpoid) or in elongate spikes; sepals 5; petals 5; ovary superior; fruit 2 or 4 nutlets or a drupe.

WILD HELIOTROPE
Heliotropium amplexicaule Vahl
[*H. anchusaefolium* Poir.]

Perennial, 10-50 cm tall; stems hairy; leaves 2-8 cm long, hairy; petals to 5 mm long, purple with yellow base; nutlets 2, each to 3 mm long. Dry sites. Infreq. CF, NF, WF. W to Tex., N to Va. Spr-sum.

SCORPION-TAIL
Heliotropium angiospermum Murr.

Annual, 0.2-1.2 m tall; stems sprawling to erect, hairy; leaves 2-8 cm long; *petals white to yellow; floral tube 1-2 mm long; *nutlets 2, to 2 mm wide. Hammocks, disturbed sites. Freq. SF, CF, NF. All yr.

SEASIDE HELIOTROPE
Heliotropium curassavicum L.

Annual or perennial, 10-60 cm long; stems prostrate to ascending, smooth, succulent; leaves 1-6 cm long; flower spikes paired; *petals white with yellow eye; floral tube to 2 mm long; *nutlets 4, 2-3 mm wide. Native to South America. Coastal marshes and dunes. Infreq. all Fla. W to Tex., N to Mass. Spr-fall.

INDIAN HELIOTROPE
Heliotropium indicum L.

Annual, to 1 m tall; stems hairy; leaves to 10 cm long, smooth on upper surface, hairy on lower surface; petals to 3 mm long, blue to purple, with white or yellow eye; *nutlets 2, each to 2 mm long. Moist to wet sites. Infreq. WF. W to Tex., N to Va. Sum-fall.

PINELAND HELIOTROPE
Heliotropium polyphyllum Lehm.
[*H. leavenworthii* (Gray) Small]

Perennial, 10-50 cm long, 0.05-1.1 m tall; stems prostrate to erect, hairy; leaves 0.5-2.5 cm long; petals white to yellow; floral tube 3-4 mm long; nutlets 4, to 1.5 mm wide. Dry to moist sites. Freq. all Fla. All yr.

SLENDER ROOTED PUCCOON
Lithospermum incisum Lehm.

Perennial, 10-50 cm tall; stems rough, covered with short stiff gray hairs; leaves 2-3 mm wide, to 5 mm long; early season flowers opening, 1.5-3.5 cm long, yellow; late season flowers not opening (cleistogamous); nutlets 3-4 mm long. Dry, open sites. Infreq. CF, NF, WF. W to Ariz., N to Canada. Spr-sum.

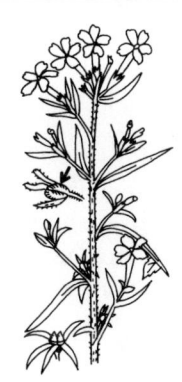

FALSE GROMWELL
Onosmodium virginianum (L.) A. DC.

Perennial, 20-80 cm tall; stems covered with short appressed hairs; leaves 2-13 cm long; flowers coiled; *petals 7-10 mm long, yellow, with lobes to 3 mm long; styles threadlike, long exserted; *nutlets to 2.5 mm long. Dry sites, woods. Infreq. CF, NF, WF. W to La., N to Mass. Spr-sum.

PINEAPPLE FAMILY   *BROMELIACEAE*

Epiphytic or terrestrial herbs; leaves mostly basal, thick, with sharp-toothed margins; cups formed by overlapping leaves accumulating water and organic matter; sepals 3; petals 3; ovary superior or inferior; fruit a capsule or berry.

NORTHERN NEEDLE-LEAF
Tillandsia bartramii Ell.

Perennial, epiphyte, 10-30 cm tall; leaves channeled above, gray and coarsely lepidote; floral bracts pink; flowers to 2.5 cm long; capsules to 2.7 cm long. Woods. Freq. CF, NF, WF. Spr-sum.

TWISTED AIR PLANT
Tillandsia flexuosa Sw.

Perennial, 20-80 cm tall; basal leaves 7-28 cm long, with *banded sheath; flower stems zigzag; *petals to 4 cm long, pink; capsules to 5 cm long. Swamps, hammocks. Infreq. SF, CF. Spr-sum.

SOUTHERN NEEDLE-LEAF
Tillandsia setacea Sw.

(not shown)

Similar to *Tillandsia bartramii* except: leaves finely lepidote, green or reddish; floral bracts green, red-tinged. Moist to wet woods. Common SF, CF. Sum.

BROAD NEEDLE-LEAF
Tillandsia simulata Small

Perennial, 20-40 cm tall; leaves pointed; bracts reddish; petals 4.5 cm long, violet; *capsules to 3 cm long. Tree trunks, swamps. Freq. CF, NF. Spr-sum.

SWOLLEN WILD-PINE
Tillandsia utriculata L.

Perennial, 0.3-2 m tall; stems light green; leaves to 80 cm long, mostly basal; flower spikes zigzag, branching, with exposed rachis; petals to 4 cm long, cream; capsules to 5 cm long. Cypress swamps, scrub. Freq. SF, CF, NF. Sum-fall. Plants die after flowering.

Illustrated Plants of Florida and the Coastal Plain

BURMANNIA FAMILY   *BURMANNIACEAE*

Herbs; most species lacking chlorophyll; leaves alternate, basal or missing; flowers mostly bisexual; sepals 3; petals 3; ovary inferior; fruit a capsule.

BLUE THREAD
Burmannia biflora L.

Annual, 5-15 cm tall; leaves 1-3 mm long; *floral tube to 5 mm long, purple, 3-winged; *capsules to 5 mm long. Moist to wet sites. Infreq. all Fla. W to Tex., E to Va. Spr-fall.

CAPITATE BURMANNIA
Burmannia capitata (Walt.) Mart.

Annual, 5-20 cm tall; leaves to 5 mm long; *flowers to 5 mm long, in terminal clusters, white to blue; *capsules to 3 mm long. Low wet sites. Infreq. all Fla. W to Tex., E to Va. Spr-fall.

WATER SHIELD FAMILY   *CABOMBACEAE*

Aquatic herbs with rhizomes; leaves opposite and alternate; flowers bisexual; sepals 3 or 4; petals 3 or 4; ovary superior; fruits nut-like.

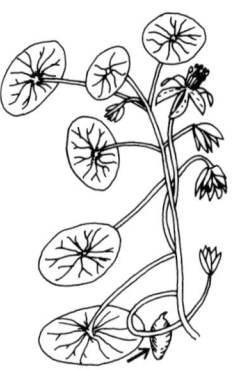

WATER SHIELD
Brasenia schreberi Gmel.

Perennial, to 2 m long; stems mucilage-coated; all leaves floating, alternate, peltate; leaf blades to 12 cm long; sepals and petals 10-15 mm long, purple; stamens 18-36; *fruits 5-8 mm long. Quiet water. Infreq. CF, NF, WF. W to Tex., N to Canada. Sum.

FANWORT
Cabomba caroliniana A. Gray

Perennial; submerged leaves opposite, dissected, with segments to 4 cm long; floating leaves 1-2 cm long, alternate, entire, attached to stems by their centers (peltate); sepals and petals 8-12 mm long, white or purple; stamens 3-6; *fruits 5-7 mm long. Quiet water. Common all Fla. W to Tex., N to Mass. Spr-fall.

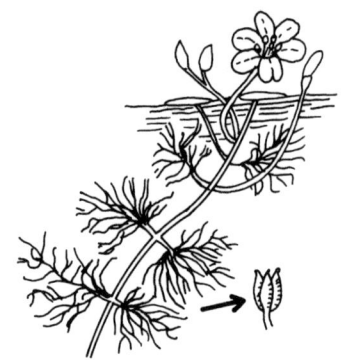

CACTUS FAMILY   *CACTACEAE*

Shrubs or trees; stems succulent, usually bearing spines; leaves usually absent or scale-like; flowers mostly bisexual; floral parts few to many; ovary inferior or superior; fruit a berry.

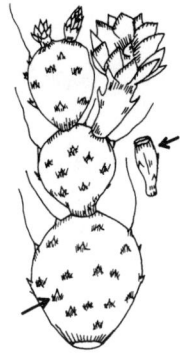

PRICKLY PEAR
Opuntia humifusa (Raf.) Raf.
[*O. compressa* (Salisb.) Macb.]
[*O. lata* Small]

Perennial, to 2 m tall; stems prostrate to ascending or erect; leaves 5-11 mm long; joints 4-16 cm long; *spines 0, 1 or 2 per cluster, white, gray or brown; flowers 4-9 cm wide, yellow; petals 2.5-4 cm long; *berries 2-6 cm long. Dry open sites. Common all Fla. W to Miss., N to Canada. Spr-sum.

COASTAL PRICKLY PEAR
Opuntia stricta Haw.

Perennial, to 3 m tall; stems erect to suberect; leaves 2-8 mm long; joints 17-25 cm long; spines yellow at maturity; flowers 5-10 cm wide, yellow to red; *berries 3.5-9.5 cm long. Coastal sites. Infreq. all Fla. N to N.C. Spr-sum, all yr S.

WATER STARWORT FAMILY   *CALLITRICHACEAE*

Aquatic or wet terrestrial herbs; leaves mostly opposite; flowers unisexual (dioecious); sepals 0; petals 0; ovary superior; fruit 4 nutlets.

SMOOTH-FRUITED WATER STARWORT
Callitriche heterophylla Pursh

(not shown)

Annual, aquatic; leaves to 25 mm long; fruits 1 mm x 1 mm, as long as wide, stalked. Moist to wet sites. Infreq. CF, NF, WF. W to La., N to Canada. Sum.

WIDE-FRUITED WATER STARWORT
Callitriche peploides Nutt.

Annual, aquatic; leaves 2-3 mm long; fruits 4.5 mm x 7 mm, wider than long, sessile. Wet sites. Infreq. all Fla. W to Tex., N to S.C. Spr-sum.

BELLFLOWER FAMILY   *CAMPANULACEAE*

Herbs or shrubs; leaves alternate, opposite or whorled; flowers bisexual; sepals 5; petals 5 or 0; ovary mostly inferior; fruit a capsule.

TALL BELLFLOWER
Campanula americana L.

Biennial, to 2 m tall; leaves 5-15 cm long; *flowers blue or white; *capsules to 12 mm long. Moist rich sites. Infreq. WF. W to Ark., N to Canada. Sum.

FLORIDA BELLFLOWER
Campanula floridana S. Wats.

Perennial, 20-40 cm tall; stems branching, ascending; leaves 1-4 cm long; sepals 7-9 mm long; petals violet; *capsules to 4 mm long. Low sites. Common all Fla. Endemic to peninsula Fla. and easternmost Fla. panhandle. All yr.

PIEDMONT LOBELIA
Lobelia amoena Michx.

Perennial, to 1.2 m tall; stems smooth; leaves to 18 cm long, to 4.5 cm wide; leaf margins variously toothed; sepals with smooth margins; *petals 18-24 mm long, blue with white center; capsules 5-6 mm wide. Low, moist to wet sites. Infreq. CF, NF, WF. W to Miss., N to Va. Sum-fall.

CARDINAL FLOWER
Lobelia cardinalis L.

Perennial, to 2 m tall; leaves 5-20 cm long; petals
to 4.5 cm long, scarlet; capsules to 1 cm long.
Low, wet sites and floating mats. Infreq. CF, NF,
WF. W to Tex., N to Canada. Sum-fall.

BAY LOBELIA
Lobelia feayana Gray

Annual, to 30 cm tall; stems smooth; leaves 0.5-
1.5 cm long, to 1 cm wide; leaf margins smooth
to slightly toothed; sepals with smooth margins;
petals 7-9 mm long, blue or purple, with white
center; capsules to 3.5 mm wide. Open, moist to
wet sites. Common all Fla. Spr-fall.

FLACCID-LEAVED LOBELIA
Lobelia flaccidifolia Small

Annual, 20-70 cm tall; leaves 3-11 cm long;
*petals to 2 cm long, blue, violet or white;
capsules to 6 mm in diam. Swamps. Rare WF.
W to Tex., E to Ga. Sum-fall.

LARGER WHITE LOBELIA
Lobelia floridana Chapm.

Perennial, 0.5-1.5 m tall; leaves 1-40 cm long;
*petals to 2 cm long, light blue; capsules 4-5 mm
in diam. Low wet sites. Freq. WF. W to Tex.
Sum-fall.

COASTAL PLAIN LOBELIA
Lobelia glandulosa Walt.

Perennial, 0.3-1.2 m tall; stems smooth; *leaves
2-15 cm long, to 5 mm wide; leaf and sepal margins with toothed glands; *petals to 3.5 cm long,
blue with white center; capsules to 8 mm wide.
Swamps, low sites. Common all Fla. W to Ala.,
N to Va. Sum-fall, all yr S.

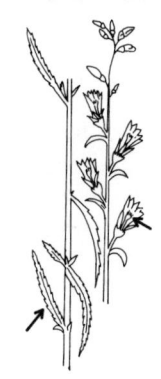

LITTLE WHITE LOBELIA
Lobelia homophylla F. E. Wimm.
[*L. cliffortania* L.]

Annual, 20-60 cm tall; stems smooth; leaves to
4 cm long, to 1 cm wide; leaf margins with blunt
teeth; sepal margins smooth; petals to 15 mm
long, blue; capsules to 6.5 mm wide. Open
moist sites. Infreq. SF, CF, NF. Spr-fall.

NUTTALL'S LOBELIA
Lobelia nuttallii Roem. & Shult.

Annual, to 70 cm tall; stems hairy on lower portion; leaves to 4 cm long, to 1 cm wide; leaf margins sparingly toothed; sepal margins smooth;
petals to 11 mm long, blue with white center;
*capsules to 3 mm wide. Low sites. Infreq. WF.
W to La., N to N.Y. Spr-fall.

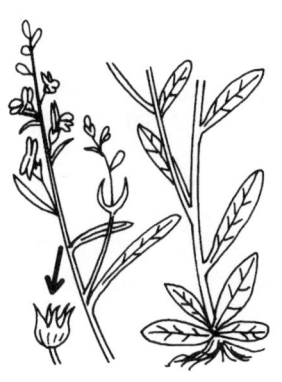

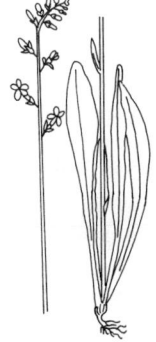

WHITE SWAMP LOBELIA
Lobelia paludosa Nutt.

Perennial, 20-80 cm tall; stems smooth; basal
leaves 3-25 cm long, to 1.5 cm wide, with
smooth to toothed margins; sepal margins smooth
to slightly toothed; petals to 2 cm long, blue or
white; capsules to 3.5 mm wide. Moist to wet
sites. Freq. all Fla. N to Ga. Spr-fall, all yr S.

DOWNY LOBELIA
Lobelia puberula Michx.

Perennial, to 1.5 m tall; stems covered with short stiff hairs; *leaves to 12 cm long, to 4 cm wide, with both sides hairy; leaf margins toothed to smooth; sepals hairy; sepal margins not glandular toothed; petals to 2 cm long, blue or purple, with white center; capsules to 10 mm wide. Moist to dry woods. Infreq. CF, NF, WF. W to Tex., N to N.J. Sum-fall.

TWO-FLOWERED VENUS' LOOKING-GLASS
Triodanis biflora (Ruiz & Pavon) Greene
[*T. perfoliata* (L.) Nieuwl.
var. *biflora* (Ruiz & Pavon) Bradley]

Annual, 10-50 cm tall; leaves sessile; petals 15-20 mm wide; capsules to 1.5 mm in diam., opening near top. Fields, disturbed sites. Infreq. CF, NF, WF. W to Tex., E to Va. Spr-fall.

VENUS' LOOKING-GLASS
Triodanis perfoliata (L.) Nieuwl.
[*Specularia perfoliata* (L.) A. DC.]

Annual, 10-50 cm tall; leaves clasping; petals 10-15 mm wide, purple; *capsules to 3 mm in diam., opening below middle. Dry sites. Freq. CF, NF, WF. W to Tex., N to Canada. Spr-fall.

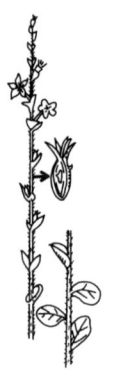

ASIATIC BELLFLOWER
Wahlenbergia marginata (Thunb.) DC.

Perennial, 10-65 cm tall; stems clumped; leaves 1-4 cm long; petals 5-8 mm long, blue to violet; capsules 3-6 mm long. Open sites. Freq. CF, NF, WF. W to Ala., N to N.C. Spr-fall.

CANNA FAMILY   *CANNACEAE*

Rhizomatous herbs; leaves alternate; flowers bisexual; sepals 3; petals 3, sepal-like; stamens 3, with 2 petal-like and 1 functional with 1 fertile anther; ovary inferior; fruit a capsule.

GOLDEN CANNA
Canna flaccida Salisb.

Perennial, to 1.3 m tall; leaves 25-60 cm long; sepals 2-3 cm long, green or purple; petals 4-6 cm long, yellow; stamens 5-7 cm long, yellow; *capsules 4-6 cm long. Wet sites. Freq. all Fla. W to Tex., E to S.C. Spr-sum.

COMMON CANNA
Canna X generalis Bailey

Perennial, to 1 m tall; stems waxy; leaves 30-50 cm long; flowers red-yellow variegated; *sepals 2-2.5 cm long; petals to 7 cm long, longer than staminodal tube; stamens to 15 cm long; staminodal tube longer than sepals; *capsules 2-2.5 cm long. Native to tropical America. Escapes from cultivation. Low, wet disturbed sites. Rare SF, CF. Spr-sum.

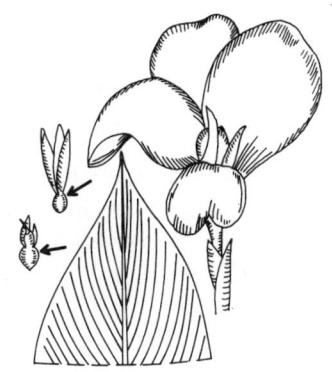

INDIAN SHOT
Canna indica L.

Perennial, to 1 m tall; *leaves 20-40 cm long; flowers red; *sepals 1-1.5 cm long; petals 3-3.5 cm long; stamens to 5 cm long; staminodal tube shorter than sepals; capsules 2.5-3.5 cm long. Native to tropical America. Escapes from cultivation. Low, wet sites. All Fla. W to Tex. Sum-fall.

CAPER FAMILY   *CAPPARACEAE*

Herbs, vines, shrubs or trees; leaves mostly alternate; flowers bisexual or unisexual (dioecious); sepals 4-8; petals 4-16 or 0; ovary superior; fruit a capsule or berry.

JAMAICA CAPER TREE
Capparis cynophallophora L.

(not shown)

Perennial shrub, to 6 m tall; upper stems with brown to red scales; leaves to 10 cm long, with brown to red scales beneath; petals to 1.5 cm long, to 3 cm wide, white; capsules to 20 cm long. Coastal sites. Infreq. SF, CF, NF. Spr-sum.

BAY-LEAVED CAPER TREE
Capparis flexuosa (L.) L.

Perennial shrub, to 8 m tall; leaves 4-8 cm long, smooth; petals 1-1.5 cm long, to 7.5 cm wide, white; capsules to 20 cm long. Coastal sites. Infreq. SF, CF, NF. Spr-sum.

SPIDER-FLOWER
Cleome spinosa L.
[*C. houtteana* Raf.]

Annual, 0.7-1 m tall; stems with glandular hairs; leaves divided into 5-7 leaflets; upper leaflets to 10 cm long; lower leaflets much smaller; *petals to 3 cm long, pink; capsules to 8 cm long. Native to South America. Disturbed, cultivated sites. Rare CF. W to Tex., N to Conn. Sum-fall.

PINELAND CATCHFLY
Polanisia tenuifolia T. & G.
[*Aldenella tenuifolia* (T. & G.) Greene]

Annual, 20-80 cm tall; leaflets 3, 1 or 2, 10-40 mm long, linear or filiform; flowers in racemes; sepals 4, reflexed; *petals 4, unequal in size, largest to 8 mm long, white; *capsules 4-6 cm long. Dry sites. Freq. all Fla. W to Miss., N to Ga. Spr-fall.

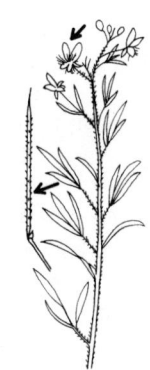

HONEYSUCKLE FAMILY   *CAPRIFOLIACEAE*

Shrubs, trees or woody vines; leaves opposite; flowers bisexual; sepals 3-5; petals 3-5; ovary inferior; fruit a berry, drupe or capsule.

JAPANESE HONEYSUCKLE
Lonicera japonica Thunb.

Perennial, trailing and/or high-climbing, semi-woody; leaves 2-8 cm long, evergreen; flowers paired, fragrant; sepals minute; petals 2.5-5 cm long, white, pink or yellow; *berries 5 mm in diam., black. Native to eastern Asia. Disturbed sites. Freq. all Fla. W to Tex., N to Mo., E to Del. Spr-sum.

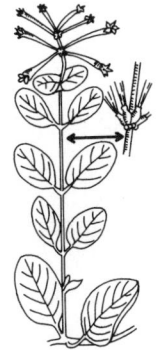

TRUMPET HONEYSUCKLE
Lonicera sempervirens L.

Perennial woody vine, to 5 m long; stems twining, trailing, smooth; leaves 3-7 cm long, with white film on lower surface; leaf pairs below flowers united around stem (connate perfoliate); *flowers in spikes, produced at leaf nodes; petals to 4.5 cm long, red or orange; berries red or orange. Various wooded sites. Freq. CF, NF, WF. W to Tex., N to Maine. Spr-sum.

ELDERBERRY or SOUTHERN ELDERBERRY
Sambucus canadensis L.
[*S. simpsonii* Rehd.]

Perennial shrub, 2-4 m tall; leaflets 3-11, 3-15 cm
long; petals 5, 3-7 mm wide, white; drupes 4-6 mm
long, purplish black, edible. Open moist to wet
sites. Common throughout Fla. W to Tex., N to
Canada. Spr-fall, all yr S.

SOUTHERN ARROWWOOD
Viburnum dentatum L.

(not shown)

Similar to *Viburnum nudum* except: leaves 3-8 cm
long, broad ovate, nearly as broad as long; leaf
margins sharply dentate; bases rounded, cordate;
petioles 10-30 mm long; drupes 5-8 mm long,
blue-black. Moist woods. Infreq. NF, WF. W to
Miss., N to Canada. Spr-sum.

POSSUM HAW
Viburnum nudum L.

Perennial shrub or small tree; leaves 5-15 cm
long; leaf margins crenate or entire; petioles 5-
15 mm long; drupes 6-10 mm long, blue.
Swamps, wet woods. W to Tex., N to Canada.
Spr-sum.

SMALL VIBURNUM
Viburnum obovatum Walt.

Perennial shrub or small tree, to 4 m tall; leaves
1.5-6 cm long; leaf margins entire or sparsely
toothed, short petiolate or sessile; petals 5, 4-7 mm
wide, white; drupes 6-8 mm long, black. Moist,
woody, swampy sites. Common all Fla. W to
Ala., E to S.C. Spr.

PINK FAMILY   *CARYOPHYLLACEAE*

Herbs; leaves mostly opposite; flowers bisexual; sepals 4 or 5; petals 4, 5 or 0; ovary superior; fruit a capsule or utricle.

CORN COCKLE
Agrostemma githago L.

Annual, 0.3-1 m tall; stems covered with long flattened hairs; leaves 4-12 cm long; sepals 5, 5-6 cm long, persistent; petals 5, to 3 cm long, red to purple; *capsules to 2 cm long. Native to Europe. Escapes from cultivation. Grain fields, disturbed sites. Rare NF. W to Tex., E to Ga. Spr-sum.

PINE-BARREN SANDWORT
Arenaria caroliniana Walt.

Perennial, 5-30 cm tall; stems prostrate, forming a basal cushion; leaves 3-12 mm long, overlapping, opposite; sepals 5, to 5 mm long; *petals 5, 5-12 mm long, white; stamens 10; *capsules 4-6 mm long, with valves entire. Dry sites. Infreq. WF. N to N.Y. Spr-sum.

WOOLLY SANDWORT
Arenaria lanuginosa (Michx.) Rohrb.

Perennial, 10-80 cm long; stems prostrate, spreading, hairy; leaves 10-30 mm long, elliptic or broad at tip; sepals 5, to 3 mm long; petals minute or 0; stamens 10; *capsules to 3.5 mm long. Shady, swampy sites. Infreq. all Fla. W to Tex., E to Va. Spr-fall.

THYME-LEAVED SANDWORT
Arenaria serpyllifolia L.

Annual, 10-40 cm tall; stems erect to lying flat;
leaves 2-7 mm long, ovate; sepals 5, to 4 mm
long; petals 5, to 2 mm long, white; stamens 10;
capsules to 2.5 mm long, with 2-lobed valves.
Disturbed sites. Rare CF, NF, WF. Throughout
U.S. Spr.

(not shown)

MOUSE-EAR CHICKWEED
Cerastium viscosum L.
[*C. glomeratum* Thuill.]

Annual, 5-30 cm tall, with sticky hairs; leaves
8-25 mm long, hairy; sepals 5; *petals 5, white,
pink or green; stamens 10; filaments smooth;
*capsules 5-9 mm long; flower stalk shorter than
capsule. Disturbed sites. Infreq. CF, NF, WF.
W to Tex., N to Canada. Spr-sum.

WEST INDIAN CHICKWEED
Drymaria cordata (L.) Willd. ex R. & S.

Annual, 10-60 cm long; stems branching, prostrate to ascending; leaves 5-20 mm long; sepals
5, 4-4.5 mm long; petals 5, to 2.2 mm long,
notched, white; stamens 5; styles 3; capsules to
3 mm long. Native to tropical America. Moist
lawns and pastures, other moist to wet sites.
Freq. all Fla. W to La. All yr.

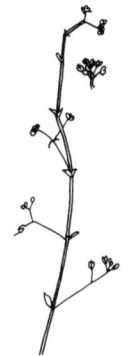

AMERICAN WHITLOW-WORT
Paronychia americana (Nutt.) Fenzl ex Walp.

Annual or perennial, to 80 cm long; stems
branching; leaves to 15 mm long; sepals 5, white,
with obscure cusp; sepals shorter than hypanthium; petals 0; utricles to 0.5 mm long. Scrub,
coastal sites. Infreq. all Fla. N to S.C. Spr-fall.

BALDWIN'S WHITLOW-WORT
Paronychia baldwinii (T. & G.) Fenzl

Annual or perennial, to 40 cm long; leaves 5-15 mm long; sepals 5, to 1 mm long, white, awned, much longer than hypanthium; petals 0; utricles to 1 mm long. Scrub, coastal sites. Infreq. all Fla. W to La., E to N.C. Spr-fall.

SAND PINE WHITLOW-WORT
Paronychia chartacea Fern.

(not shown)

Annual or biennial, 5-20 cm long; stems prostrate, branching; leaves 1.5-3 mm long; leaf margins revolute; sepals 4-5, to 0.5 mm long, hooded; petals 0; utricles to 0.4 mm long. Scrub. Infreq. CF. Sum.

HAIRY WHITLOW-WORT
Paronychia herniarioides (Michx.) Nutt.

Annual, 5-20 cm long; stems prostrate, branching or matting; leaves 0.3-1 cm long, elliptic, spine-tipped, hairy; flowers to 1.5 mm long; utricles 0.5-0.7 mm long. Scrub, sandhills. Infreq. CF, NF. N to N.C. Spr-fall.

(not shown)

SPREADING WHITLOW-WORT
Paronychia patula Shinners

Annual or perennial, 10-60 cm long; stems prostrate, branching; leaves 5-10 mm long; sepals 5, to 1.5 mm long, white, awned, cuspidate, longer than hypanthium; petals 0; utricles to 1.5 mm long. Dry sites. Infreq. CF, NF, WF. W to La., N to Ga. Spr-fall.

SAND SQUARES
Paronychia rugelii Shuttlew. ex Chapm.

Annual or biennial, 10-50 cm tall; stems erect,
branching; leaves 1-3 cm long, narrow; *flowers
to 2.5 mm long, pink; petals absent; *utricles less
than 1 mm long. Dry sites. Infreq. CF, NF, WF.
N to Ala., E to Ga. Sum-fall.

CHINESE SANDWORT
Polycarpaea corymbosa (L.) Lam.

(not shown)

Annual, 6-11 cm tall; stems corymbosely branched
above; flowers cymose, white; sepals to 3 mm
long, scarious. Dry sites. Rare (freq. locally) CF.
Spr-fall.

PEARLWORT
Sagina decumbens (Ell.) T. & G.

Annual, 2-15 cm tall; stems erect to ascending;
leaves 2-15 mm long; sepals 5 or 4, 1-2.5 mm
long; petals 0, 4 or 5, as long as sepals, white;
capsules 2-3 mm long. Open moist, disturbed
sites. Infreq. CF, NF, WF. W to Tex., E to Vt.,
W to Mo. and Ill. Spr-sum.

(not shown)

SLEEPY CATCHFLY
Silene antirrhina L.

Annual, 30-90 cm tall; stems smooth except for
sticky areas between leaf nodes; leaves 2.5-8 cm
long; sepals 5; *petals 5, 6-8 mm long, white or
pink; *capsules to 8 mm long. Open, disturbed
sites. Infreq. CF, NF, WF. W to Tex., N to
Canada. Spr-sum.

SMALL-FLOWERED CATCHFLY
Silene gallica L.

Annual, 10-45 cm tall; stems covered with shaggy hairs; leaves 15-30 mm long; sepals to 1 cm long, united; petals 5, 8-12 mm long, white or pink; stamens 10; *capsules 6-9.5 mm long. Disturbed sites. Rare NF, WF. N to Canada. Spr-fall.

SAND SPURREY
Spergularia marina (L.) Griseb.

Annual, 10-20 cm tall, branching; leaves 1-3 cm long; sepals 5; *petals 5, white or pink; stamens 2-5; *capsules 3-5 mm long. Salty flats. Rare CF. W to Tex., N to Canada. Spr-fall.

COMMON CHICKWEED
Stellaria media (L.) Cyrillo

Annual, 10-30 (or occasionally up to 80) cm long; stems lying flat; leaves 5-25 mm long; sepals 5, longer than petals; *petals 5, appearing 10 due to deep notch, 2.5-3.5 mm long, white; capsules to 5 mm long. Disturbed sites. Freq. all Fla. Throughout U.S. Spr, all yr southern peninsula Fla.

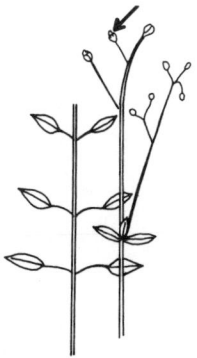

GULF CHICKWEED
Stellaria prostrata Baldw.
[*S. cuspidata* Willd.]

Annual, 10-60 cm long; stems lying flat, branching; leaves 5-20 mm long; sepals 2-3 mm long; petals to 5 mm long, white; stamens 10; *capsules 3-4 mm long. Moist sites. Rare CF, NF. W to Tex., N to Ga. Spr-sum.

WIRE WEED
Stipulicida setacea Michx.
[*S. filiformis* Nash]

Perennial, 5-22 cm tall; stems smooth; basal
leaves 4-15 mm long; upper leaves smaller; sepals 5, to 1.5 mm long; *petals 5, 1.5-2 mm long,
white; stamens 3-5; *capsules to 1 mm long. Dry
sites, scrub. Freq. all Fla. W to Miss., E to N.C.
Spr-fall.

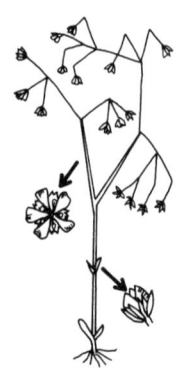

STAFF-TREE FAMILY   *CELASTRACEAE*

Shrubs or trees; leaves alternate or opposite; flowers bisexual
or unisexual; sepals 3-5; petals 3-5 or 0; ovary superior; fruit
a capsule, samara, drupe or berry.

HEARTS-A-BUSTIN'
Euonymus americanus L.

Perennial, to 2 m tall; leaves 2-10 cm long,
opposite; sepals 5; petals 5; flowers green-purple;
capsules 1-2 cm in diam., red, splitting when
mature. Rich woods, hammocks. Freq. CF, NF,
WF. W to Tex., E to N.Y. Spr-fall.

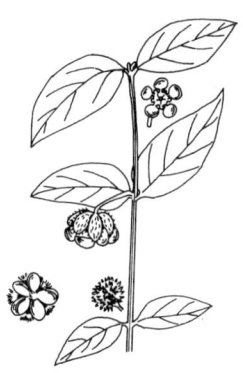

HORNWORT FAMILY   *CERATOPHYLLACEAE*

Aquatic herbs; leaves whorled; flowers unisexual (monoecious); sepals 0; petals 0; flowers enclosed by 8-12 bracts; ovary superior; fruit an achene.

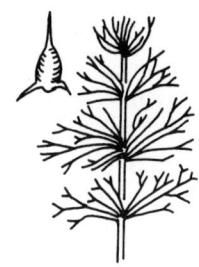

COON TAIL
Ceratophyllum demersum L.

Perennial, to 1 m long; stems several, branching; leaves 5-12 at a node, to 3 cm long, 1-, 2- or 3-forked, toothed; flowers small; achenes to 5 mm long, with 2 basal spines. Quiet water. Freq. all Fla. W to Tex., N to Canada. Sum, all yr S.

SPINY HORNWORT
Ceratophyllum muricatum Cham.
[*C. echinatum* Gray]

Perennial, 20-50 cm long; *leaves to 3 cm long, 2- or 3-forked, with *inconspicuous segment teeth; *achenes to 7 mm long, with 2 basal spines, also with lateral spines. Wet sites. Infreq. WF. W to Tex., N to Maine. All yr.

MUSK-GRASS FAMILY   *CHARACEAE*

Freshwater algae; roots, stems and leaves 0; short branches occurring at joints; branches not forked; sporangia borne on branches.

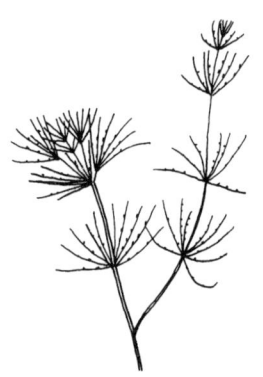

MUSK-GRASS
Chara species

Annual above frost line, perennial below; appearance like a flowering plant with cell-like divisions; plant submersed, to 18 inches tall, erect; stems usually rough to touch, often lime-encrusted; branches whorled; plants with musky odor. Hard water lakes, spring runs, ponds and ditches. Freq. all Fla. Throughout U.S. Spores found during spr and sum above frost line, all yr below.

GOOSEFOOT FAMILY   *CHENOPODIACEAE*

Herbs, shrubs or trees; leaves alternate, opposite or missing; flowers unisexual (monoecious); sepals 1, 3, 5 or 0; petals 0; ovary superior; fruit a utricle or nutlet.

LAMB'S-QUARTERS
Chenopodium album L.

Annual, 0.5-1.5 m tall; entire plant covered with mealy powder (farinose); leaves to 7 cm long, to 5 cm wide; margins wavy or entire; flowers and fruits in spikes; seeds to 1.5 mm in diam., shiny, black. Disturbed sites. Infreq. all Fla. Throughout U.S. Sum-fall.

MEXICAN-TEA
Chenopodium ambrosioides L.

Annual, 0.4-1.5 m tall; leaves 1.5-15 cm long, glandular-dotted, having pungent odor; sepals to 1 mm long, white. Dry disturbed sites. Common all Fla. W to Tex., N to Canada. Sum-fall.

NETTLE LEAVED GOOSEFOOT
Chenopodium murale L.

(not shown)

Annual, 10-60 cm tall; plant smooth to farinose; leaves to 8 cm long, to 5 cm wide, shiny above; margins coarsely toothed; flowers in panicles or cymes; seeds to 1.5 mm in diam., dull, black. Disturbed sites. Infreq. CF, NF, WF. W to Tex., N to Canada. Spr-fall.

ANNUAL GLASSWORT
Salicornia bigelowii Torr.

Annual, 10-40 cm tall; stems erect to ascending; leaves scale-like; flowers in spikes to 6 mm in diam. Brackish marshes. Infreq. all Fla. N to Canada, W to Calif. Sum-fall.

(not shown)

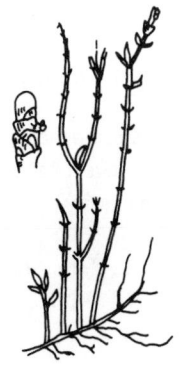

PERENNIAL GLASSWORT
Salicornia virginica L.

Perennial, 10-30 cm long; stems creeping or matting; leaves mere scales; flowers in spikes to 3 mm in diam. Brackish marshes. Freq. all Fla. W to Tex., N to N.H. Sum-fall.

COCO PLUM FAMILY — *CHRYSOBALANACEAE*

Trees or shrubs; leaves alternate; stipules present; flowers mostly bisexual; sepals 5; petals 5; ovary superior; fruit a drupe.

GOPHER APPLE
Licania michauxii Prance
[*Chrysobalanus oblongifolia* Michx.]

Perennial small shrub, 10-40 cm tall, with underground stems; leaves 3-12 cm long, evergreen, with tips retuse to mucronate; flowers to 2 mm long, white; drupes 2-3 cm long. Dry sites. Freq. all Fla. W to Miss., E to Ga. Spr, all yr S.

ROCK-ROSE FAMILY — *CISTACEAE*

Herbs or shrubs; leaves opposite, alternate or whorled; flowers bisexual; sepals 5; petals 3, 5 or 0; ovary superior; fruit a capsule.

CAROLINA ROCK-ROSE
Helianthemum carolinianum (Walt.) Michx.

Perennial, 5-25 cm tall; stems covered with long soft hairs; leaves 2-4 cm long, often forming basal rosette; both leaf surfaces covered with soft coarse hairs; flowers few; petals 5, to 3 cm long, yellow; *capsules to 1 cm long. Dry sites. Infreq. CF, NF, F. W to Tex., N to N.C. Spr-sum.

CLUSTERED ROCK-ROSE
Helianthemum corymbosum Michx.

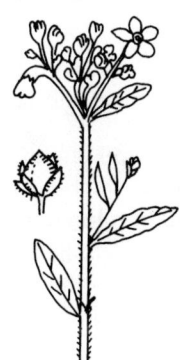

Perennial, 15-20 cm tall; stems covered with gray hairs; leaves 1.5-4 cm long, dark green on upper surface, pale green on lower surface; flowers of 2 kinds (closed with petals lacking or poorly developed; and opened with petals 5, 16-20 mm wide, yellow), in cymes; capsules 4-6 mm wide. Dry to moist sites. Common all Fla. N to N.C. Spr-sum, all yr S.

SCRUB ROCK-ROSE
Helianthemum nashii Britt.

(not shown)

Perennial, to 40 cm long; stems covered with gray hairs or smooth; leaves 1-3 cm long, pale green on both surfaces; flowers of 2 kinds (closed with petals lacking or poorly developed; and opened with petals 5, 16-20 mm wide, yellow), in panicles or racemes; capsules 3-3.5 mm long. Scrub. Freq. SF, CF, NF. Spr-sum.

DROOPING PINWEED
Lechea cernua Small

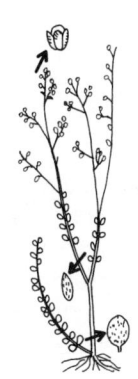

Perennial, 30-60 cm tall; stems smooth or covered with flattened hairs; leaves to 1 cm long, with soft gray hairs on upper and lower surfaces; flowers and fruits in groups of 2 or 3; outer sepals shorter than inner; capsules to 2 mm long, slightly exserted. Dry sites. Infreq. SF, CF. Endemic. Sum-fall.

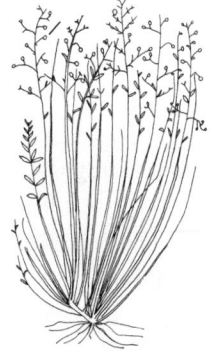

WOODY PINWEED
Lechea deckertii Small

Perennial, to 30 cm tall; stems woody below, branching, smooth or covered with flattened hairs; leaves 1.5-5 mm long, smooth above; flowers and fruits solitary; sepals hairy to smooth, with outer sepals shorter than inner; capsules longer than sepals; capsules to 1.5 mm long, exserted. Dry sites. Infreq. all Fla. N to Ga. Sum-fall.

HAIRY SHORT-STALKED PINWEED
Lechea divaricata Shuttlw. ex Britt.

Perennial, to 30 cm tall; stems branching,
ascending to erect, covered with spreading hairs;
leaves 4-8 mm long; outer sepals shorter than
inner; capsules longer than sepals; capsules to
2 mm long, exserted, clustered. Dry sites. Infreq.
SF, CF. Sum-fall.

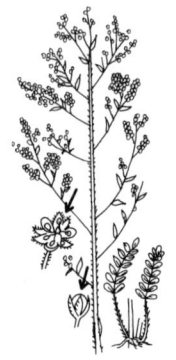

THYME-LEAVED PINWEED
Lechea minor L.

Perennial, 15-70 cm tall; stems smooth or covered
with flattened hairs; leaves 6-12 mm long, hairy;
outer sepals longer than inner; sepals equal in
length to or longer than capsules; *capsules to
1.5 mm long, not exserted. Dry sites. Infreq. all
Fla. W to La., N to Mass. Sum.

HAIRY LONG-LEAVED PINWEED
Lechea mucronata Raf.
[*L. villosa* Ell.]

Perennial, 30-90 cm tall; stems covered with
spreading hairs; leaves 1-3 cm long, with gray
hairs; outer sepals equal to or longer than inner;
sepals and capsules equal in length; capsules to
2 mm long, exserted, clustered. Dry sites. Infreq.
CF, NF, WF. W to Tex., N to Canada. Sum.

LEGGETT'S PINWEED
Lechea pulchella Raf.

Perennial, 30-80 cm tall; stems smooth or covered with flattened hairs; leaves 1-2.5 cm long;
outer sepals shorter than inner; capsules longer
than sepals; capsules to 1.5 mm long, slightly
exserted. Dry sites. Infreq. NF, WF. W to La.,
N to Mass. Sum-fall.

NARROW-LEAVED PINWEED
Lechea sessiliflora Raf.
[*L. patula* Legg.]

Perennial, 10-60 cm tall; stems smooth or covered with flattened hairs; leaves 4-9 mm long; flowers clustered; outer sepals nearly as long as inner; *capsules longer than sepals; capsules to 1.5 mm long, exserted. Dry sites. Common all Fla. N to Va. Sum-fall.

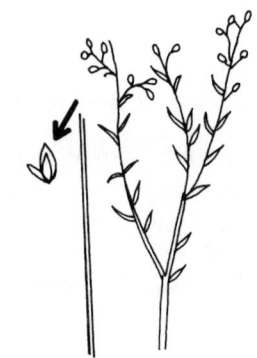

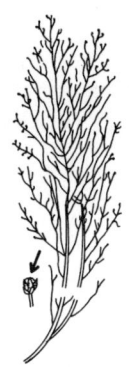

COMPACT PINWEED
Lechea torreyi Legg. ex Britt.

Perennial, 15-50 cm tall; stems branching, smooth or covered with flattened hairs; leaves 5-20 mm long, 0.5-1 mm wide, hairy on lower surface; flowers and fruits solitary; sepals hairy, with outer sepals shorter than inner; *capsules to 1.5 mm long, slightly exserted. Dry sites. Freq. all Fla. N to N.C. Spr-fall, all yr S.

WHITE MANGROVE FAMILY *COMBRETACEAE*

Trees, shrubs or vines, with woody stems; leaves alternate or opposite; flowers bisexual or unisexual; sepals 4 or 5; petals 4, 5 or 0; ovary inferior; fruits winged or wingless.

BUTTON WOOD
Conocarpus erectus L.

Perennial shrub, to 20 m tall; leaves 2-10 cm long, evergreen, smooth or covered with dense soft silver hairs; stamens 5-10; *flowers in spike-like heads, 9-14 mm in diam., green; *drupes 4-10 mm long, flattened. Coastal sites. Freq. SF, CF, WF. All yr.

WHITE MANGROVE
Laguncularia racemosa (L.) Gaertn. f.

Perennial shrub or small tree, to 20 m tall; leaves
2-7 cm long, opposite, evergreen, with *2 glands
on petiole; *flowers in spikes, 3-6 cm long, fragrant; sepals 5; petals 5, white; *drupes to 2 cm
long. Tidal swamps. Freq. SF, CF, NF. Spr.

SPIDERWORT FAMILY   *COMMELINACEAE*

Herbs; leaves alternate, sheathing; flowers bisexual; sepals 3;
petals 3; ovary superior; fruit a capsule.

BASKET PLANT
Callisia repens L.

Perennial, to 1 m long; stems creeping or ascending; leaves 15-35 mm long; sepals to 3 mm long;
petals shorter than sepals, white; capsules to
1.5 mm long. Native to tropical America.
Rarely escapes from cultivation. Disturbed sites.
All Fla. W to Tex. Sum-fall.

TROPICAL SPIDERWORT
Commelina benghalensis L.

Perennial, 25-40 cm high; stems sprawling;
leaves to 12 cm long; some leaf sheath hairs red;
petals, 2 blue and 1 white, each to 3 mm long.
Disturbed sites. Troublesome garden and crop
weed. Infreq. CF, NF. Sum-fall.

ASIATIC DAY-FLOWER
Commelina communis L.

Similar to *Commelina diffusa* except: may be
erect; petals, 2 blue and 1 white. Disturbed
areas, moist woods. Rare NF. W to Tex.,
E to Mass. Spr-sum.

(not shown)

SPREADING DAY-FLOWER
Commelina diffusa Burm. f.

Annual, 0.2-1 m long; stems lying flat, creeping,
branching, smooth; spathe margins free at base;
petals 3, blue; seeds 2.5 mm long. Disturbed sites.
Common all Fla. W to Tex., E to Del. Spr-fall.

ERECT DAY-FLOWER
Commelina erecta L.

Perennial, 20-90 cm tall; stems lying flat; leaves
3-15 cm long; petals, 2 blue and 1 white and
smaller; spathe 1.5-3 cm long, with fused base;
capsules 4-5 mm long. Dry sites, woods.
Common all Fla. W to Ariz., E to N.Y. Spr-fall.

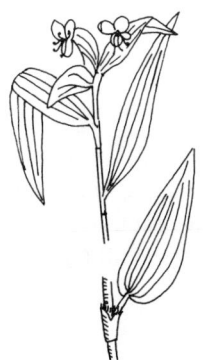

WOODS DAY-FLOWER
Commelina virginica L.

Perennial, 0.2-1.2 m tall; leaves to 20 cm long,
with rust-colored hairs on sheaths; spathes 1.5-
2.5 cm long, with base edges united; petals 3,
blue; capsules to 7.5 mm long. Low woods.
Infreq. NF, WF. W to Tex., E to N.J. Sum-fall.

GRASS-LEAF ROSELING
Cuthbertia graminea Small

Similar to *Cuthbertia rosea* except: leaves 1-4 mm
wide; *flowers to 2.5 cm wide. Dry sites, woods.
Infreq. CF, NF, WF. E to Ga. Spr-sum.

BIG ROSELING
Cuthbertia ornata Small

Perennial, to 60 cm tall; leaves to 30 cm long, to
2 mm wide; *flowers 2.5-3.5 cm wide, rose; capsules to 4 mm in diam. Scrub. Infreq. SF, CF,
NF. Endemic Fla. Spr-fall.

WIDE-LEAF ROSELING
Cuthbertia rosea (Vent.) Small

Perennial, 20-40 cm tall; leaves to 30 cm long, to
6 mm wide; *flowers to 4 cm wide, rose; *capsules to 4 mm in diam. Sandy sites. Infreq. WF.
N to N.C. Spr-sum.

DOVE WEED
Murdannia nudiflora (L.) Brenan

Annual, to 15 cm tall; leaves 5-10 cm long;
*terminal cymes bracted; petals blue, lavender or
purple; stamens 2-3; *capsules to 5 mm long.
Disturbed sites, turf. Common all Fla. E to N.C.
Spr-fall.

CREEPING WHITE SPIDERWORT
Phyodina cordifolia (Sw.) Rohw.
[*Leiandra cordifolia* (Sw.) Raf.]

Perennial, creeping, forming mats; leaves 1-2 cm
long; tepals 2-3 mm long, white; stamens 6;
**bracts of the cymules scale-like; capsules
enclosed by sepals. Moist shady sites. Infreq. SF,
CF, NF. Spr-fall.

WHITE-FLOWERED SPIDERWORT
Tradescantia fluminensis Vell.

Perennial, to 1 m long; stems spreading, ascending or erect; leaves 2-9 cm long, ovate-oblong;
*flowers cymose, lasting for one day or less;
sepals to 6 mm long; petals 8-10 mm long, white;
capsules to 3 mm long. Native to South America.
Cultivated and locally common as escape.
Disturbed sites, moist to wet woods. Infreq. CF,
NF, WF. N to N.C. Spr-fall.

HAIRY SPIDERWORT
Tradescantia hirsuticaulis Small

Similar to *Tradescantia ohiensis* except: 20-50 cm
tall; branches with white hairs; leaves 7-25 cm
long; sepals hairy; petals blue, rose. Dry sites.
Infreq. WF. W to Tex., E to N.C. Spr.

(not shown)

COMMON SPIDERWORT
Tradescantia ohiensis Raf.

Perennial, 20-80 cm tall; stems smooth, glaucous;
leaves 15-40 cm long; sepals hairy near apex
only; petals blue; capsules to 5 mm long. Open
sites, woods. Common CF, NF, WF. W to Tex.,
E to Mass. Spr-sum.

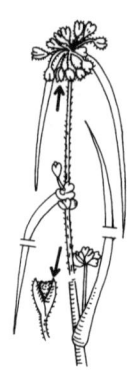

SCRUB SPIDERWORT
Tradescantia roseolens Small

Perennial, 30-60 cm tall; stems with glandular
hairs; leaves to 30 cm long; sepals 9-10 mm long,
having only glandular hairs; petals 1-2 cm long;
*capsules to 6 mm long. Dry sandy sites. Infreq.
CF, NF. N to S.C. Spr-sum.

ZIG ZAG SPIDERWORT
Tradescantia subaspera Ker.

Perennial, 0.3-1 m tall; stems often zigzag; leaves
10-35 cm long, linear; leaf sheaths white-ciliate;
sepals 6-8 mm long, hairy; petals 1-1.5 cm long,
white or blue; *capsules 4-5 mm long. Dry, rich
woods. Rare WF. W to La., N to Mo., E to Va.
Spr-sum.

WANDERING JEW
Zebrina pendula Schnizl.

Perennial, diffusely branching, creeping; leaves 3-
8 cm long, often purple-striped, mottled; flowers
in cymes, arising from a leaf-like structure; petals
5-8 mm long, purple, rose or white; capsules to
1.5 mm long. Native to tropical America.
Cultivated and escapes from cultivation. Moist
disturbed sites. Infreq. SF, CF, NF. Spr-fall.

(not shown)

DAISY FAMILY   *COMPOSITAE*

Herbs, shrubs or trees; leaves alternate, opposite or whorled; flowers bisexual or unisexual, in heads; involucre of bracts surrounding each head; flowers disc, ray or both types together; fruit tip with or without long scales or bristles; fruit an achene.

PARAGUAY STARBUR
Acanthospermum australe (Loefl.) Kuntze

Annual, 30-60 cm long; stems prostrate, slightly hairy; leaves 1-2.5 cm long; petioles slightly winged; heads consisting of white ray and disc flowers; involucral bracts prickly; achenes to 1 cm long, star-like. Disturbed sites. Freq. CF, NF, WF. W to La., N to Va. Spr-fall.

BRISTLY STARBUR
Acanthospermum hispidum DC.

Annual, to 1 m tall; stems erect, covered with dense fuzzy hairs; leaves 3-6 cm long; heads consisting of yellow ray and disc flowers; involucral bracts with 2 spines at tip; *achenes to 1 cm long, star-shaped, spiny. Disturbed sites. Infreq. CF, NF, WF. N to Ala., E to Ga. Spr-fall.

YARROW
Achillea millefolium L.

Perennial, 0.2-1 m tall; stems hairy to smooth; leaves 2-15 cm long, finely divided; ray flowers 5-10, 1-3 mm long, white; disc flowers to 2 mm long, yellowish white; achenes to 2 mm long. Dry sites. Infreq. CF, NF, WF. W to Ark., N to Canada. Spr-fall.

BUTTON OF GOLD
Acmella repens (Walt.) L. C. Rich.
[*Spilanthes americana* (Mutis ex L. f.) Hieron.]

Perennial, 20-80 cm long; stems lying flat, creeping to ascending; leaf blades 2-6 cm long; ray
and disc flowers yellow; achenes 1.5 mm long,
ray achene 3-angled, disc achene flattened pappus 0. Wet sites. Common all Fla. W to Tex.,
N to Mo., E to N.C. Spr-fall, all yr S.

COMMON RAGWEED
Ambrosia artemisiifolia L.

Annual, 0.2-2 m tall; stems erect, covered with
short hairs; leaves 5-15 cm long, once or twice
pinnatified; flowers monoecious; staminate
flowers 2-4 mm wide; achenes 3-4 mm long.
Disturbed sites. Common all Fla. W to Calif.,
N to Canada. Spr-fall.

COASTAL RAGWEED
Ambrosia hispida Pursh

Perennial, to 3 m long; stems prostrate, creeping,
covered with white densely packed upright hairs;
leaves tripinnatified; flowers monoecious; male
flowers 1-2 mm high; achenes 3.5-4 mm long.
Beaches, dunes. Infreq. SF, CF. Sum, all yr S.

(not shown)

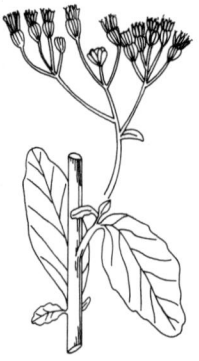

INDIAN PLANTAIN
Arnoglossum floridanum (A. Gray) H. Robins
[*Cacalia floridana* Gray]

Perennial, 0.7-1.2 m tall; leaves to 18 cm long,
mostly basal; involucral bracts 5, 10-12 mm long,
winged on midrib; disc flowers only, 5 per head;
petals white or yellow; achenes to 5 mm long;
pappus of many bristles. Dry sites. Freq. CF, NF,
WF. Spr-fall.

WETLAND INDIAN PLANTAIN
Arnoglossum ovatum (Walt.) H. Robins.

Perennial, 0.5-3 m tall; leaves to 30 cm long, mostly basal; involucral bracts 5, 8-10 mm long, not winged; disc flowers only, 5 per head; petals white; achenes 6-6.5 mm long; pappus of many bristles. Wet sites. Freq. all Fla. W to Tex., N to Ga. Sum-fall.

(not shown)

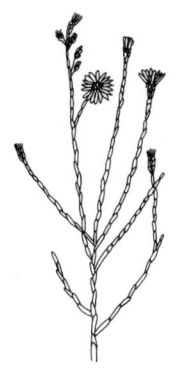

TIGHT-LEAVED ASTER
Aster adnatus Nutt.

Perennial, 20-80 cm tall; stems hairy; basal leaves to 2.5 cm long, oval; upper leaves reduced, ascending, occurring very close together on stem giving stem jointed appearance; ray flowers to 1.5 cm long, violet, blue or purple; disc flowers yellow; achenes to 3 mm long, smooth. Dry sites. Freq. all Fla. W to Miss., N to Ga. Fall.

SMALL-HEADED COASTAL ASTER
Aster bracei Britt. ex Small
[*A. tenuifolius* var. *aphylla* R. W. Long]

Similar to *Aster tenuifolius* except: to 1 m tall; lower leaves filiform, often absent; *ray flowers 5-7 mm long. Salt marshes. Infreq. CF, SF. W to Miss., N to Mass. All yr.

CLIMBING ASTER
Aster carolinianus Walt.

Perennial, 1-4 m long; stems woody, arching or climbing; leaves 2-10 cm long, clasping at base; ray flowers to 2 cm long, pink or lavender; achenes to 3 mm long. Wet sites. Freq. all Fla. N to S.C. Spr, fall.

CHAPMAN'S ASTER
Aster chapmanii T. & G.

Perennial, to 80 cm tall; stems smooth; leaves to
30 cm long, linear; ray flowers 1-2 cm long, violet, purple or blue; disc flowers yellow; achenes
to 4.5 mm long, smooth to hairy. Wet sites.
Infreq. NF, WF. N to Ala. Fall.

(not shown)

SILVER ASTER
Aster concolor L.

Perennial, 30-70 cm tall; stems hairy; leaves
0.8-4 cm long, oval to linear, with silky hairs on
upper and lower surfaces; *bracts with white,
appressed hairs; margins with stiff erect hairs; ray
flowers to 12 mm long, blue or pink; disc flowers
yellow; achenes to 2.5 mm long, hairy. Dry sites.
Infreq. all Fla. W to La., N to Mass. Sum-fall.

BUSHY ASTER
Aster dumosus L.
[*A. pinifolius* Alex.]

Perennial, 0.3-1.5 m tall; stems smooth; leaves to
12 cm long, linear, mostly deciduous below; *bracts
4-7 mm long, with white margins; ray flowers 4-11
mm long, white, lavender, purple or blue; disc flowers yellow or red; achenes 1.2-2.3 mm long, slightly
hairy. Dry to moist disturbed sites. Common all
Fla. W to Tex., N to Canada. Sum-fall.

ELLIOTT'S ASTER
Aster elliottii T. & G.

Perennial, 0.5-2.5 m tall; leaves to 25 cm long;
ray flowers 25-45, to 10 mm long, purple or
lavender. Swampy sites. Common all Fla.
N to Va. Sum-fall.

THISTLE-LEAF ASTER
Aster eryngiifolius T. & G.

Perennial, 30-80 cm tall; stems hairy; leaves persistent; basal leaves to 25 cm long, linear, grass-
like; upper leaves reduced; ray flowers 1-2 cm
long, white, lavender or blue; disc flowers yellow; achenes 1-2 mm long, smooth to hairy.
Low, wet sites. Freq. WF. N to Ala., E to Ga.
Sum.

(not shown)

SOUTHERN ASTER
Aster hemisphaericus Alex.

Perennial, 10-70 cm tall, colonial; *leaves to
20 cm long, scabrous; ray flowers to 2 cm long,
violet or purple; achenes to 2.5 mm long; pappus
to 5 mm long, tan. Dry sites. Rare WF. W to
Tex., N to Mo. Sum-fall.

CALICO ASTER
Aster lateriflorus (L.) Britt.

Perennial, to 1.2 m tall; leaves to 9 cm long, hairy
on midrib below; *ray flowers 9-14, to 6.5 mm
long, white to purple; disc flowers yellow; achenes to 2 mm long. Moist to wet sites. Infreq.
NF, WF. W to Tex., N to Canada. Sum-fall.

STIFF-LEAVED ASTER
Aster linariifolius L.

(not shown)

Perennial, 20-70 cm tall, tufted; leaves 1.5-3 cm
long, scabrous; ray flowers to 15 mm long, violet;
achenes to 3 mm long; pappus to 6 mm long,
tan. Dry sites. Infreq. NF, WF. W to Tex., N to
Maine. Sum-fall.

SWAMP ASTER
Aster paludosus Ait.

Perennial, 20-80 cm tall; stems smooth to hairy; leaves to 20 cm long, grass-like; ray flowers 1-2 cm long, blue or violet; disc flowers yellow; achenes 3-4 mm long, smooth to hairy. Moist sites. Ga., N to N.C. Fall.

(not shown)

CLASPING-LEAF ASTER
Aster patens Ait.

Perennial, 0.3-1.2 m tall; leaves 3-13 cm long, covered with short coarse, rough-to-the-touch hairs; ray flowers 18-30, 8-15 mm long, blue or pink. Sandhills, dry sites. Infreq. WF. W to Tex., N to Minn., E to Mass. Sum-fall.

FROST ASTER
Aster pilosus Willd.

Perennial, to l m tall, bushy; stems hairy; leaves 2-8 cm long, hairy, scabrous above; *ray flowers 16-35, to 1.5 cm long, white; disc flowers yellow; achenes to 1.3 mm long. Dry sites. Infreq. WF. W to Miss., N to Pa. Sum-fall.

YELLOW-HEADED WHITE-TOP ASTER
Aster reticulatus Pursh

Perennial, to 90 cm tall; stems hairy; leaves to 8 cm long; ray flowers 8-20, to 1.5 cm long, white; disc flowers yellow or red; achenes to 3 mm long; pappus to 7 mm long. Moist sites. Common all Fla. N to S.C. Spr-sum.

ARROW LEAVED ASTER
Aster sagittifolius Wedem. ex Willd.

Perennial, 0.6-1.5 m tall; leaves to 15 cm long;
*ray flowers to 8 mm long, light blue or purple;
achenes to 2 mm long; pappus to 4 mm long,
white. Dry sites. Infreq. WF. W to Tex.,
N to N.Y. Sum-fall.

BROAD LEAVED WHITE-TOP ASTER
Aster sericocarpoides (Small) K. Schum.
[*A. humilis* Willd. - misapplied;
A. umbellatus Mill. var. *latifolius* Gray]

Perennial, 1-2 m tall; stems smooth; leaves 5-
10 cm long; ray flowers 2-7, to 1 cm long; disc
flowers yellow; achenes to 3 mm long; *pappus
in 2 series. Swamps. Infreq. WF. W to Tex.,
N to N.C. Sum-fall.

LARGE-HEADED BUSHY ASTER
Aster simmondsii Small

Perennial, to 1.2 m tall; stems hairy; leaves to
12 cm long, linear to oval; ray flowers 7-11 mm
long, white, lavender or purple; disc flowers yellow; achenes to 2 mm long, hairy. Moist to dry
sites. Infreq. all Fla. Fall-wint.

ANNUAL MARSH ASTER
Aster subulatus Michx.
var. *ligulatus* Shinners
[*A. exilis* Ell.; *A. inconspicuous* Less.]

Annual, 0.3-1.2 m tall; stems smooth, fleshy;
leaves to 20 cm long, linear; ray flowers 0.1-1 cm
long, white, lavender or blue; disc flowers yellow;
achenes 1.5-2 mm long, hairy. Moist to wet sites,
salt and brackish marshes. Freq. all Fla. W to
Calif., N to Maine. Sum-fall.

PERENNIAL SALT MARSH ASTER
Aster tenuifolius L.

Perennial, 30-60 cm tall; stems smooth; leaves
4-15 cm long; ray flowers 10-25, to 1 cm long,
white to violet; disc flowers yellow; achenes to
4 mm long. Salt and brackish marshes. Freq. all
Fla. W to Tex., N to N.H. Sum-fall.

WHITE-TOP ASTER
Aster tortifolius Michx.
[*Sericocarpus bifoliatus* (Walt.) Porter]

Perennial, 0.3-1 m tall; stems hairy; leaves 1-3 cm
long, with upper surfaces hairy; ray flowers 3-7,
to 8 mm long, white; disc flowers white; achenes
to 3 mm long; pappus to 8 mm long. Dry, scrub
sites. Freq. all Fla. W to La., N to Va. Sum-fall.

WAVYLEAF ASTER
Aster undulatus L.

Perennial, 0.3-1 m tall; stems hairy; leaves to
14 cm long; ray flowers 6-12 mm long, violet;
achenes to 3 mm long. Dry sites. Infreq. CF, NF,
WF. W to Ark., N to Canada. Sum-fall.

SMALL WHITE ASTER
Aster vimineus Lam.

Perennial, 0.5-1.5 m tall; leaves 6-11 cm long; ray
flowers 17-30, to 6 mm long, white to purple;
disc flowers yellow; achenes to 2 mm long. Moist
sites. Infreq. WF, NF. N to Maine. Sum-fall.

DROOPING-LEAF ASTER
Aster walteri Alex.
[*A. squarosus* Walt.]

Perennial, 20-80 cm tall; stems rough-to-the-touch; leaves 2-3 cm long; basal leaves 2-9 mm long; upper leaves reflexed; ray flowers 18-24, 7-15 mm long, lilac to purple. Dry sites. Infreq. CF, NF, WF. N to N.C. Fall.

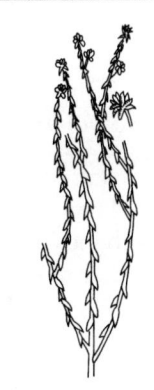

FALSE WILLOW
Baccharis angustifolia Michx.

Perennial, 0.5-3 m tall; stems smooth, branching; leaves 1-5 cm long, 1-5 mm wide, linear; male and female flowers on separate plants; involucres 4-5 mm high; flowers white to yellow; achenes 1-1.5 mm long. Coastal hammocks, beaches, flats. Freq. all Fla. N to N.C. Fall, all yr S.

STALKLESS GROUNDSEL BUSH
Baccharis glomeruliflora Pers.

Perennial shrub, to 2 m tall; leaves 3-4 cm long, 1-2 cm wide; flowers dioecious, in clusters spreading over entire branch, with 2-3 nearly stalkless heads per cluster; achenes to 1.5 mm long. Wet woods, marshes. Common all Fla. N to N.C. Sum-wint.

GROUNDSEL BUSH
Baccharis halimifolia L.

Perennial shrub to small tree, 1-4 m tall; leaves 3-7 cm long, 1-2 cm wide, with glands on lower side; male and female flowers on different plants (dioecious); flowers in terminal clusters, with 1-3 stalked heads per cluster; achenes to 1.5 mm long. Disturbed, low sites. Common all Fla. W to Tex., N to Mass. Sum-fall.

YELLOW BUTTONS
Balduina angustifolia (Pursh) Robins.

Annual, to 1 m tall; leaves linear, to 5 cm long; ray flowers to 2 cm long, yellow; disc flowers 4-5 mm long; achenes to 2 mm long; *pappus with round scales. Dry sites. Common all Fla. W to Miss., N to Ga. Sum-fall.

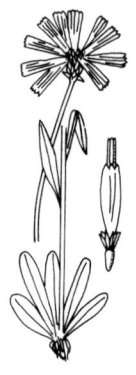

HONEYCOMB HEAD
Balduina uniflora Nutt.

Annual or biennial, 30-90 cm tall; stems erect, branching, hairy to smooth; basal and lower leaves to 10 cm long, deciduous; upper stem leaves reduced; ray and disc flowers yellow; receptacle honeycombed; pappus scales several, rounded. Low sites. Infreq. NF, WF. W to La., E to N.C. Sum-fall.

RED-CENTERED GREEN-EYES
Berlandiera pumila (Michx.) Nutt.

Perennial, 0.2-1.1 m tall; stems covered with woolly hairs, leafy; leaves 4-10 cm long, with gray woolly hairs beneath; ray flowers 1-1.5 cm long, yellow; disc flowers red; achenes to 6 mm long. Dry sites, open woods. Infreq. CF, NF, WF. W to Tex., E to S.C. Sum-fall.

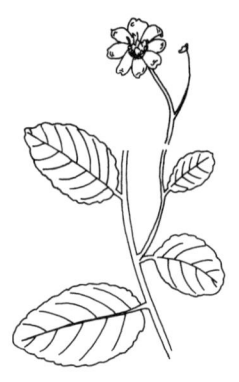

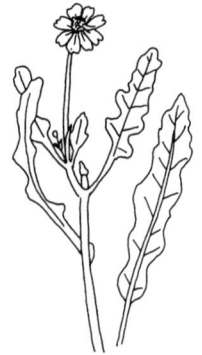

COMMON GREEN-EYES
Berlandiera subacaulis Nutt.

Perennial, 10-50 cm tall, scapose; leaves 4-12 cm long, pinnatifid; ray flowers 1-1.5 cm long, yellow; disc flowers yellow or green; achenes 5-6 mm long. Dry sites. Common SF, CF, NF. Endemic. Sum-fall, all yr S.

COMMON BEGGAR-TICK
Bidens alba (L.) DC.
[*B. pilosa* L.]

Annual, 0.1-1.2 m tall; leaves simple to compound, with segments 2-10 cm long; ray flowers
5, 1-2 cm long, white; achenes to 1.2 cm long;
pappus of 2-3 awns. Disturbed sites. Common
all Fla. W to Tex., N to N.C. Spr-fall, all yr S.

SPANISH NEEDLES
Bidens bipinnata L.

Annual, 0.3-1.7 m tall; leaves to 20 cm long,
divided 1-3 times; ray flowers to 5 mm long, yellowish; achenes to 1.3 cm long; *pappus of 3-4
awns. Moist disturbed woods and hammocks.
Freq. CF, NF, WF. W to Ariz., E to Va. Sum-fall.

SOUTHERN TICKSEED-SUNFLOWER
Bidens coronata (L.) Britt.

Annual, to 1.5 m tall; stems smooth to hairy;
leaves to 15 cm long, 3- to 7-divided, linear; ray   (not shown)
flowers 8, 1.5-3 cm long, yellow; disc flowers
yellow to orange; achenes to 7 mm long, ciliate,
with 2 scale-like awns. Brackish marshes. S.C.,
N to Wis., E to Maine. Sum-fall.

BUR MARIGOLD
Bidens laevis (L.) BSP.

Annual, 0.5-1.5 m tall; leaves to 12 cm long,
simple, sessile, with perfoliate base; ray flowers
8, 1.5-3 cm long, yellow; disc flowers yellow to
orange; achenes to 7 mm long; awns 4, 2 long
and 2 short. Wet sites. Freq. SF, CF, NF. W to
Ariz., E to N.H. Fall, all yr S.

MARSH BEGGAR-TICK
Bidens mitis (Michx.) Sherff

Annual, 0.3-1 m tall; stems smooth; leaves 7-12 cm long, simple to 3- to 5-divided, petioled; ray flowers 8, 1-3 cm long, yellow with purple or brown marks; disc flowers yellow to orange; bracts in 2 series; achenes 2-4 mm long, smooth, with minute awns. Moist to wet sites. Common CF, NF, WF. W to Tex., N to Md. Spr-fall.

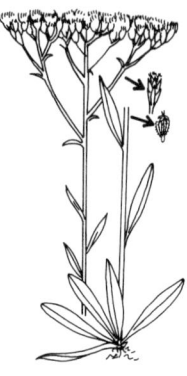

RAYLESS GOLDENROD
Bigelowia nudata (Michx.) DC.
[*Chondrophora nudata* (Michx.) Britt.]

Perennial, 20-80 cm tall; leaves 6-15, to 14 cm long, 4-14 mm wide; *flowers 2-5 per head, yellow; achenes to 1.7 mm long. Low wet sites. Infreq. all Fla. W to La., N to N.C. Sum-fall.

NARROW LEAF RAYLESS GOLDENROD
Bigelowia nuttallii L. C. Anderson

Similar to *Bigelowia nudata* except: leaves 12-23, 1-2 mm wide; achenes to 3.5 mm long. Dry rocky outcrops, sand pine scrub. Rare SF, CF, WF. Sum.

(not shown)

(not shown)

ASTER DAISY
Boltonia asteroides (L.) L'Her.

Perennial, to 2 m tall; leaves to 12 cm long; ray flowers to 15 mm long, pink or purple; disc flowers 7-10 mm wide; achenes to 2.5 mm long, winged; pappus with awns. Wet places. Rare WF. W to La., N to Minn. Sum-fall.

DOLL'S DAISY
Boltonia diffusa Ell.

Perennial, 1-2 m tall; leaves to 8 cm long; ray
flowers to 8 mm long, white or lilac; discs 4-5 mm
wide; achenes to 2.5 mm long, *winged; pappus
with awns. Low wet sites. Infreq. all Fla. W to
Tex., N to N.C. Fall.

SEA OXEYE
Borrichia aborescens (L.) DC.

(not shown)

Perennial, to 1.2 m tall; leaves 3-6 cm long,
opposite; heads with pointed scales, not spiny,
outer scales appressed; ray flowers yellow; discs
brown; achenes to 3 mm long. Coastal sites.
Infreq. SF. All yr.

SEA OXEYE
Borrichia frutescens (L.) DC.

Perennial, 0.15-1 m tall; leaves 2-6 cm long,
opposite; heads with spiny scales, outer scales
spreading; ray flowers yellow; discs 1-2 cm wide,
brownish yellow; achenes to 3.5 mm long.
Coastal sites. Common all Fla. W to Tex.,
E to Va. Sum-fall, all yr S.

Thick Leaves

FALSE BONESET
Brickellia eupatorioides (L.) Shinners

Perennial, 0.3-1.5 m tall; flower stems and bracts
hairy; leaves 1-10 cm long; flower heads 8-10 mm
high; disc flowers only, 7-33 per head; petals
white or yellow; achenes 3.5-5.5 mm long; pappus of 20 or more bristles. Open, wooded sites.
Infreq. all Fla. W to Miss., E to N.J. Sum-fall.

SPRAWLING HORSEWEED
Calyptocarpus vialis Less.

Perennial, 10-40 cm long; stems prostrate; leaves
1-5 cm long; ray flowers to 3 mm long, yellow;
achenes to 5 mm long. Disturbed areas. Infreq.
all Fla. W to Tex. All yr.

BALD-HEADED CARPHEPHORUS
Carphephorus carnosus (Small) James

Perennial, 30-80 cm tall; stems hairy; leaves 3-7 cm
long; petals violet; involucres to 6 mm long; achenes to 2.5 mm long. Low sites. Infreq. SF, CF.
Sum-fall.

LARGE-HEADED CARPHEPHORUS
Carphephorus corymbosus (Nutt.) T. & G.

Perennial, 30-90 cm tall; stem solitary, hairy;
leaves 5-15 cm long; flowers 15-20 per head;
heads over 1 cm long; petals to 9 mm long, pink
to purple, smooth; achenes 3-4 mm long. Dry,
moist to wet sites. Common all Fla. N to Ga.
Sum-fall.

DEER'S TONGUE
Carphephorus odoratissimus (J. F. Gmel.) Hebert
[*Trilisa odoratissima* (Walt.) Cass.]

Perennial, to 1.5 cm tall; stems smooth; basal
leaves 10-30 cm long, 1-12 cm wide; flowers 4-
10 per head, in corymbes; bracts oblong; petals
purple; achenes to 3.5 cm long; pappus barbed.
Moist sites. Freq. all Fla. W to La., N to N.C.
Sum-fall.

HAIRY TRILISA
Carphephorus paniculatus (J. F. Gmel.) Hebert
[*Trilisa paniculata* (J. F. Gmel.) Cass.]

Perennial, 0.2-1.7 m tall; stems with sticky hairs; basal leaves 5-25 cm long, to 3 cm wide, in rosettes; flowers 4-10 per head, in panicles; bracts rounded; petals purple, 4-5 mm long; achenes to 2 mm long; pappus barbed. Moist sites. Freq. all Fla. W to La., N to N.C. Fall-wint.

DENSE TOPPED CARPHEPHORUS
Carphephorus pseudoliatris Cass.

Perennial, 30-90 cm tall; stems hairy; leaves 12-34 cm long; flowers 20-35 per head, petals 6-9 mm long, lavender, pink or white; *achenes to 3 mm long. Moist to wet sites. Infreq. WF. W to La., E to Ga. Fall.

HAIRY INDIAN PAINTBRUSH
Carphephorus tomentosus (Michx.) T. & G.

Perennial, to 80 cm tall; stems 1 to many, hairy; leaves to 20 cm long; flowers 20-30 per head; petals to 9 mm long, pink to purple, with fine glandular hairs; achenes to 4 mm long. Pinelands, sandy woods. Ga., N to Va. Sum-fall.

(not shown)

PINELAND DAISY
Chaptalia dentata (L.) Cass.

Perennial; leaves to 11 cm long, with woolly hairs beneath; flower stems 10-30 cm tall, woolly; flower heads to 2 cm long, white; achenes to 4.5 mm long. Pinelands. Infreq. SF. Spr.

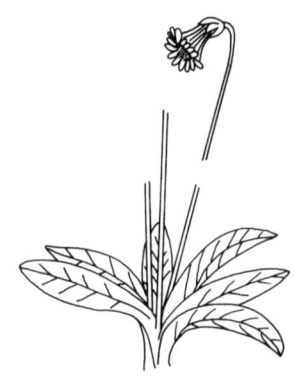

PINELAND DAISY
Chaptalia tomentosa Vent.

Perennial; leaves 4-15 cm long, with white felty
hairs beneath; flower stems 5-40 cm tall, hairy;
ray flowers 13-21, 0.7-1 cm long, pink to purple
below, white above; achenes to 6 mm long; pappus tawny. Moist, wet flatwoods. Freq. all Fla.
W to Tex., N to N.C. Spr.

FLORIDA GOLDEN ASTER
Chrysopsis floridana Small

Biennial or perennial, 30-70 cm long; stems erect
or ascending, branching, with white hairs; leaves
with white hairs; basal leaves 1-10 cm long; stem
leaves reduced; bracts lightly glandular; involucre 6-8 mm high; achenes to 2.5 mm long.
Scrub, sandhills. Rare CF. Endemic. Sum-fall.

HAIRY GOLDEN ASTER
Chrysopsis gossypina (Michx.) Ell.
subsp. *gossypina*
[*C. pilosa* (Walt.) Britt., misapplied]

Annual, 30-80 cm tall; stems erect, woolly; basal
leaves to 10 cm long; stem leaves to 1 cm long;
peduncles and bracts woolly; involucre 6-9 mm
high. Scrub, dry sites. Infreq. CF, NF, WF.
W to Ala., E to Va. Sum-fall.

FRINGED LEAF GOLDEN ASTER
Chrysopsis gossypina (Michx.) Ell
subsp. *hyssopifolia* (Nutt.) Semple
[*C. mixta* Dress]

Perennial or biennial, 20-70 cm tall; stems erect
or ascending, smooth or hairy; basal leaves 3-
10 cm long; stem leaves reduced; bracts eglandu-
lar; involucre 8-11 mm high; achenes 2.5-3 mm
long. Dry sites. Infreq. CF, NF, WF. W to Ala.
Sum-fall.

GOLDEN ASTER
Chrysopsis latisquamea Pollard

Biennial or perennial, 30-70 cm tall; stems erect,
unbranched to branched, with silvery hairs;
leaves with silvery hairs; basal leaves to 8 cm
long; stem leaves reduced; bracts stipate-glandu-
lar; involucre 8-11 mm high; achenes 2-3 mm
long. Pinelands. Infreq. CF, NF, WF. Sum-fall.

(not shown)

GOLDEN ASTER
Chrysopsis linearifolia Semple

(not shown)

Biennial, 0.3-2 m tall; stems erect, smooth; basal
leaves to 10 cm long; stem leaves reduced; bracts
smooth; involucre 6-12 mm high; achenes 2-2.5
mm long. Dry sites. Infreq. CF, NF, WF. Sum-fall.

MARYLAND GOLDEN ASTER
Chrysopsis mariana (L.) Ell.

Perennial, to 1 m tall; stems with cobweb-like
hairs; leaves to 18 cm long, with cobweb-like
hairs; involucre 8-10 mm high; ray flowers to 1 cm
long, yellow; achenes to 2 mm long. Dry sites,
woods. Infreq. CF, NF, WF. W to Tex., N to Ohio
and N.Y. Sum-fall.

ROUGH LEAF GOLDEN ASTER
Chrysopsis scabrella T. & G.

Biennial or perennial, 0.4-1 m tall; stems erect or
ascending, with glandular hairs; leaves with rough
glandular hairs; basal rosette leaves 4-10 cm long,
densely hairy; stem leaves reduced; bracts glandular; involucre 6-9 mm high; achenes 2-3 mm
long. Scrub, pinelands. Infreq. SF, CF, NF.
W to Ala. Sum-fall.

LEAFY GOLDEN ASTER
Chrysopsis subulata Small

Perennial, 0.3-1.1 m tall; stems with cobweb-like hairs; leaves to 7 cm long; involucre 6-8 mm high; ray flowers to 1 cm long, yellow; achenes to 2 mm long. Dry pinelands, scrub. Infreq. CF, NF. Fall.

ROADSIDE THISTLE
Cirsium altissimum (L.) Spreng.

Perennial, 1-4 m tall; stems hairy; lower leaves to 60 cm long, white tomentose beneath; involucres 25-35 mm long, spiny; petals lilac to purple. Various dry sites. WF. W to La., N to Minn., E to Mass. Sum-fall.

HORRIBLE THISTLE
Cirsium horridulum Michx.

Biennial, 0.2-1.5 m tall; leaves spiny, to 30 cm long; heads involucrate with spiny bracts; inner involucre 40-45 mm long; petals usually purple in Fla., but can be lavender, cream or yellow. Open sites. Common all Fla. W to Tex., E to Maine. Spr-sum.

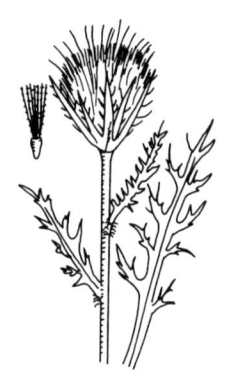

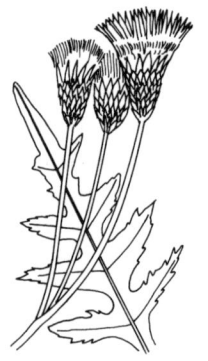

SWAMP THISTLE
Cirsium muticum Michx.

Biennial, 1-3 m tall; stems angled; leaves 17-40 cm long, spiny; bracts not spiny; involucres 25-35 cm wide; flowers purple or white. Swamps, wet sites. Infreq. WF, NF. W to Tex., N to Canada. Sum-fall.

NUTTALL'S THISTLE
Cirsium nuttallii DC.

Biennial, 1-3.5 m tall; stems winged below;
leaves to 25 cm long, often decurrent; involucre
20-25 mm long; petals lavender to pink. Dry to
moist sites. Common all Fla. W to Miss., E to
Va. Spr-fall, all yr S.

VIRGINIA THISTLE
Cirsium virginianum (L.) Michx.

Biennial, 0.5-2 m tall; stems with cobweb-like hairs;
leaves to 15 cm long, with gray felty hairs beneath;
leaf margins revolute; involucres 12-20 mm long;
heads corymbose; petals lilac to rose, purple.
Woods, low sites. Ga., N to N.J. Spr-fall.

HORSEWEED
Conyza canadensis (L.) Cronq.
var. *canadensis*

Annual, 0.1-2 m tall; stems hairy to smooth;
leaves 2.5-10 cm long, bristly ciliate; involucres
2-5 mm high; bract tips pale; ray flowers white or
yellow; disc flowers yellow; achenes 2-2.5 mm
long; pappus of many bristles. Disturbed sites.
Infreq. NF, WF. N to most of U.S. Sum-fall.

DWARF HORSEWEED
Conyza canadensis (L.) Cronq.
var. *pusilla* (Nutt.) Cronq.

Similar to *Conyza canadensis* var. *canadensis*
except: to 1.5 m tall; stems smooth to hairy;
leaves to 8 cm long, with ciliate margins at base;
bracts purple-tipped; ray flowers white or pink;
disc flowers white. Disturbed sites. Common all
Fla. N to most of U.S. Sum-fall.

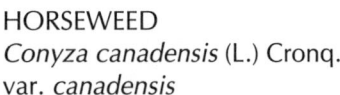

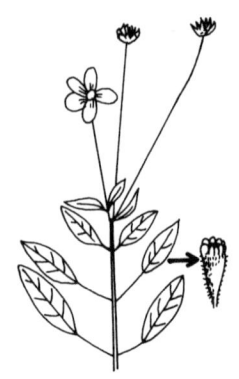

TICKSEED
Coreopsis auriculata L.

Perennial, 20-60 cm tall; stems hairy; leaves to
7 cm long; ray flowers to 3 cm long, yellow; disc
flowers to 1.5 cm wide, yellow; *achenes to 2.5
mm long, winged; pappus of 2 scales that fall off.
Dry sites. Miss., E to Tenn. Spr-sum.

DYE FLOWER
Coreopsis basilis (Otto & Dietr.) Blake
[*C. drummondii* (D. Don) T. & G.]

Annual or biennial, 20-70 cm tall; stems erect to
lying flat, hairy to smooth; leaves pinnatifid or
pinnately compound; leaf segments 1-5 cm long;
ray flowers yellow; disc flowers red to purple;
pappus scales 2, or absent. Native to Texas.
Cultivated and escapes. Dry sites. Infreq. CF,
NF, WF. W to Tex., N to Minn. Spr-sum.

FLORIDA TICKSEED
Coreopsis floridana E. B. Sm.

Perennial, to 1.2 m tall; leaves to 15 cm long,
alternate; ray flowers 2-3 cm long, yellow; disc   (not shown)
flowers to 6 mm long, yellow and purple-brown;
achenes to 5 mm long, with lobed wing; pappus
with 2 awns. Low, wet sites. Infreq. all Fla.
Endemic. Fall-wint.

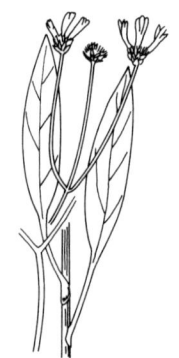

LONG-LEAF COREOPSIS
Coreopsis gladiata Walt.
[*C. longifolia* Small]

Perennial, 0.6-1.2 m tall; leaves 6-15 cm long,
alternate; ray flowers to 2.5 cm long, yellow; disc
flowers to 4 mm long, purple; achenes to 4.5 mm
long, winged or lobed; pappus with 2 awns. Low
wet sites. Common all Fla. W to Miss., N to N.C.
Sum-fall.

LARGE-FLOWER TICKSEED
Coreopsis grandiflora Hogg

Perennial, 0.3-1 m tall; stems smooth; leaves 5-8 cm long, dissected with linear segments; ray flowers 1.3-2.5 cm long, yellow; disc flowers 5-lobed, yellow; achenes to 2.5 mm long, winged, having tip with 2 awns. Dry sites. Rare NF, WF. W to Tex., E to Ga. Spr-sum.

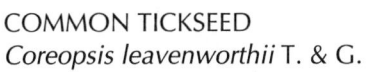

LANCE-LEAF TICKSEED
Coreopsis lanceolata L.

Perennial, 20-60 cm tall; leaves to 20 cm long, mostly entire; bracts to 1 cm long; ray flowers to 3 cm long, yellow; *achenes to 4 mm long, winged. Dry to moist sites. Rare WF. W to La., N to Canada. Spr-fall.

COMMON TICKSEED
Coreopsis leavenworthii T. & G.

Annual, to 1.5 m tall; leaves to 15 cm long, opposite; ray flowers 1-2 cm long, yellow; disc flowers to 3 mm long, black or purple; achenes to 3 mm long, broadly winged; pappus with 2 awns. Wet sites. Common all Fla. Endemic to Fla. Spr.

WHORLED-LEAF COREOPSIS
Coreopsis major Walt.

Perennial, 50-90 cm tall; leaf blades 3-9 cm long, *palmately lobed, sessile; bracts 5-6 mm long; ray flowers to 2.5 cm long, yellow; *achenes to 7 mm long. Dry woods. Infreq. WF. W to Miss., E to Va. Sum-fall.

YERBA-DE-TAGO
Eclipta alba (L.) Hassk.

Annual, to 1 m tall; stems lying flat, hairy; leaves
3-10 cm long; ray flowers many, 1-2 mm long,
white; achenes to 2.2 mm long; pappus minute.
Open, low wet sites. Common all Fla. W to Tex.,
N to Wis. Sum-fall, all yr S.

LEAFY ELEPHANT'S-FOOT
Elephantopus carolinianus Willd.

Perennial, to 70 cm tall; leaves to 25 cm long;
petals violet; *involucres 8-10 mm long; *achenes 3-4 mm long. Low woods. Infreq. CF, NF,
WF. W to Tex., N to N.J. Sum-fall.

FLORIDA ELEPHANT'S-FOOT
Elephantopus elatus Bertol.

Perennial, 0.7-1.2 m tall; stems hairy, subscapose;
rosette leaves 10-25 cm long, densely hairy on
lower surface; involucre 7-9 mm long, silky; achenes 3-4 mm long; pappus to 5 mm long. Dry
sites. Freq. all Fla. W to Miss., E to S.C. Sum-fall.

PURPLE ELEPHANT'S-FOOT
Elephantopus nudatus A. Gray

(not shown)

Perennial, 20-80 cm tall; plant sparsely hairy;
leaves to 20 cm long; petals purple; involucres to
8 mm long; achenes 2-3 mm long. Moist sites.
Freq. CF, NF, WF. W to Tex., N to Del. Sum-fall.

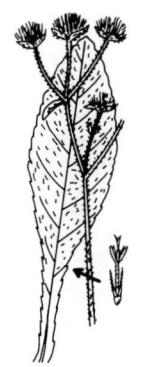

SOUTHEASTERN ELEPHANT'S-FOOT
Elephantopus tomentosus L.

Perennial, 20-70 cm tall; upper stems hairy;
*leaves 10-32 cm long, densely hairy beneath;
petals white to lavender; involucres 10-14 mm
long; achenes 4-5 mm long. Dry sites. Freq. WF.
W to Tex., E to Va. Sum-fall.

TASSEL FLOWER
Emilia fosbergii Nicols.
[*E. coccinea* of auth.]

Annual, 10-80 cm tall; stems hairy; leaves to
12 cm long, toothed; petals red, orange or rose;
bracts 10-12 mm long; achenes to 4 mm long.
Native to Old World Tropics. Disturbed sites.
Freq. SF, CF, NF. All yr.

RED TASSEL FLOWER
Emilia sonchifolia (L.) DC.

Annual, 10-50 cm tall; leaves to 10 cm long,   (not shown)
lobed, toothed; petals 7-8 mm long, purple or
lilac; bracts 8-10 mm long; achenes to 3 mm
long. Disturbed sites. Infreq. SF, CF, NF. All yr.

PHILADELPHIA FLEABANE
Erigeron philadelphicus L.

Biennial or perennial, to 70 cm tall; stems hairy;
leaves to 15 cm long; ray flowers 150-400, 5-10
mm long, pink to purple or white; bracts to 6 mm
long; achenes to 1 mm long. Low sites. Infreq.
to rare WF. Most of U.S. Spr-sum.

ROBIN'S PLANTAIN
Erigeron pulchellus Michx.

Biennial or perennial, 10-60 cm tall; stems hairy; leaves to 8 cm long; ray flowers 50-100, to 1 cm long, blue to pink or white; bracts to 7 mm long; achenes to 1.5 mm long. Disturbed, moist sites. Rare WF. W to Tex., E to Ga., N to Minn. Spr.

(not shown)

SOUTHERN FLEABANE
Erigeron quercifolius Lam.

Perennial, 10-80 cm tall; stems hairy; leaves to 14 cm long; ray flowers 100+, to 3.5 mm long, bluish white; disc flowers yellow; achenes to 7 mm long. Disturbed areas. Common all Fla. W to Tex., N to Md. Spr-sum.

DAISY FLEABANE
Erigeron strigosus Muhl.
[*E. ramosus* (Walt.) BSP.]

Annual, 0.2-1 m tall; leaves to 15 cm long; ray flowers 50-100, to 6 mm long, white, pink or bluish; achenes to 1 mm long; pappus with short bristles. Disturbed open sites. Freq. CF, NF, WF. W to Tex, N to Wis. Spr-sum.

MARSH FLEABANE
Erigeron vernus (L.) Torr. & Gray

(not shown)

Perennial, 10-60 cm tall; stems smooth, scape-like; leaves to 15 cm long, mostly basal; ray flowers 20-40, to 8 mm long, white or pink; achenes to 1.5 mm long; pappus with long bristles. Moist sites. Common all Fla. W to La., E to Va. Spr.

WHITE THOROUGHWORT
Eupatorium album L.

Perennial, 30-90 cm tall; stems with rough hairs;
leaves to 12 cm long, elliptic, lanceolate to
ovate; petioles to 6 mm long; heads 8-10 mm
high, with 5 flowers per head; bracts white,
smooth; petals white. Pinelands, woods. Infreq.
all Fla. W to La., N to Ohio, E to N.Y. Sum-fall.

WILD HOARHOUND
Eupatorium aromaticum L.

(not shown)

Perennial, 0.3-1.5 m tall; leaves 2-10 cm long,
sparsely hairy, sessile or short-petioled; margins
crenate; flowers 13-14 per head; bracts equal in
1 row (a few short), 4-5 mm long, ciliate; petals
4.5-5 mm long, white; achenes 2.5-3 mm long.
Dry sites. Infreq. WF. W to Miss., E to Pa. and
Mass. Sum-fall.

DOG FENNEL
Eupatorium capillifolium (Lam.) Small

Perennial, 0.9-4 m tall; leaves 1-1.5 mm wide,
sparsely punctate, not sticky; upper leaf segments
filiform; flower bracts smooth; petals to 2 mm
long, white; achenes 1-1.5 mm long. Disturbed,
moist sites. Freq. all Fla. W to Tex., E to N.J.
Sum-fall.

MIST FLOWER
Eupatorium coelestinum L.
[*Conoclinum coelestinum* (L.) DC.]

Perennial, 0.2-1.1 m tall; stems hairy; leaves 3-
12 cm long; *heads bearing only disc flowers,
purple to blue; *achenes to 1.5 mm long. Low
sites. Common all Fla. W to Tex., N to Ind.
Sum-fall.

DOG FENNEL
Eupatorium compositifolium Walt.

Perennial, 0.6-1.5 m tall; stems hairy; leaves 5-8 cm long, 1-2 mm wide, hairy, abundantly glandular-punctate, sticky, dissected; leaf segments nearly linear; flower bracts punctate, sticky; petals to 3 mm long, white; achenes to 1.5 mm long. Dry sites. Freq. all Fla. W to Tex., E to N.C. Sum-fall.

SMALL LEAF THOROUGHWORT
Eupatorium jucundum Greene
[*Ageratina jucunda* (Greene) Clewell & Wooten]

Perennial, 0.4-1.2 m tall; leaves 1.5-6 cm long, smooth; leaves on upper part of plant with distinct stalks; leaf margins sharp-toothed; flowers 10-19 per head; bracts to 3.5 mm long; petals to 4 mm long, white; achenes 2-3 mm long. Moist, shady sites. Common all Fla. Fall-wint.

DOG FENNEL
Eupatorium leptophyllum DC.

Perennial, to 2 m tall; stems smooth; leaves 1-2 mm wide; flower heads on 1 side of branch; branches recurved, ascending; petals to 3 mm long; achenes to 1.5 mm long. Dry to moist, wet sites. Freq. all Fla. W to Miss., E to N.C. Sum-fall.

WHITE BRACTED THOROUGHWORT
Eupatorium leucolepis (DC.) Torr. & Gray

Perennial, 0.3-1 m tall; stems with rough hairs; leaves 3-10 cm long, linear to lanceolate, smaller at lower end of stem, increasing in size up to midstem and becoming smaller again past midstem; petioles to 0.2 mm long; heads 6-8 mm high, with 5 flowers per head; bracts white, hairy; petals white. Pinelands, bogs. Freq. NF, WF. W to Tex., E to Mass. Sum-fall.

SEMAPHORE EUPATORIUM
Eupatorium mikanioides Chapm.

Perennial, 0.5-1 m tall; stems smooth or with short
hairs; leaf blades 2-4 cm long, vertical, fleshy, glandular-punctate; margins crenate; flowers 3-5 per
head; bracts 4-5 mm long, hairy, glandular; petals
to 4 mm long, white or pink; achenes to 1 mm
long. Low coastal areas. Freq. all Fla. Sum-fall.

(not shown)

MOHR'S EUPATORIUM
Eupatorium mohrii Greene
[*E. recurvans* Small]

Perennial, 0.4-1.5 m tall; stems hairy; leaves 1.5-
4 cm long, reflexed, hairy, glandular-punctate;
margins entire to blunt-toothed; flowers 3-5 per
head; bracts 3-4 mm long, with glandular hairs;
petals 3-4 mm long, white; achenes to 2 mm
long. Pinelands. Freq. all Fla. W to Miss.,
E to Va. Sum-fall.

BITTER BUSH
Eupatorium odoratum L.

Perennial, 0.6-3 m tall; stems woody, hairy;
leaves 4-12 cm long, lanceolate to ovate; flowers
20-35 per head; petioles 3-12 mm long; involucre
9-10 mm high, chartaceous; bracts green-tipped;
petals white to purple. Hammocks. Infreq. SF,
CF. All yr.

(not shown)

BONESET
Eupatorium perfoliatum L.

Perennial, 0.3-1.5 m tall; stems hairy; leaves 5-
25 cm long, lanceolate, fused together at stem
base; flowers 7-15 per head; involucre 3.5-6 mm
high; bracts white, hairy; petals white. Moist to
wet sites. Infreq. CF, NF, WF. W to Tex., N to
Canada. Sum-fall.

HAIRY THOROUGHWORT
Eupatorium pilosum Walt.

Perennial, 0.5-1.5 m tall; stems hairy; leaves 2-12 cm long, sparsely hairy; leaf margins irregularly toothed; flowers 5 per head; bracts 3-6 mm long, overlapping in several series; petals to 3 mm long, white; achenes to 3 mm long. Dry to moist, wet sites. Infreq. NF, CF, WF. W to Miss., E to Pa. and N.J. Sum-fall.

FALSE HOARHOUND
Eupatorium rotundifolium L.

(not shown)

Perennial, 0.4-1.2 m tall; stems densely hairy; leaves to 8 cm long, sessile, hairy, glandular; margins crenate-dentate; bracts to 6 mm long, hairy, glandular, overlapping in several series; petals to 3 mm long, white; achenes 2-2.5 mm long. Dry sites, bogs, swamps. Freq. CF, NF, WF. W to Tex., E to Maine. Sum-fall.

WHITE SNAKEROOT
Eupatorium rugosum Houtt.

Perennial, 0.2-1.5 m tall; leaves to 18 cm long, acuminate; petioles to 6 cm long; flowers 9-34 per head; bracts 4-6 mm long; petals white. Woods, pastures, floodplains. Common NF. W to Tex., N to Canada. Sum-fall. Poisonous to grazing animals.

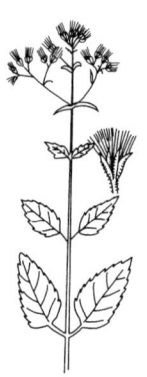

LATE BONESET
Eupatorium serotinum Michx.

Perennial, 0.8-3 m tall; stems hairy; leaves to 25 cm long, lanceolate; petioles 1-4 cm long; heads 4-6 mm high, with 10-15 flowers per head, white, hairy; petals white. Moist to wet, natural and disturbed areas. Freq. all Fla. W to Tex., N to Wis., E to N.Y. Sum-fall.

FLAT-TOPPED GOLDENROD
Euthamia tenuifolia (Pursh) Greene
[*E. minor* (Michx.) Greene]

Perennial, to 1.2 m tall; leaves to 8 cm long, 3-5 mm wide, with 1-5 veins; inflorescence flat-topped; involucre 3.5-6 mm high; ray flowers 7-16, to 2 mm long, yellow. Pinelands. Freq. all Fla. W to Miss., N to Mass. Sum-fall.

FLORIDA YELLOW-TOP
Flaveria floridana J. R. Johnston

(not shown)

Similar to *Flaveria linearis* except: leaves to 1.5 cm wide; flowers 10-19 per head. Coastal areas. Infreq. SF, CF, NF. Sum-fall.

YELLOW-TOP
Flaveria linearis Lag.

Perennial, to l m tall; stems smooth; leaves 2.5-10 cm long, to 6 mm wide; bracts 5-6, 3-4 mm long; flowers 5-10 per head, yellow; ray flower 1, 2.5-3 mm long, yellow. Coastal areas. Infreq. all Fla. Spr-fall.

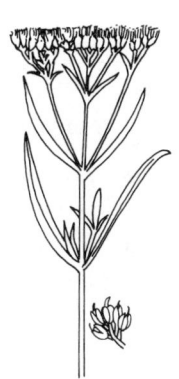

STALKLESS YELLOW-TOP
Flaveria trinervia (Spreng.) C. Mohr

Annual, 0.2-1.1 m tall, branching; leaves 3-10 cm long, to 2.5 cm wide; bracts 1 or 2, to 4 mm long; flower 1 per head, ray or disc, yellow; ray flower 1, to 1 mm long; achenes to 2 mm long. Native to tropical America. Pinelands and disturbed sites. Infreq. SF, CF. Sum-fall.

SMOOTH-HEADED BLANKET FLOWER
Gaillardia aestivalis (Walt.) H. Rock
[*G. lanceolata* Michx.]

Annual or biennial, 30-70 cm tall; stems branching, erect, hairy; basal leaves 5-10 cm long; stem leaves smaller; ray flowers to 2 cm long, reddish yellow; disc flowers purple to reddish brown; pappus scales 7-10; achenes to 2 mm long. Dry sites. Infreq. CF, NF, WF. W to Tex., N to Ill., E to N.C. Spr-fall.

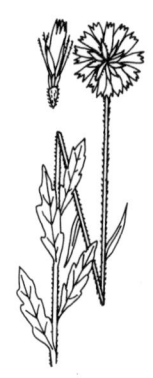

BLANKET FLOWER
Gaillardia pulchella Foug.

Annual or perennial, to 60 cm tall; stems lying flat, hairy; leaves 3-8 cm long; ray flowers to 2 cm long, red to purple, with or without yellow tips; achenes to 2 mm long. Dry sites, often along coasts. Infreq. all Fla. W to Tex., N to N.C. Spr-fall.

GARBERIA
Garberia heterophylla (Bartr.) Merr. & Harp.

Perennial shrub, 1-2 m tall; leaves 2-3 cm long, evergreen; flower heads to 1 cm high, rose-purple; achenes 7-8 mm long, surrounded at tip by brown bristles (pappus). Sand pine scrub. Freq. CF, NF. Endemic. Spr-fall.

SWEET EVERLASTING
Gnaphalium obtusifolium L.

Annual, 20-80 cm tall; stems covered with woolly hairs; leaves to 8 cm long, with white woolly hairs beneath; flower heads 6-7 mm long, white. Dry, disturbed, open sites. Freq. all Fla. W to Tex., N to Canada. Sum-fall.

WANDERING CUDWEED
Gnaphalium pensylvanicum Willd.

Annual, 20-40 cm tall; stems covered with cob-web-like hairs; leaves 2-8 cm long, thin, with gray woolly hairs beneath; flower heads to 2 cm long, white, woolly. Disturbed sites. Common all Fla. W to Tex. Spr-sum.

PURPLE CUDWEED
Gnaphalium purpureum L.

Annual, to 50 cm tall; stems covered with white woolly, felty hairs; leaves to 10 cm long, with white hairs beneath; flower heads to 1 cm long, white. Disturbed sites. Common all Fla. W to Tex., N to Maine. Spr-sum.

SCRATCH DAISY
Haplopappus divaricatus (Nutt.) Gray

Annual or biennial, 0.3-1.2 m tall; stems branching, hairy; leaves spine-tipped; basal leaves 2-10 cm long; stem leaves reduced; ray and disc flowers yellow; *pappus 1 series of bristles. Dry sites. Infreq. CF, NF, WF. W to Tex., E to N.C. Sum-fall.

CAMPHOR DAISY
Haplopappus phyllocephalus DC.
[*H. megacephalus* (Nash) Small]

Biennial or perennial, 0.8-1.2 m tall; leaves 2-7 cm long; bracts in 2 to several series; ray flowers to 1.5 cm long, yellow; achenes to 2 mm long. Sandy disturbed and coastal sites. Infreq. SF, CF, NF. W to Tex. Sum-fall.

BITTERWEED
Helenium amarum (Raf.) H. Rock

Annual, 20-50 cm tall; stems smooth; leaves 2-
7 cm long, wingless; ray flowers to 1.2 cm long,
yellow; disc flowers 0.6-1.2 cm wide, yellow;
achenes to 1 mm long; pappus scales 6-8.
Disturbed sites. Common all Fla. W to Tex.,
E to Va. Spr-fall.

FALL SNEEZEWEED
Helenium autumnale L.

(not shown)

Perennial, to 1.5 m tall; leaves to 15 cm long,
forming wings at attachment to stem; ray flowers
to 2.5 cm long, yellow; disc flowers to 2 cm
wide, yellow; achenes to 2 mm long, hairy; pappus scales 5-10. Moist sites. Infreq. WF. W to
Tex., N to Canada. Spr-fall.

BOG SNEEZEWEED
Helenium brevifolium (Nutt.) Wood

Perennial, to 80 cm tall; basal leaves to 17 cm
long; stem leaves reduced, slightly winged; ray
flowers to 2 cm long, yellow; disc flowers to 2 cm
wide, red to brown or purple to brown; achenes
to 1.5 mm long, hairy; pappus scales 5-10. Bogs.
Common WF. W to La., N to N.C. Spr-sum.

(not shown)

WINGED SNEEZEWEED
Helenium flexuosum Raf.

Perennial, 0.4-1 m tall; rosette and lower leaves
to 14 cm long, usually shed; ray flowers to 2 cm
long, yellow to brown, with purple bases; disc to
15 mm wide, reddish brown or purplish; achenes
to 1 mm long. Moist to wet sites. Infreq. CF, NF,
WF. W to Tex., N to Mich. Spr-fall.

SWAMP SNEEZEWEED
Helenium pinnatifidum (Nutt.) Rydb.

Perennial, 0.3-1 m tall; stems hairy; rosette leaves 3-20 cm long; stem leaves shorter, forming wings at attachment to stem; wings to 5 mm long; ray flowers to 1.5 cm long, yellow; disc flowers 1-2.5 cm wide, yellow; achenes to 1.5 mm long, hairy; pappus scales 5-10. Wet sites. Common all Fla. W to La., N to N.C. Spr-sum.

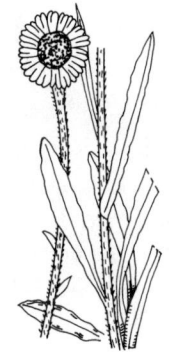

SMOOTH SWAMP SNEEZEWEED
Helenium vernale Walt.

(not shown)

Similar to *Helenium pinnatifidum* except: stems smooth; rosette leaves to 30 cm long; stem leaves forming wings to 2 cm long; achenes smooth. Wet sites. Common SF, CF, NF. Spr.

ANNUAL WILD SUNFLOWER
Helianthus agrestis Pollard

Annual, to 2 m tall; stems smooth; leaves to 15 cm long, to 4 cm wide; flower heads 1-3 cm in diam.; ray flowers to 2 cm long, yellow; achenes to 4 mm long. Scrub or moist sites. Infreq. SF, CF, NF. N to Ga. Sum-fall.

NARROW-LEAVED SUNFLOWER
Helianthus angustifolius L.

Perennial, to 2 m tall; stems with short to long hairs; leaves to 20 cm long, to 1 cm wide; upper leaves alternate; lower leaves opposite; leaf margins rolled under; ray flowers 10-12, 1-4 cm long, yellow; discs to 1.5 cm in diam., purple-red; achenes to 2.7 mm long. Disturbed, moist to wet sites. Freq. all Fla. W to Tex., N to Ohio, E to N.Y. Sum-fall. Hybridizes with *H. floridanus*.

ANNUAL SUNFLOWER or COMMON SUNFLOWER
Helianthus annuus L.

Annual, 0.5-3 m tall; stems with rough hairs; leaves 4-40 cm long, to 15 cm wide; flower heads 2-10 cm in diam.; ray flowers to 5 cm long, yellow; achenes 0.5-1.5 cm long. Escapes from cultivation. Disturbed sites. Infreq. all Fla. W to Tex., N to Canada. Sum-fall.

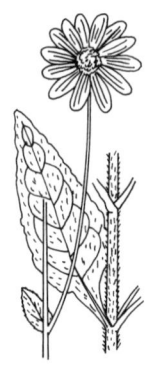

HAIRY WOOD SUNFLOWER
Helianthus atrorubens L.

Perennial, 0.5-2 m tall; lower portion of stem hairy; mainly stem leaves 6-20 cm long, 3-10 cm wide, opposite, ovate-elliptic, rough; basal petioles winged, as long as blade; other leaves reduced up stem; ray flowers 1-3 cm long, yellow; disc flowers dark red; achenes 3-3.5 mm long. Shady, low sites. Infreq. WF. N to Ga., W to La., N to Va. Sum-fall.

SMOOTH BEACH SUNFLOWER
Helianthus debilis Nutt.
subsp. *cucumerifolius* (Torr. & Gray) Heiser

Annual, 0.5-2 m tall; stems smooth, erect to lying flat, red- to purple-mottled; leaves 4-10 cm long; leaf margins entire to toothed; ray flowers 1-3 cm long, yellow; discs 15-20 mm wide, red to purple; phyllaries 1-3 mm wide; achenes 3-7 mm long. Disturbed and coastal sites. Rare CF, NF, WF. W to Tex., E to N.C. Sum-fall.

BEACH SUNFLOWER
Helianthus debilis Nutt.
subsp. *debilis*

Annual, 1-1.5 m tall, or to 10 cm long; stems lying flat; branches ascending to erect, red-, brown-, purple-mottled, covered with short hairs; leaves irregularly toothed; ray flowers to 1.5 cm long, yellow; discs to 15 mm wide, red to purple. Coastal sites. Freq. CF, NF, WF. N to Ga. All yr.

HAIRY BEACH SUNFLOWER
Helianthus debilis Nutt.
subsp. *vestitus* (E. E. Wats.) Heiser

Annual, to 1 m long; stems lying flat, ascending
to erect, covered with long shaggy white hairs;
leaf blades 5-8 cm long, with strongly toothed
margins; ray flowers 10-20, 1-3 cm long, yellow;
discs to 15 mm wide, red-purple; achenes 3.5-
4 mm long. Dry, coastal sites. Infreq. CF.
W to Tex., N to Ga. All yr.

FLORIDA SUNFLOWER
Helianthus floridanus Gray

Perennial, 1-2 m tall; stems covered with short
hairs; mostly all stem leaves 5-8 cm long, 1-3 cm
wide, with wavy margins; heads 5-6 cm wide;
ray flowers yellow; disc flowers with red-purple
or yellow lobes; achenes to 3 mm long. Moist
sites. Infreq. CF, NF, WF. W to La., E to S.C.
Sum-fall.

BOG SUNFLOWER
Helianthus heterophyllus Nutt.

Perennial, 0.4-1.2 m tall; lower portion of stem
rough, hairy; basal leaves 6-30 cm long, 1-4 cm
wide, opposite, oblanceolate to linear; stem leaves
0.5-2 cm wide, alternate, linear; ray flowers 3-5
cm long, yellow; disc flowers dark purple; heads
6-8 cm wide; achenes 4-5 mm long. Moist to wet
sites. Freq. WF. W to La., E to N.C. Sum-fall.

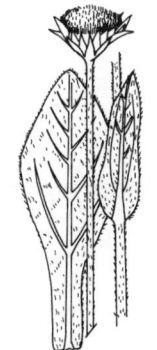

RAYLESS SUNFLOWER
Helianthus radula (Pursh) Torr. & Gray

Perennial, 0.5-1 m tall; stems hairy; rosette leaves
4-12 cm long; stem leaves to 7 cm long; ray flowers when present to 1.5 mm long, dark purple; disc
to 3 cm wide, dark purple; achenes to 4.5 mm
long. Flatwoods. Freq. CF, NF, WF. W to La.,
N to S.C. Sum-fall.

CAMPHOR WEED
Heterotheca subaxillaris (Lam.) Britt. & Rusby

Annual or biennial, 0.3-1 m tall; stems ascending
to erect, covered with dense short hairs; basal
leaves 1-7 cm long; stem leaves reduced; *ray
and disc flowers yellow; pappus of disc flowers
double, with outer series shorter than inner; ray
flowers lacking pappus. Open dry sites. Freq. all
Fla. W to Tex., E to Del. Sum-fall, all yr S.

HAWKWEED
Hieracium gronovii L.

Perennial, 0.2-1.2 m tall; stems hairy; leaves to
18 cm long, simple, hairy; flower heads consisting of ray flowers only, yellow, orange or white;
*achenes 3-3.5 mm long, with white bristles
(pappus) covering tip. Dry sites. Freq. CF, NF,
WF. W to Tex., N to Canada. Sum-fall.

LARGE-HEAD HAWKWEED
Hieracium megacephalon Nash

Perennial, 20-70 cm tall; stems hairy; leaves to
17 cm long; bracts in 2-3 series; ray flowers to
1 cm long, yellow; achenes to 5 mm long.
Pinelands. Freq. all Fla. N to Ga. Spr-fall.

MARSH ELDER
Iva frutescens L.

Perennial shrub, to 3.5 m tall; leaves to 10 cm
long, with toothed margins; petals 1-2 mm long,
white; achenes to 2.5 mm long. Salt marshes.
Freq. all coastal Fla. W to Tex., E to Va. Sum-fall.

DUNE ELDER
Iva imbricata Walt.

Perennial shrub, 0.3-1 m tall; leaves 2-5 cm long, with almost entire margins; petals to 1.5 mm long, white; achenes 3.5-5 mm long. Coastal dunes. Freq. all coastal Fla. W to Tex., E to Va. Sum-fall.

(not shown)

NARROW-LEAVED ELDER
Iva microcephala Nutt.

Annual, 0.3-1 m tall; stems smooth to slightly hairy; leaves 2-6 cm long, linear, glandular-punctate; male and female flowers separate but on same plant (monoecious); heads less than 5 mm long; flowers green; bracts 4-5, to 2 mm long, distinct; achenes to 1 mm long. Low moist to wet sites. Freq. all Fla. N to Ala., E to S.C. Sum-wint.

SERINIA
Krigia cespitosa (Raf.) Chambers
[*K. oppositifolia* Raf.]

Annual, 5-30 cm tall; leaves 2-12 cm long; bracts in 1 series; ray flowers to 1 cm long, yellow to orange; achenes to 1.5 mm long. Floodplains, moist disturbed sites. Rare WF. W to Miss, N to N.C. Spr.

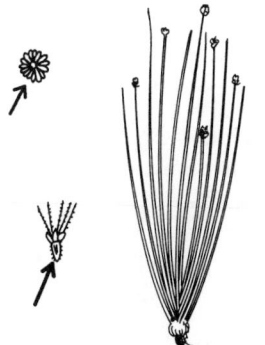

DWARF DANDELION
Krigia virginica (L.) Willd.

Annual; leaves 2-12 cm long, pinnatifid; flower stems 3-40 cm tall with glandular hairs; *flower heads 1-1.2 cm long, terminal, consisting of ray flowers only, yellow or orange; *achenes to 2 mm long, with pappus of 5 scales and 5 to many bristles. Moist woods, flatwoods, disturbed sites. Freq. CF, NF, WF. W to Tex., N to Wis., E to Maine. Spr-sum.

WILD LETTUCE
Lactuca floridana (L.) Gaertn.

Annual or biennial, 1-3 m tall; leaves 10-20 cm
long, to 10 cm wide; ray flowers mostly blue;
achenes 6-7 mm long, usually beakless. Moist,
open sites. Infreq. CF, NF, WF. W to Tex.,
N to N.Y. Sum-fall.

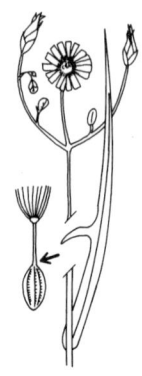

COMMON WILD LETTUCE
Lactuca graminifolia Michx.

Biennial, 0.6-1.5 m tall; leaves to 30 cm long, to
5 cm wide; ray flowers mostly blue; *achenes to
9 mm long, long-beaked. Fields, woods. Freq.
all Fla. N to Ga. and S.C. Spr-sum.

ROUGH BLAZING-STAR
Liatris aspera Michx.

Perennial, 0.3-1.5 m tall; stems hairy; leaves to
20 cm long, to 3 cm wide; flowers 10 or more
per head; bracts obtuse, rounded; petals to 7 mm
long, purple; achenes to 5 mm long; pappus to
8 mm long, barbed. Sandy disturbed areas and
woods. Infreq. WF. W to Tex., N to Canada.
Sum-fall.

(not shown)

SLENDER-HEADED BLAZING-STAR
Liatris chapmanii Torr. & Gray

Perennial, 20-60 cm tall; stems hairy; leaves to
15 cm long, to 1 cm wide, with punctate glands;
bracts acuminate; petals to 11 mm long, purple;
achenes to 6 mm long; pappus to 11 mm long,
barbed. Dry sites. Freq. all Fla. N to Ala. and
N.C. Sum-fall.

HANDSOME BLAZING-STAR
Liatris elegans (Walt.) Michx.

Perennial, 0.3-1.5 m tall; stems hairy; leaves to
14 cm long, to 7 mm wide; flowers 5 per head;
bracts petal-like; petals to 5 mm long, purple,
pink or white; achenes to 6 mm long; pappus to
8 mm long, feather-like. Dry sites. Infreq. CF,
NF, WF. W to Tex., N to Va. Sum-fall.

(not shown)

GARBER'S BLAZING-STAR
Liatris garberi Gray

(not shown)

Perennial, 20-80 cm tall; stems hairy, from a cluster of tuberous roots; basal leaves 7-30 cm long,
linear; involucral bracts 7-14 mm high, sticky,
glandular; petals 6-7 mm long, red to purple; achenes to 3 mm long; pappus bristles to 5 mm long.
Pinelands. Freq. SF, CF, NF. Endemic. Sum-fall.

COMMON BLAZING-STAR
Liatris gracilis Pursh

Perennial, 0.4-1 m tall; stems hairy; *leaves to
35 cm long, to 1 cm wide; *flowers 3-6 per head;
bracts rounded; petals to 5 mm long, purple;
achenes to 3 mm long; pappus to 5 mm long,
barbed. Dry to moist sites. Freq. all Fla.
W to Miss., N to S.C. Sum-fall.

GRASS-LEAVED LIATRIS
Liatris graminifolia (Walt.) Willd.

(not shown)

Perennial, to 1.2 m tall; stems smooth to hairy;
basal rosette of leaves absent; leaves to 30 cm long,
to 7 mm wide; flower heads to 12 mm long; flowers 10-14 per head; petals to 5 mm long, purple,
hairy; achenes to 5 mm long. Flatwoods. Infreq.
CF, NF, WF. W to Miss., N to N.J. Sum-fall.

LONG-LEAF BLAZING-STAR
Liatris laevigata (Nutt.) Small

Perennial, 0.6-1.2 m tall; stems smooth; leaves
to 20 cm long, 2-8 mm wide; flowers 5 per head;
bracts acuminate; petals to 7 mm long, purple;
achenes to 4 mm long; pappus to 5 mm long,
barbed. Disturbed sites. Common all Fla.
N to Ga. Sum.

LARGE-HEADED BLAZING-STAR
Liatris ohlingerae (Blake) Robins.

Perennial, to 1 m tall; stems hairy; leaves to 15 cm
long; flower heads to 7 cm long, rose-purple; petals
to 2 cm long, rose to purple; achenes to 1 cm long.
Scrub. Rare CF. Endemic. Sum.

LOPSIDED BLAZING-STAR
Liatris pauciflora Pursh

Perennial, 20-90 cm tall; leaves to 20 cm long;
flower heads on curved stalks to 1.5 cm long;
flowers 3-6 per head; petals to 16 mm long, with
few hairs on lower portion; achenes 3-4 mm
long. Dry sites. Infreq. CF, NF, WF. N to S.C.
Sum-fall.

DENSE BLAZING-STAR
Liatris spicata (L.) Willd.

Perennial, 0.6-2.5 m tall; stems smooth to hairy;
leaves to 40 cm long; *flower heads to 1 cm
long; petals 6.5-9 mm long, rose to purple; *achenes to 5 mm long. Low sites. Common all Fla.
W to Tex. Sum-fall.

FINE LEAF BLAZING-STAR
Liatris tenuifolia Nutt.

Perennial, 0.6-1.2 m tall; stems smooth; basal leaves to 40 cm long, to 2 mm wide; flower heads 6-9 mm long; flowers 3-6 per head; bracts sharp-pointed; petals 6-8 mm long, purple; achenes to 4 mm long. Dry to moist sites. Freq. all Fla. W to Ala., N to N.C. Sum-fall.

ROSERUSH
Lygodesmia aphylla (Nutt.) DC.

Perennial, 30-80 cm tall; leaves if any, few, to 30 cm long, basal; heads to 2 cm long; ray flowers to 2 cm long, pink, purple or white; achenes l0-12 mm long. Dry sites. Freq. all Fla. N to Ga. Spr-fall.

BLACK ANTHERS or CAT-TONGUE
Melanthera nivea (L.) Small [*M. aspera* (Jacq.) Small]

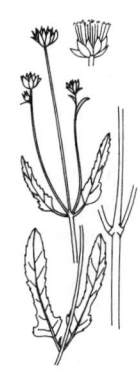

Perennial, 0.1-1.8 m tall; stems 4-angled, mottled, covered with rough hairs; leaves 1-16 cm long, ovate, triangular or narrow; involucral bracts acuminate, acute or obtuse, in 2-3 series; disc flowers only; petals white; anthers black-tipped; achenes 2-3 mm long; pappus of few teeth or bristles. Moist to dry sites. Freq. all Fla. W to La., E to S.C. Spr-fall, all yr S.

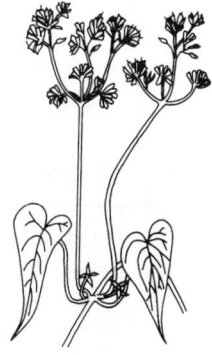

CLIMBING HEMP WEED
Mikania scandens (L.) Willd.
[*M. batatifolia* DC.]

Perennial, low-climbing matted vine; stems smooth to slightly hairy; leaves 2-10 cm long; bracts 3.5-6 mm long, white to pink to purple; achenes to 2.5 mm long. Moist to wet sites. Freq. all Fla. N to Canada. Sum, all yr S.

PALAFOXIA
Palafoxia feayi A. Gray

Perennial herb or shrub, 1-3 m tall; leaves 2-7 cm long, evergreen, with rough hairs; lower leaves ovate-lanceolate; disc flowers only, 17-30 per head; petals to 7.5 mm long, white, pink or lavender, with throat longer than tube; achenes 5-6 mm long; pappus scales to 1 mm long. Dry sites. Common SF, CF, NF. Fall.

MANY WINGS
Palafoxia integrifolia (Nutt.) T. & G.
[*Polypteris integrifolia* Nutt.]

Perennial, to 1.5 m tall; leaves 5-7 cm long; lower leaves linear; disc flowers only, 13-26 per head; petals 9-14 mm long, white, with throat shorter than tube; achenes 4-5 mm long; pappus scales 4-5 mm long. Dry sites. Freq. all Fla. N to Ga. Fall.

SANTA MARIA
Parthenium hysterophorus L.

Annual, 30-70 cm tall; leaves to 20 cm long, divided into leaflets; ray flowers 5, minute, white; disc flowers white; achenes to 2.5 mm long. Disturbed sites. Infreq. all Fla. W to Tex., N to Pa. Spr-fall.

(not shown)

WILD QUININE
Parthenium integrifolium L.

(not shown)

Perennial, to 1.2 m tall; leaves to 20.5 cm long, toothed; ray flowers 5, to 2 mm long, white; disc flowers white; achenes to 3.7 mm long. Dry sites. Ga., W to Tex., N to Mo. Spr-sum.

STALKED CHICKEN WEED
Pectis leptocephala (Cass.) Urban

Perennial, to 40 cm tall; stems spreading; leaves to 3 cm long, less than 2 mm wide, with glands on lower surface in rows on margins; flower heads stalked; ray flowers to 3 mm long, yellow; achenes to 3 mm long. Pinelands, beaches. Infreq. SF, CF. Sum-fall.

(not shown)

LEMON CHICKEN WEED
Pectis linearifolia Urban

(not shown)

Annual, to 40 cm tall; stems erect to ascending; leaves to 3 cm long, less than 2 mm wide, with glands on lower surface in rows on margins; heads sub-sessile; ray flowers 2-3 mm long, yellow; achenes to 3 mm long. Pinelands, beaches. Infreq. SF, CF. Sum-fall.

CHICKEN WEED
Pectis prostrata Cav.

Annual, to 20 cm long; stems prostrate; leaves to 3 cm long, 2-5 mm wide, with scattered glands on lower surface; heads sub-sessile; ray flowers to 2 mm long, yellow; achenes to 4 mm long. Disturbed sites. Infreq. all Fla. W to Ariz. Sum-fall.

PHOEBANTHUS
Phoebanthus grandiflora (Torr. & Gray) Blake

Perennial, 0.6-1.3 m tall; leaves 2-6 cm long; ray flowers 10-20, 2-4 cm long, yellow; discs 1.5-2 cm in diam., yellow; achenes to 5.5 mm long, with 1 tooth per side. Sandhills. Freq. CF, NF, WF. Spr-fall.

SILK-GRASS
Pityopsis graminifolia (Michx.) Nutt. [*Chrysopsis graminifolia* (Michx.) Shinners; *Pityopsis tracyi* Small]

Perennial, 30-90 cm tall; stems erect with silvery-silky hairs (floccose); leaves grass-like, with silvery-silky hairs; basal leaves 8-25 or 40 cm long; stem leaves reduced; bracts lightly glandular; involucre 5-12 mm high; achenes 2.5-4.5 mm long. Dry sites. Common all Fla. W to Tex., N to Ohio, E to Del. Sum-fall, all yr S.

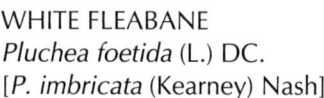

MARSH FLEABANE
Pluchea camphorata (L.) DC.

Annual or perennial, 0.3-1.5 cm tall; stems hairy; leaves 3-15 cm long, glandular; flower heads to 5 mm high; petals pink to purple; achenes to 1 mm long. Moist to wet sites. Infreq. CF, NF, WF. W to Tex., N to Mass. Sum-fall.

WHITE FLEABANE
Pluchea foetida (L.) DC.
[*P. imbricata* (Kearney) Nash]

Perennial, 0.3-1 m tall; stems solitary or several; leaves to 13 cm long, sessile, smooth to hairy; flower heads to 1.5 cm long; petals creamy-white; achenes to 1 mm long. Low, wet sites. Common all Fla. W to Tex., N to N.J. Spr-sum, all yr S.

TALL WHITE FLEABANE
Pluchea longifolia Nash

Perennial, 0.6-2.5 m tall; stems hairy; leaves 5-20 cm long, sessile; heads 10-12 mm long; petals creamy white; achenes to 7 mm long. Moist to wet sites. Infreq. CF, NF, WF. Sum-fall.

SALTMARSH FLEABANE
Pluchea odorata (L.) Cass.
[*P. purpurascens* (Sw.) DC.]

Annual, 0.3-1.4 m tall; stems hairy; petioles to
2 cm long; leaves 5-20 cm long; heads to 7 mm
long, with glandular hairs; petals pink to purple;
achenes to 1.1 cm long. Moist to wet sites. Freq.
all Fla. W to Calif., N to Maine. Sum-fall.

GODFREY'S FLEABANE
Pluchea rosea Godfrey
var. *rosea*

Perennial, 0.4-1.1 m tall; stems solitary or several,
covered with fine glandular, sticky hairs; leaves 2-
7 cm long, sessile; flower heads 4-6 mm long;
bracts and petals rose-purple; achenes to 1 mm
long. Wet flatwoods. Common all Fla. W to
Tex., N to N.C. Spr-sum.

BLACK ROOT
Pterocaulon pycnostachyum (Michx.) Ell.
[*P. virgatum* (L.) DC.]

Perennial, 20-80 cm tall; stems with soft white
wool-like hairs, winged; roots black; leaves 3-
10 cm long, with white wool-like hairs beneath;
heads to 10 cm tall, cymose in a terminal spike;
flowers white; involucres to 4 mm high; achenes
minute. Dry to moist sites. Infreq. all Fla. W to
La., E to N.C. Spr-fall.

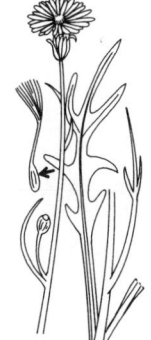

FALSE DANDELION
Pyrrhopappus carolinianus (Walt.) DC.

Annual, 0.3-1 m tall; stems erect, smooth; leaves
8-25 cm long, entire to lobed; flower heads 1.5-
4 cm in diam.; rays yellow or white; achenes 4-
6 mm long. Disturbed sites. Common all Fla.
W to Tex., N to Del. Spr-fall.

FALL CONEFLOWER
Rudbeckia fulgida Ait.

Perennial, to 1 m tall; stems hairy; leaves 3-10 cm
long; ray flowers to 2 cm long, orange to yellow;
disc flowers to 3.5 mm long, purple-brown;
achenes to 2.5 mm long; pappus crowned. Dry
to wet sites. Infreq. WF. W to Tex., N to N.J.
Sum-fall.

(not shown)

BLACK-EYED SUSAN
Rudbeckia hirta L.

Perennial, biennial or annual, 0.3-1 m tall; stems
hairy; leaves 5-18 cm long; ray flowers to 3.5 cm
long, orange or yellow; disc flowers to 2 cm long,
purple-black or brown; achenes to 2.7 mm long;
pappus absent. Disturbed sites. Common all Fla.
W to Tex., N to Canada. Sum-fall.

CUT-LEAVED CONEFLOWER
Rudbeckia laciniata L.

Perennial, 0.3-3 m tall; basal leaves deeply
lobed; petioles 5-30 cm long, with length
decreasing up stem; ray flowers 5-12, 2-6 cm
long, yellow; discs 1-1.5 cm in diam., yellow;
achenes 2-4 cm long; pappus a crown. Some
forms cultivated. Moist to wet sites. Rare CF,
NF, WF. W to Ariz., N to Canada. Sum-fall.

SANDHILL CONEFLOWER
Rudbeckia mollis Ell.

(not shown)

Annual, 0.3-1 m tall; stems hairy; leaves 3-7 cm
long; ray flowers to 4 cm long, yellow; disc flowers to 2 cm in diam., brown or black; achenes to
2.7 mm long; pappus absent. Dry sites. Rare CF,
NF. N to S.C. Sum-fall.

ST. JOHN'S-SUSAN
Rudbeckia nitida Nutt.

Perennial, 0.5-1.3 m tall; stems smooth; leaves
to 30 cm long; ray flowers to 5 cm long, yellow;   (not shown)
disc flowers 3-4.5 mm long, purple-black; ach-
enes to 7 mm long; pappus crowned. Low, wet
sites. Rare CF, NF, WF. W to Tex., E to Ga.
Sum-fall.

SOUTHERN RAGWORT
Senecio anonymus A. Wood
[*S. smallii* Britt.]

Perennial, 30-60 cm tall; leaves 3-10 cm long;
flower heads 30-150, 0.8-1 cm broad; ray flowers
to 8 mm long, yellow to orange; achenes to 1.5 mm
long. Dry to moist sites. Rare WF. W to Miss.,
N to Pa. Spr-sum.

GOLDEN RAGWORT
Senecio aureus L.

Perennial, 20-80 cm tall, smooth or with cottony
hairs in leaf axils on flower branches; *basal
leaves to 12 cm long, purple beneath; upper
leaves pinnatifid; *ray flowers 5-15 mm long,
orange to yellow; disc flowers yellow; achenes
2.5-3 mm long. Moist to wet sites. Rare WF.
E to Ga., W to Ark., N to Canada. Spr-sum.

MEXICAN FLAME VINE
Senecio confusus Britt.

Perennial, high climbing herbaceous or woody
vine, to 4 m long; leaves 3-10 cm long; petals
to 2.5 cm long, orange to reddish. Native to
Mexico. Cultivated and escapes. Disturbed
areas. Freq. SF, CF, NF. All yr.

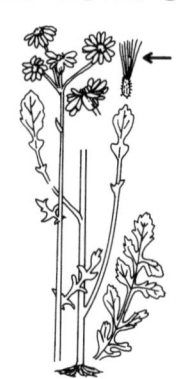

BUTTER WEED
Senecio glabellus Poir.

Annual, 0.15-1.5 m tall; stems branching, smooth; basal leaves 3-20 cm long; stem leaves reduced; ray and disc flowers yellow; *pappus of disc flowers 1 series of bristles; ray flowers lacking pappus. Wet sites. Freq. all Fla. W to Tex., N to Mo., E to N.C. Spr.

BALSAM GROUNDSEL
Senecio pauperculus Michx.

Perennial, 20-60 cm tall; basal leaves 7-10 cm long; involucre 5-7 mm high; ray flowers to 1 cm long, yellow to orange; achenes to 3 mm long. Dry to wet sites. Rare WF. N to Canada. Spr-sum.

WOOLLY RAGWORT
Senecio tomentosus Michx.

Perennial, 20-70 cm tall; base and petioles covered with cottony hairs; basal leaves 3-10 cm long, with cottony hairs beneath; ray flowers 5-10 mm long, yellow; disc flowers orange-yellow; *achenes 2-2.5 mm long. Moist to wet, open sites. Infreq. NF, WF. W to Tex., N to N.J. Spr.

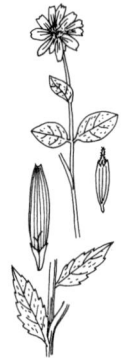

STARRY ROSIN-WEED
Silphium asteriscus L.

Perennial, 0.6-1.5 m tall; stems with rough hairs; leaves 4-15 cm long; bracts in 2 to several series; ray flowers to 3 cm long, yellow; achenes to 9 mm long. Dry sites. Infreq. all Fla. W to Miss., E to Va. Sum-fall.

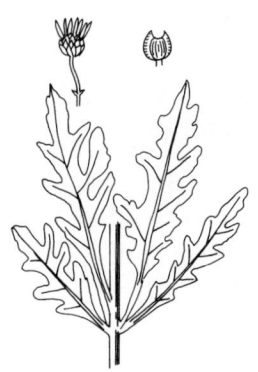

ROSIN-WEED
Silphium compositum Michx.
[*S. ovatifolium* (Torr. & Gray) Small]

Perennial, to 30 cm tall; stems smooth; basal
leaves to 50 cm long, pinnately cut; bracts in 2 to
several series; ray flowers 1-1.5 cm long, yellow;
achenes to 9 mm long. Dry, woody sites. Infreq.
CF, NF, WF. E to Ga. Spr-sum.

(not shown)

TOOTH ROSIN-WEED
Silphium dentatum Ell.

Similar to *Silphium asteriscus* except: stems
smooth; leaves to 25 cm long. Dry woods.
Ala., E to N.C. Sum-fall.

SHARP GOLDENROD
Solidago arguta Ait.

Perennial, 0.6-1.2 m tall; stems smooth except for
few hairs on flower branches; basal leaves 8-15 cm
long, having sharply toothed margins with long
petioles; upper leaves reduced, with entire margins;
flower heads 3-4.5 mm high, yellow; ray flowers 2-
8; disc flowers 8-20; achenes to 3 mm long, hairy.
Wet to dry and disturbed sites. Rare CF, NF, WF.
W to Ala., N to Maine. Sum-fall.

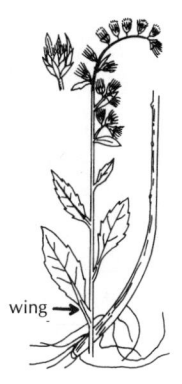

wing →

BLUESTEM GOLDENROD
Solidago caesia L.

Perennial, 0.3-1 m tall; leaves 5-10 cm long;
*flower heads to 4.5 mm high; ray flowers to 3 mm
long, yellow; achenes to 2 mm long. Dry to moist
woods. Rare WF. N to Canada. Sum-fall.

TALL GOLDENROD
Solidago canadensis L.
[*S. altissima* L.]

Perennial, to 1 m tall; stems hairy; leaves to 15 cm
long; *flower heads to 5 mm high; ray flowers to
4 mm long, yellow; achenes to 1.5 mm long.
Disturbed woods. Freq. NF, WF. N to Canada.
Sum-fall.

CHAPMAN'S GOLDENROD
Solidago chapmanii Torr. & Gray

Perennial, to 1.5 m tall; stems hairy; lower leaves
3-6 cm long, with rolled under margins; upper
leaves reduced, with entire margins; flower heads
6-7 mm high, yellow; ray and disc flowers 3-5;
achenes to 1.5 mm long, hairy; pappus exceeding bracts. Pinelands, sandy sites. Common all
Fla. N to Ga. Sum-fall.

ELLIOTT'S GOLDENROD
Solidago elliottii Torr. & Gray

Perennial, to 2 m tall; stems smooth except for few
hairs on flower branches; lower leaves 6-15 cm
long, with toothed margins; upper leaves smaller,
with toothed margins; *flower heads to 5 mm high;
ray flowers 6-12, yellow; disc flowers 4-7; achenes
1-2 mm long, hairy. Moist woody sites.
Occasional CF. N to Canada. Sum-fall.

HOLLOW GOLDENROD
Solidago fistulosa Mill.

Perennial, 0.6-2 m tall; stems with white hairs;
leaves to 12 cm long; *flower heads to 7 mm
high; ray flowers to 4 mm long, yellow; achenes
to 1.5 mm long. Low sites. Freq. all Fla. W to
La., N to N.J. Sum-fall.

GIANT GOLDENROD
Solidago gigantea Ait.

Perennial, 0.8-2.5 cm tall; stems smooth; lower leaves 6-15 cm long, with sharply toothed margins; upper leaves reduced, with entire margins; flower heads 4.5-6 mm high, yellow; ray flowers 10-17; disc flowers 6-10; achenes to 1.5 mm long, hairy. Moist sites. Infreq. SF, CF, NF. W to Tex., N to Canada. Sum-fall.

(not shown)

LEAVENWORTH'S GOLDENROD
Solidago leavenworthii Torr. & Gray

Perennial, 0.5-1.2 m tall; upper part of stem smooth; lower portion of stem with curved hairs; lower leaves to 18 cm long, with toothed margins; upper leaves reduced, with toothed margins; flower heads 4-6 mm high, yellow; rays 10-15; discs 6-10; achenes to 1 mm long, hairy. Moist sites. Freq. all Fla. N to N.C. Sum-fall.

DWARF GOLDENROD
Solidago nemoralis Ait.

Perennial, 10-90 cm tall; stems prostrate, hairy; lower leaves 4-25 cm long; *flower heads to 6 mm high; ray flowers to 4 mm long, yellow; achenes to 1.3 mm long. Sunny dry sites. Rare WF. N to Canada. Sum-fall.

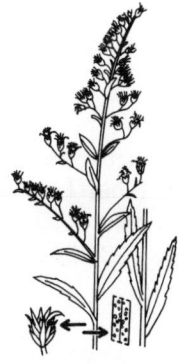

SWEET GOLDENROD *Solidago odora* Ait.

Perennial, 0.5-1.5 m tall; hairs on flower branches and in lines on leaf bases; leaves having odor of anise when bruised, 1-veined, with *pellucid-punctate dots; lower leaves 3-16 cm long, with finely hairy margins; upper leaves reduced, with entire margins; *flower heads 4-5.5 mm high, yellow; ray and disc flowers 3-5; achenes 1.5-1.8 mm long, hairy; pappus not exceeding bracts. Dry sites. Freq. CF, NF, WF. W to Tex., N to N.H. Sum-fall.

SPREADING GOLDENROD
Solidago patula Muhl. ex Willd.

Perennial, 0.7-2 m tall; stems smooth; leaves 15-20 cm long, 6-8 cm wide, scabrous above; involucre to 4 mm high; ray flowers 5-12, yellow. Low sites. Rare WF. W to La., N to Wis. Sum-fall.

DOWNY GOLDENROD
Solidago puberula Nutt.

Perennial, 0.3-1.1 m tall; stems hairy; leaves 3-10 cm long, to 3.5 cm wide; involucre to 5 mm high; ray flowers 9-16, yellow. Dry woods, wet sites. Rare WF. W to Miss., N to Canada. Sum-fall.

ROUGH GOLDENROD
Solidago rugosa Mill.

Perennial, 0.4-2 m tall; stems hairy; leaves 2-10 cm long; flower heads to 5 mm high; ray flowers to 4 mm long, yellow; achenes to 1.5 mm long. Low sites. Rare WF. N to Canada. Sum-fall.

SEASIDE GOLDENROD
Solidago sempervirens L.

Perennial, 0.4-2 m tall; stems smooth; leaves to 25 cm long; flower heads to 6 mm high; ray flowers to 5 mm long, yellow; achenes to 2 mm long, hairy. Low sites, coastal areas. Freq. all Fla. N to Canada. Sum-fall.

SMOOTH GOLDENROD
Solidago stricta Ait.

Perennial, to 2 m tall; stems smooth; basal leaves
6-30 cm long, with entire margins; stem leaves
numerous and abruptly reduced, with entire margins; flower heads 4-7 mm high, yellow; ray
flowers 3-7; disc flowers 8-12; achenes to 2 mm
long, hairy. Wet and coastal sites. Freq. all Fla.
W to Tex., N to N.J. Sum-fall.

TWISTED LEAF GOLDENROD
Solidago tortifolia Ell.

Perennial, 0.3-1 m tall; stems with short hairs;
lower leaves 1.5-8 cm long, with toothed margins; upper leaves reduced, with entire margins;
flower heads to 3.5 mm high, yellow; ray and
disc flowers 2-6; achenes to 1 mm long, hairy.
Dry sites. Infreq. all Fla. W to Tex., N to Md.
Sum-fall, all yr S.

SPINY-LEAVED SOW THISTLE
Sonchus asper (L.) Hill

Annual, 0.3-2 m tall; stems smooth; leaves 6-
30 cm long, spiny; leaf basal portion round;
flower heads 1.5-2.5 cm in diam.; rays yellow;
*achenes to 2.5 mm long, 3-ribbed. Disturbed
areas. Common all Fla. All U.S. Spr-sum.

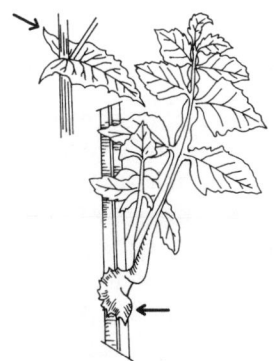

COMMON SOW THISTLE
Sonchus oleraceus L.

Annual, to 2 m tall; stems smooth; leaves to
30 cm long, spiny, lobed; *leaf basal portion
pointed/arrow-shaped; flower heads 1.5-2.5 cm in
diam.; rays yellow; achenes to 3 mm long, wrin-
kled, covered with projections. Disturbed sites.
Common all Fla. N to Pa., W to Tex. Spr-sum.

NODEWEED
Synedrella nodiflora (L.) Gaertn.

Annual, 15-80 cm tall; leaves 2-10 cm long, with
petiole-like bases; *flowers yellow, with bracts in
1 series; *achenes 4-5 mm long, bristly, having
2 awns. Sandy, wet sites. Rare SF. All yr.

COMMON MARIGOLD
Tagetes erecta L.

Annual, 0.2-1.5 m tall; stems branching or lying
somewhat flat; leaves divided into leaflets; each
leaflet to 5 cm long, strongly scented; involucre
to 1.8 cm high; ray flowers to 3 cm long, orange
or yellow; *achenes to 1 cm long. Native to
Mexico. Disturbed sites. Escapes from cultivation in Fla. Warm months.

COMMON DANDELION
Taraxacum officinale Weber

Perennial, low herb; leaves 8-40 cm long, entire
to lobed; flower stems to 50 cm tall; flower heads
3-5 cm in diam.; rays orange to yellow; *achenes
to 3 mm long. Lawns, disturbed sites. Freq. all
Fla. Nearly all U.S. Spr-fall.

MEXICAN SUNFLOWER
Tithonia diversifolia (Hemsl.) A. Gray

Annual, to 5 m tall; lower portion of plant
woody; leaves 8-20 cm long, hairy beneath;
bracts in 2 series, with the inner 18-25 mm long
and the outer shorter; ray flowers 7-21, 4-6 cm
long, yellow; disc flowers yellow; *achenes to
6 mm long, with 2 long and several short scales.
Escapes from cultivation. Disturbed sites. Rare
all Fla. Fall.

MEXICAN DAISY
Tridax procumbens L.

Perennial, to 40 cm long; stems lying flat, branching, hairy; leaves 2-5 cm long, hairy; flower stems (peduncles) to 30 cm long; ray flowers 3-8, to 4 mm long, yellow; achenes to 2 mm long; pappus of 20 long bristles. Native to tropical America. Open disturbed sites. Freq. SF, CF, NF. All yr.

SKUNK DAISY
Verbesina encelioides (Cav.) Benth. & Hook

Annual, 30-70 cm tall, hairy; upper leaves 4-12 cm long, 2-10 cm wide, alternate; lower leaves opposite, ovate-deltoid, with gray hairs on lower side; petioles shorter than blades; involucre spreading, flattish; ray flowers 1-2.5 cm long, yellow; discs 13-20 mm in diam., yellow; achenes to 5 mm long, winged. Native to southwest U.S. Moist, disturbed sites. Infreq. SF, CF. N to N.C. Sum.

SMALL YELLOW CROWNBEARD
Verbesina occidentalis (L.) Walt.

Perennial, 0.9-2 m tall; stems smooth to slightly hairy, 4-winged; leaves to 20 cm long, to 10 cm wide, opposite, ovate; petiole winged, half as long as blade; ray flowers 1-5, 1-2 cm long, yellow; discs 3-7 mm in diam., yellow; achenes 4-5 mm long, nearly wingless. Disturbed and wet sites. Infreq. WF. W to Miss., E to Va. Sum-fall.

(not shown)

FROSTWEED
Verbesina virginica L.

Perennial, 0.6-2.8 m tall; stems hairy; leaves 5-22 cm long, simple, with winged petioles; ray flowers 1-5, to 1 cm long, white; achenes to 5 mm long; pappus of 2 awns. Dry to wet wooded sites. Common all Fla. W to Tex., N to Va. Sum-fall.

CUT-LEAF FROSTWEED
Verbesina virginica L.
var. *laciniata* (Poir.) Small

Very similar to *Verbesina virginica* except:
leaves divided into sections or lobed. Dry to wet
wooded sites. Infreq. SF, CF, NF. N to S.C.
Sum-fall.

NARROW-LEAVED IRONWEED
Vernonia angustifolia Michx.

Perennial, 0.6-1.8 m tall; leaves 4.5-12 cm long,
scabrous above; heads 5-7 mm long; achenes to
2 mm long; pappus white, pink or purple. Dry
sites. Freq. CF, NF, WF. W to Miss., E to N.C.
Sum-fall.

BLODGETT'S IRONWEED
Vernonia blodgettii Small

Perennial, 20-50 cm tall; leaves 2-4.5 cm long,   (not shown)
with smooth upper surface; petals purple; pappus
yellow. Pinelands. Infreq. SF, CF. Endemic.
Sum-fall.

ASIAN IRONWEED
Vernonia cinera (L.) Less.

Annual, to 1 m tall; stems hairy; leaf blades 2-8 cm
long; involucre 4-5 mm long; flowers pale purple;
*achenes 1 mm long; pappus in 2 series, with
inner series deciduous. Disturbed sites. Infreq. SF,
CF. All yr.

TALL IRONWEED
Vernonia gigantea (Walt.) Trel.
[*V. altissima* Nutt.]

Perennial, 1-3 m tall; stems smooth; leaves 10-
30 cm long, 3-5 cm wide; heads 12- to 30-flow-
ered; petals 5-6 mm long, purple; achenes to 3 mm
long; pappus purple. Low, wet sites. Freq. all Fla.
W to La., E to S.C. Sum-fall, all yr S.

NEW YORK IRONWEED
Vernonia noveboracensis (L.) Michx.

(not shown)

Similar to *Vernonia gigantea* except: bracts
broad-based, filiform-tipped; petals 8-11 mm
long; pappus brown to purple. Low, wet sites.
Infreq. NF, WF. W to Ala., E to Mass. Fall.

CREEPING OXEYE
Wedelia trilobata (L.) A. Hitchc.

Perennial, to 40 cm tall; stems creeping to erect;
leaves 3-10 cm long; bracts in 2-3 series; ray flowers to 1.5 cm long, yellow; achenes to 5 mm long.
Native to West Indies. Escapes from cultivation.
Disturbed sites. Infreq. SF, CF, NF. All yr.

ASIATIC HAWKS BEARD
Youngia japonica (L.) DC.

Annual, 0.1-1 m tall; stem smooth to hairy at
base; leaves to 15 cm long, mostly all basal,
pinnatifid; flower heads consisting of ray flowers
only, to 5 mm long, yellow; achenes 1.5-2.5 mm
long, with many white long bristles (pappus)
covering tip. Native to SE Asia. Disturbed sites.
Common all Fla. W to La., N to Pa., E to Va.
Sum, all yr S.

MORNING-GLORY FAMILY *CONVOLVULACEAE*

Herbs, vines, shrubs or trees; leaves alternate or absent; flowers bisexual; sepals 4 or 5; petals 4 or 5; ovary superior; fruit a capsule.

BIG BLUE PROSTRATE MORNING-GLORY
Bonamia grandiflora (A. Gray) Heller

Perennial, 0.5-3 m long; stems trailing; leaves 2.5-5 cm long; flowers 7-10 cm long, solitary, blue; capsules to 2 cm long. Dry sites. Infreq. CF, NF. Spr-sum.

HEDGE BINDWEED
Calystegia sepium (L.) R. Brown

Perennial, 6-20 cm long; stems trailing or twining; leaf blades to 8 cm long, ovate or narrowly sagittate; petals 4-7 cm long and broad, white to pink or purple; capsules to 1.5 cm long. Roadsides, cultivated and disturbed sites. Infreq. all Fla. Naturalized locally in temperate North America. Sum-fall.

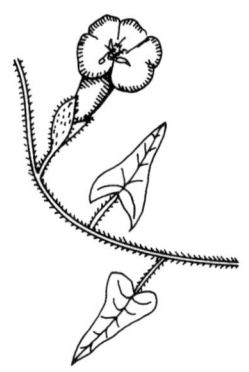

LOW BINDWEED
Calystegia spithamea L.

Perennial, 10-70 cm long; stems erect or ascending, hairy; leaf blades 4-10 cm long, oval to sagittate, with velvety hairs; petals 4-6 cm long, white; capsules to l cm long. Disturbed, calcareous sites. Rare WF. N to Canada. Spr-sum.

COMPACT DODDER
Cuscuta compacta Juss.

Annual, parasitic vine; leaves scale-like; flowers
to 4.5 mm long, with 2 to several appressed
bracts, greenish white, cymose; petal lobes bent
backwards; capsules 3-5 mm wide. Dry and wet
sites, on woody and herbaceous plants. Infreq.
all Fla. W to Tex., N to Ill., E to N.H. Sum-fall.

(not shown)

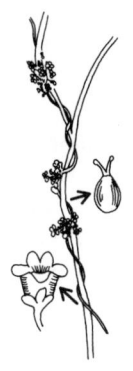

DODDER
Cuscuta gronovii Willd.

Annual parasitic vine on herbs or shrubs; leaves
mere scales; *flowers to 2.5 mm long, white to
yellow; *capsules oval. Low wet sites. Infreq.
CF, NF, WF. N to Canada. Sum-fall.

PRETTY DODDER
Cuscuta indecora Choisy

Annual, parasitic vine; leaves scale-like; flowers
2-3 mm long, white to green, cymose; *petal
lobes bent inwards, persisting; *capsules to 5 mm
in diam. On herbaceous plants and low shrubs.
Infreq. CF, NF, WF. W to Calif., N to Idaho,
E to Mich. Sum-fall, all yr S.

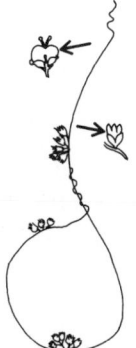

DODDER
Cuscuta obtusiflora Kunth in HBK.

Similar to *Cuscuta gronovii* except: *flowers to
2 mm long; *capsules rounded. Parasitic on
herbaceous plants in open areas. Rare CF, NF,
WF. W to Tex. and N.Mex., E to Ga. Sum-fall.

DODDER LOVE VINE
Cuscuta pentagona Engelm.

Annual, parasitic vine; leaves scale-like; flowers
to 1.5 mm long, white, cymose; petal lobes bent
backwards; capsules 3-4 mm wide. Sandy sites,
on herbaceous plants. Infreq. all Fla. W to Calif.
and Mont. Sum-fall, all yr S.

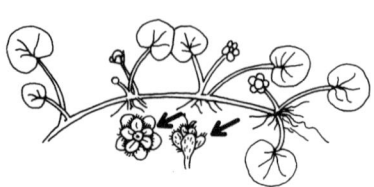

PONY-FOOT
Dichondra carolinensis Michx

Perennial, 15-60 cm long, creeping; leaves
5-20 mm wide, rounded to kidney-shaped;
*sepals 2-3 mm long; petals greenish white;
*capsules to 2.5 mm long. Moist sites and
lawns. Freq. all Fla. W to Tex., E to Va.
Spr-fall, all yr S.

MOON FLOWER
Ipomoea alba L.

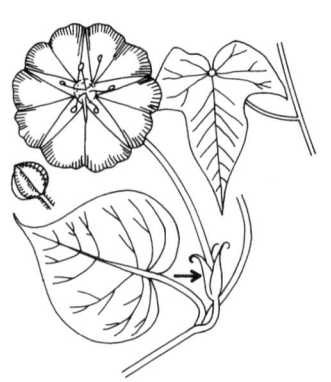

Perennial, twining, often high in trees; leaf blades
to 20 cm long, heart- to oval-shaped; *sepals to
1.5 cm long, with pointed tips; petals to 12 cm
long, white; capsules to 3 cm long. Hammocks,
lakesides, often prevalent after fire. Freq. SF, CF,
NF. All yr.

CAIRO MORNING-GLORY
Ipomoea cairica (L.) Sweet

(not shown)

Annual; stems smooth, twining; leaf blades divided like a hand into 5 leaflets, each 4-9 cm long;
petals 5-8 cm long, purple; capsules to 1.3 cm
wide. Disturbed areas. Infreq. all Fla. All yr.

IVY-LEAF MORNING-GLORY
Ipomoea hederacea (L.) Jacq.

Annual twining vine, to 3 m long; stems covered
with dense hairs; leaf blades 5-12 cm long,
3-lobed to heart-shaped; sepals covered with
brown hairs; petals 2.5-4.5 cm long, blue, later
changing to rose-purple; *capsules to 1.2 cm
wide. Disturbed areas. Infreq. all Fla.
W to Ariz., N to Maine. Sum-fall.

RED MORNING-GLORY
Ipomoea hederifolia L.

Annual twining vine, smooth; leaves 2-10 cm long,
3-lobed or heart-shaped; petals to 3.5 cm long,
scarlet, orange or white; *capsules to 5.5 mm
wide. Disturbed areas. Infreq. all Fla. W to Tex.,
N to Mass. Sum-fall.

BLUE MORNING-GLORY
Ipomoea indica (Burm. f.) Merr.

Perennial, twining vine; leaves 4-16 cm long;
sepals hairy; petals 6-9 cm long, 6-8 cm wide,
blue, purple or white; capsules 3-celled; seeds 6.
Thickets, disturbed sites. Freq. SF, CF, WF.
W to Tex. All yr.

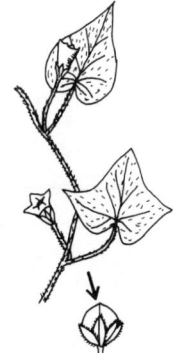

PITTED MORNING-GLORY
Ipomoea lacunosa L.

Annual, twining; leaves 2-10 cm long, simple,
3- or 5-lobed; petals to 2 cm long, white to pink;
*capsules to 6 mm in diam. Low sites. Infreq.
SF, NF, WF. W to Tex., N to N.J. Sum-fall.

BEACH MORNING-GLORY
Ipomoea macrantha Roem. & Schult.
[*I. tuba* (Schlecht.) G. Don]

Perennial, trailing; leaf blades to 16 cm long,
heart- to oval-shaped; sepals to 2.5 cm long, with
rounded tips; petals to 12 cm long, white; capsules to 2.5 cm long. Disturbed sites. Infreq. SF,
CF. All yr.

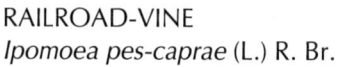

WILD POTATO VINE
Ipomoea pandurata (L.) G. F. W. Meyer

Perennial, twining or trailing; leaf blades to 10 cm
long, lobed or heart- to oval-shaped; sepals to
1.5 cm long, with rounded tips; petals to 8 cm
long, white with lavender center; *capsules to 1 cm
long. Dry sites. Freq. CF, NF, WF. W to Tex., N to
Canada. Sum-fall.

RAILROAD-VINE
Ipomoea pes-caprae (L.) R. Br.

Perennial; stems creeping, trailing, smooth,
fleshy; leaf blades 3-10 cm long; petals 4-6 cm
long, lavender to purple, funnel-shaped; capsules
to 15 mm in diam. Coastal sites. Freq. all Fla.
W to Tex., N to Ga. Sum-fall, all yr S.

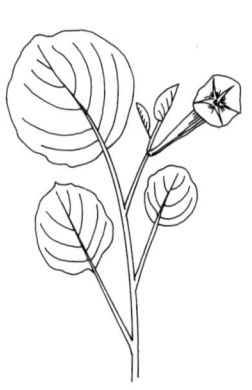

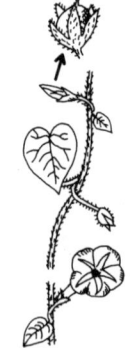

TALL MORNING-GLORY
Ipomoea purpurea (L.) Roth

Perennial, climbing, trailing; leaf blades 4-16 cm
long, heart-shaped; flowers 4.5-7 cm long, blue,
purple, pink or white; stigma 3-lobed; *capsules
to 1 cm long. Disturbed sites. Rare CF, NF, WF.
W to Tex., N to Canada. Sum-fall.

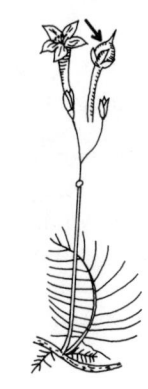

CYPRESS-VINE MORNING-GLORY
Ipomoea quamoclit L.

Annual; stems twining, smooth; leaves 2-6 cm
long, divided (like cypress tree leaves) into several
leaflets; petals 2-3 cm long, red or white, tubular-
shaped; capsules 5-10 mm long. Native to
Mexico. Cultivated. Disturbed areas. Infreq. all
Fla. W to Tex., N to Va. Sum-fall.

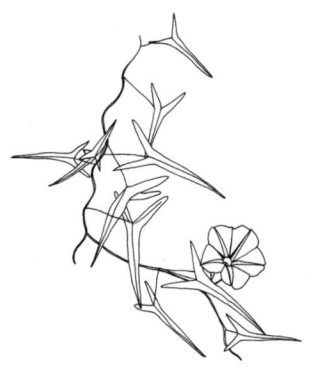

GLADES MORNING-GLORY
Ipomoea sagittata Poir.

Perennial; stems smooth, twining or trailing;
leaves 4-10 cm long, 3-lobed, arrow-shaped;
petals 6-8 cm long, pink, rose, lavender or purple;
capsules to 1 cm wide. Low, wet and coastal
sites. Freq. all Fla. W to Tex., N to N.C. Spr-fall.

BEACH MORNING-GLORY
Ipomoea stolonifera (Cyrillo) J. F. Gmel.

Perennial, to 50 m long; stems prostrate, creeping
or trailing; leaves 2-12 cm long, leather-like;
petals 3.5-7 cm long, white with yellow center.
Coastal sites. Freq. all Fla. W to Tex., N to N.C.
Sum-fall.

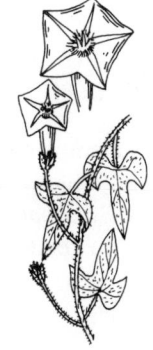

SHARP-POD MORNING-GLORY
Ipomoea trichocarpa Ell.

Perennial twining vine; leaves 2-10 cm long,
heart-shaped to lobed; petals 3-5 cm long, pink
or purple; capsules to 7 mm wide. Low sites.
Freq. all Fla. W to Tex., N to S.C. Sum-fall.

SMALL-FLOWERED MORNING-GLORY
Jacquemontia tamnifolia (L.) Griseb.

Annual; stems erect to prostrate or twining, hairy;
leaves 5-12 cm long; inflorescence cymose,
head-like, tawny hairy; flowers 1-2 cm wide,
blue; capsules to 6 mm in diam. Weed in cultivated fields and floodplains. Infreq. all Fla.
W to Tex., E to Va. Sum-wint.

ALAMO VINE
Merremia dissecta (Jacq.) Hall. f.

Perennial, trailing or high climbing; stems frequently hairy; leaves 3-10 cm long, 5- to 7-lobed;
petals 3-5 cm long, white with dark center; capsules to 3 cm long, to 1.5 cm in diam. Disturbed
sites. Infreq. all Fla. W to Tex., E to Ga. Spr-fall.

WATER STYLISMA
Stylisma aquatica (Walt.) Raf.

Perennial; stems prostrate or twining, hairy; leaves
to 3.5 cm long; petals to 1.5 cm long, rose, purple
or lavender; *styles 2, joined at bases; capsules to
5 mm long. Wet places, ponds. Infreq. WF.
W to Tex., N to N.C. Spr-fall.

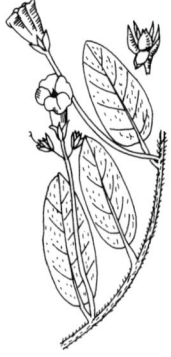

SPREADING STYLISMA
Stylisma humistrata (Walt.) Chapm.

Perennial; stems lying flat, hairy; leaves 2.5-5 cm
long; flowers 1-3 or rarely up to 7, on 1 peduncle,
15-20 mm long, white; styles 2, partially united.
Dry sites. Infreq. NF, WF. W to Tex., E to Va.
Sum-fall.

TRAILING STYLISMA
Stylisma patens (Desr.). Myint
ssp. *angustifolia* (Nash) Myint

Perennial; leaf blades to 3 cm long, linear to
lanceolate, hairy; petals 1.3-2 cm long, white;
*capsules to 2.5 mm long. Dry pinewoods.
Freq. CF, NF, WF. N to N.C. Sum.

DOGWOOD FAMILY   *CORNACEAE*

Trees, shrubs or herbs; leaves alternate or opposite; flowers
bisexual or unisexual (dioecious); sepals 4-5; petals 4-5;
ovary inferior; fruit a drupe or berry.

PAGODA DOGWOOD
Cornus alternifolia L.

Perennial small tree, to 9 m tall; twigs smooth;
leaves to 12 cm long, alternate, elliptic to oval,
crowded at twig tips, smooth above, with flattened to erect hairs below; petals white; drupes
to 8 mm in diam., blue. Woody sites. Rare WF.
N to Canada. Spr.

ROUGH LEAF CORNEL
Cornus asperifolia Michx.

(not shown)

Perennial shrub, to 5 m tall; twigs with long hairs;
leaves to 8 cm long, opposite, ovate, with long,
forked, erect hairs beneath and rough hairs
above; petals white; drupes to 8 mm in diam.,
blue. Dry woods. Infreq. CF. W to Tex.,
N to Canada. Spr.

WILD DOGWOOD
Cornus foemina Mill.

Perennial shrub, to 6 m tall; twigs smooth; leaves
4-10 cm long, opposite, elliptic to oval, with few,
short, flattened hairs beneath; petals white; drupes
to 6 mm in diam., blue. Rich, swampy sites. Freq.
all Fla. W to Tex., N to Ind. and Del. Spr-sum.

ORPINE FAMILY   *CRASSULACEAE*

Succulent herbs or shrubs; leaves in rosettes; flowers bisexual
or unisexual (dioecious); sepals 3-30; petals 3-30; ovary superior; fruit a group of follicles.

LIFE PLANT
Kalanchoe pinnata (Lam.) Pers.

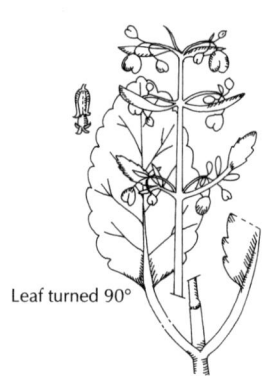

Perennial, 40-60 cm tall; stems glaucous; leaves
simple or pinnately compound; leaflets 10-30 cm
long, sometimes producing new plants at each
notch; sepals 4, to 3 cm long; petals 4, to 4 cm
long, red; follicles to 1.5 cm long. Native to
Asia. Escapes from cultivation. Disturbed sites.
Freq. SF; infreq. CF, NF. All yr.

Leaf turned 90°

CHANDELIER PLANT
Kalanchoe tubiflora (Harv.) Ham.
[*K. verticillata* Elliot]

Perennial, to 1 m tall; leaves 3-15 cm long, purple-
mottled, producing small plantlets at tips; flowers
to 2.5 cm long, red, pink or orange; follicles 4.
Native to Africa. Disturbed sites. Infreq. SF, CF,
NF. All yr.

MUSTARD FAMILY   *CRUCIFERAE*

Herbs or shrubs, with watery sap; leaves mostly alternate; flowers bisexual; sepals 4; petals 4 or 0; ovary superior; fruit a capsule (silique or silicle).

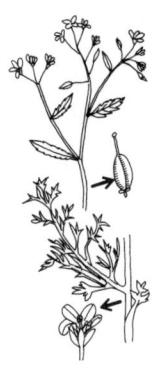

LAKE CRESS
Armoracia aquatica (A. A. Eaton) Wieg.

Perennial; stems submerged; leaves to 7 cm long; submersed leaves often narrow, divided into 1-3 segments; other leaves simple to divided; *petals to 8 mm long, white; *siliques to 8 mm long. Wet sites. Rare WF. W to Tex., N to Canada. Sum.

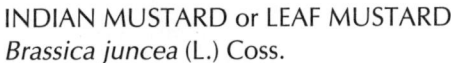

INDIAN MUSTARD or LEAF MUSTARD
Brassica juncea (L.) Coss.

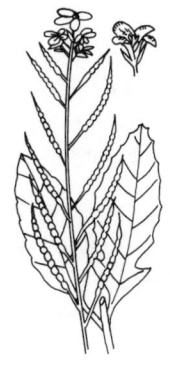

Annual, 0.3-2 m tall; stems glaucous; lower leaves 5-40 cm long, coarsely toothed or divided into segments (pinnatifid); upper leaves smaller; margins entire; petals 9-12 mm long, yellow; siliques 3-5 cm long. Native to Asia. Disturbed sites. Infreq. all Fla. Cultivated throughout U.S. N to Mich., E to N.H. Spr-sum.

WILD MUSTARD
Brassica kaber (DC.) L. C. Wheeler

Annual, 30-80 cm tall; stems with piercing bristles; leaves 3-17 cm long; basal leaves with lobes or margins toothed; *petals 6-9 mm long, yellow; *siliques (capsules) 2.1-4 cm long. Disturbed sites. Infreq. SF, CF, NF. All North America. Spr-fall.

FIELD MUSTARD
Brassica rapa L.
[*B. campestris* L.]

Annual or biennial, 0.3 to 1 m tall; stems glaucous;
leaves 5-27 cm long; lower leaves pinnatifid;
upper leaves auriculate-clasping, entire to toothed;
petals 6-10 mm long, yellow; siliques 3-7 cm
long. Native to Europe. Cultivated, persisting.
Disturbed sites. Infreq. CF, NF, WF. Most of U.S.
Spr-sum.

SEA ROCKET
Cakile constricta Rod.

(not shown)

Annual, to 30 cm tall; stems erect to prostrate;
leaves more or less entire; petals to 8 mm long,
white to lavender; fruits 14-24 mm long, with
constriction at joint. Coastal sites. Infreq. CF,
NF, WF. W to Tex. Spr-sum.

NORTHERN SEA ROCKET
Cakile edentula (Bigel.) Hook.
subsp. *harperi* (Small) Rodman
[*C. harperi* Small]

Annual, succulent, to 80 cm tall; stems lying flat;
leaves 2-10 cm long; *petals to 1 cm long, white,
pink or lavender; *pods 2-2.5 cm long. Coastal
sands. Rare CF, NF. W to Tex., N to N.C. Spr-fall.

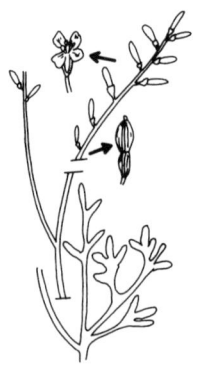

SOUTHERN SEA ROCKET
Cakile lanceolata (Willd.) Schulz
[*C. fusiformis* Greene]

Annual, to 60 cm long; stems erect to prostrate,
smooth; leaves 3-15 cm long, entire to lobed or
pinnatifid; *petals to 3 mm wide, white to lavender; *fruits 13-30 mm long, 2-segmented, without
constriction at joint. Coastal sites. Freq. SF, CF.
Spr-fall, all yr S.

SHEPHERD'S PURSE
Capsella bursa-pastoris (L.) Medicus

Annual, 30-90 cm tall; basal leaves 3-14 cm long, lobed; upper leaves to 4 cm long, arrow-shaped; petals to 4 mm long, white; siliques to 8 mm long. Disturbed sites. Infreq. NF, WF. All U.S. Spr.

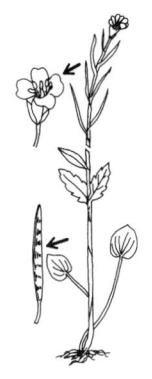

SPRING CRESS
Cardamine bulbosa (Schreb.) BSP.

Perennial, 20-60 cm tall; base bulb-like; stems erect, smooth; leaves to 8 cm long, simple; *petals to 12 mm long, white or pinkish; stamens 6; *siliques 15-30 mm long. Wet sites. Infreq. CF, NF, WF. W to Tex., N to Canada. Spr.

HAIRY BITTER CRESS
Cardamine hirsuta L.

Annual or biennial, 5-30 cm tall; stems covered with short stiff hairs or smooth; basal leaves 3-8 cm long; stem leaves shorter than basal leaves; petioles with fine hairs; *petals 2-3 mm long, white; stamens 4; siliques to 2.5 cm long. Moist to wet sites. Infreq. all Fla. W to Tex., N to N.Y. Spr.

SMALL-FLOWER BITTER CRESS
Cardamine parviflora L.

Annual, 5-30 cm tall; stems smooth; leaves 1.5-6 cm long, divided into several leaflets; *petals to 3.5 mm long, white; *siliques to 2.5 cm long. Moist to wet sites. Infreq. NF, WF. W to Tex., N to Canada. Wint-spr.

PENNSYLVANIA BITTER CRESS
Cardamine pensylvanica Muhl.

Biennial or perennial, terrestrial, 10-80 cm tall;
stems smooth, not rooting at nodes; leaves 4-8 cm
long; lower and basal leaves divided into 3-13
leaflets; petals 2-4 mm long, white; stamens 6;
*siliques 1-3 cm long, flattened. Wet sites.
Infreq. but locally common all Fla. W to Okla.,
N to Canada. Spr.

SWINE WART CRESS
Coronopus didymus (L.) J. E. Smith

Annual, 5-37 cm long; stems lying flat to ascending; lower leaves 3-10 cm long, divided into several leaflets; upper leaves to 4 cm long, divided
into several leaflets; flowers minute, green to
white; silicles to 1.5 mm long. Disturbed sites
often in turf. Infreq. all Fla. W to Calif., N to
Canada. Spr.

PINNATE TANSY MUSTARD
Descurainia pinnata (Walt.) Britt.

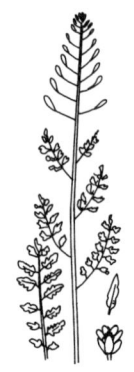

Annual, 20-80 cm tall; stems and leaves sparsely
to densely covered with short gray hairs; leaves
1-10 cm long, divided 2-3 times; petals 1-1.5 mm
long; siliques 5-8 mm long, straight. Disturbed
sites. Common all Fla. W to Tex., N to Pa. and
Va. Wint-spr.

FLIX WEED
Descurainia sophia (L.) Webb

(not shown)

Annual, 30-70 cm tall; leaves 3-10 cm long, 2- to
3-pinnate; petals 2-3 mm long, yellow; siliques
15-25 mm long, curved upward. Disturbed
areas. N.Y., W to Tex., N to Nebr. and Canada.
Spr-sum.

CRUCIFERAE 143

SHORT-FRUIT WHITLOW-GRASS
Draba brachycarpa Nutt.

Annual, 4-20 cm tall; stems hairy; leaves 5-15 mm
long; sepals purple-pigmented; petals white; capsules 2-4 mm long. Open disturbed sites. Rare
WF. W to Tex., N to Mo., E to Va. Spr.

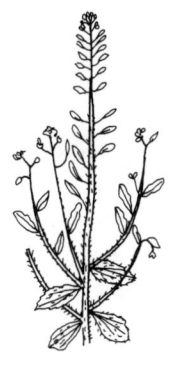

WEDGE-LEAF WHITLOW-GRASS
Draba cuneifolia Nutt. ex Torr. & Gray

Annual, 10-20 cm tall; basal leaves to 5 cm long;
stem leaves 0.8-3.5 cm long; petals 4-5 mm long,
white; siliques 8-12 mm long. Dry sites. Rare in
Fla. W to Tex., N to N.C., Central U.S. Spr.

PEPPER WEED
Lepidium virginicum L.

Annual or biennial, 20-90 cm tall; stems smooth
to hairy; lower leaves 2-10 cm long, deciduous;
petals to 1 mm long, white; stamens 2 or 4;
siliques 3-4 mm long. Disturbed sites. Freq. all
Fla. W to Tex., N to Canada. Spr-fall, all yr S.

WATER-CRESS
Nasturtium officinale R. Br.

Perennial aquatic, 10-70 cm long; stems smooth,
rooting at nodes in substrate, floating; leaves 2-
15 cm long, divided into 3-11 leaflets; petals 4-
5 mm long, white; stamens 6; siliques 10-20 mm
long, curved. Shallow streams. Infreq. CF, NF,
WF. Scattered throughout U.S. Spr-fall.

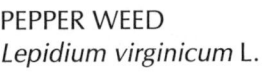

Illustrated Plants of Florida and the Coastal Plain

WILD RADISH
Raphanus raphanistrum L.

Annual, 30-80 cm tall; stems covered with short stiff hairs; basal leaves 10-30 cm long, dissected or lobed; upper leaves smaller, lobed; petals 12-20 mm wide, yellow, white, purple or pink; stamens 6; siliques 2-6 cm long. Native to Eurasia. Disturbed sites. Freq. all Fla. E to Va., N to Pa. and Canada. Spr-sum.

GARDEN RADISH
Raphanus sativus L.

Annual, to 1.1 m tall; leaves to 25 cm long, divided or lobed; petals to 1.5 cm long, white or pink; *siliques to 5 cm long. Native to Europe. Cultivated and persisting. Disturbed sites. Infreq. CF, NF, WF. N to Canada. Spr.

MARSH YELLOW-CRESS
Rorippa palustris (L.) Besser

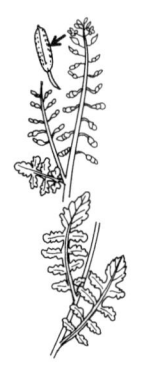

Annual or biennial, to 1.4 m tall; leaves 2-20 cm long; margins toothed, deeply cut or divided but not to midrib; lower leaves with winged stalks; upper leaves sessile; petals 1-3 mm long, yellow; *siliques 2-3 mm long. Disturbed areas. Rare all Fla. W to Tex., N to Canada. Spr-sum.

TERETE YELLOW-CRESS
Rorippa teres (Michx.) Stuckey

Annual or biennial, 10-40 cm tall; stems lying flat; leaves 2-10 cm long, appearing twice pinnate, cut to midrib; lower leaves petiolate; upper leaves sessile; petals to 1 mm long, yellow; *siliques 0.8-2 cm long. Moist to wet sites. Freq. all Fla. W to Tex., E to N.C. Wint-spr.

VIRGINIA ROCK-CRESS
Sibara virginica (L.) Rollins
[*Arabis virginica* (L.) Poir.]

Annual, 10-40 cm tall; stems branching, lying flat
to ascending, hairy at base; leaves 2-7 cm long, pinnatifid; *petals white; *siliques (capsules) 1-3 cm
long; seeds winged. Open disturbed sites. Rare
WF. W to Tex., E to Va. Spr.

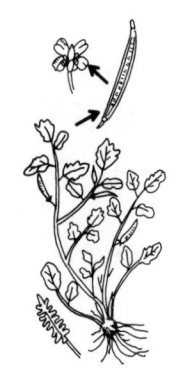

WIDE LEAF WAREA
Warea amplexifolia (Nutt.) Small

Annual, 30-70 cm tall; leaves 1-3 cm long, lanceolate, ovate, clasping; petals to 8 mm long, white
or purple; siliques (capsules) 3-5 cm long. Dry
sites. Rare CF, NF. Spr-sum.

NARROW LEAF WAREA
Warea carteri Small

Annual, 0.4-1.5 m tall; leaves 1-3 cm long;
flower stems (pedicels) to 1.2 cm long; petals to
6 mm long, white, sub-obicular; siliques to 6 cm
long. Dry sites. Rare SF, CF. All yr.

(not shown)

COASTAL PLAIN WAREA
Warea cuneifolia (Muhl.) Nutt.

Annual, to 1.2 m tall; leaves to 2.5 cm long;
pedicels to 6 mm long; petals to 6.5 mm long,
white, or pink to purple, obovate, cuneate;
siliques to 4 cm long. Dry sites. Infreq. WF.
N to N.C. Spr-sum.

GOURD FAMILY   *CUCURBITACEAE*

Herbaceous or woody vines; leaves alternate; flowers unisexual (monoecious or dioecious); sepals 3-6; petals 3-6; ovary inferior; fruit a berry, pepo or capsule.

4 sepals

CITRON
Citrullus lanatus (Thunb.) Mats. & Nakai
[*C. vulgaris* Schrad.]

Annual, to several m long; stems creeping, hairy; leaves 10-18 cm long, 7- to 9-lobed; petals 5, 1.5-2.5 cm long, yellow; berries (watermelons), 10-40 cm long. Disturbed sites. Infreq. all Fla. W to Tex., N to N.C. Spr-fall.

CREEPING CUCUMBER
Melothria pendula L.

Perennial vine, creeping, climbing or trailing; leaves 2-8 cm long, 3- to 5-lobed; petals 5, to 2.7 mm long, yellow; berries 10-25 mm long. Moist, disturbed sites. Common all Fla. W to Tex., N to Va. Spr-fall.

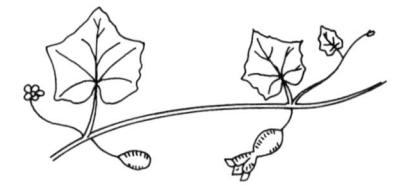

WILD BALSAM-APPLE
Momordica charantia L.

Annual or perennial vine, creeping, climbing, to several m long; leaves 5-7 cm long, 4-12 cm wide, 5- to 7-lobed; petals 1.5-2.5 cm wide, yellow; berries 4-12 cm long, orange, with soft spines. Thickets, disturbed sites. Freq. SF, CF, NF. W to Tex. All yr.

SEDGE FAMILY   *CYPERACEAE*

Herbs resembling grasses; stems mostly 3-angled; leaves alternate, occurring in 3 levels; flowers bisexual or unisexual (monoecious); flowers in spikelets; floral parts minute or absent; ovary superior; fruit an achene with somewhat persistent style (tubercle).

FINE LEAVED WHITE TOP SEDGE
Dichromena colorata (L.) Hitchc.

Perennial, to 0.5 m tall; involucral bracts 4-6, to 3 mm wide at base, white; apex of achenes with truncate tubercle. Moist sites. Common all Fla. W to Tex., E to Va. Spr-fall.

GIANT WHITE TOP SEDGE
Dichromena latifolia Baldw. ex Ell.

Similar to *Dichromena colorata* except: 0.5-1 m tall; involucral bracts 7-10, 8-10 mm wide at base; achenes with decurrent tubercle. Moist sites. Common all Fla. W to Tex., E to Va. Spr-fall.

CYRILLA FAMILY   *CYRILLACEAE*

Shrubs or trees; leaves alternate; flowers bisexual; sepals 5; petals 5; ovary superior; fruit a drupe or capsule.

LEATHERWOOD or TITI
Cyrilla racemiflora L.
[*C. arida* Small]

Perennial shrub, to 10 m tall; leaves 1-10 cm long; sepals 5; *petals 5, 2-5 mm long, white; *drupes 1.5-3 mm long. Swampy sites or scrub. Common CF, NF, WF. W to Tex., E to Va. Spr-sum.

YAM FAMILY   *DIOSCOREACEAE*

Climbing herbs or shrubs, arising from rhizomes or tubers; leaves alternate; flowers bisexual or unisexual (monoecious) or (dioecious); floral parts 6; ovary inferior; fruit a capsule, samara or berry.

AIR POTATO
Dioscorea bulbifera L.

Perennial, high climbing vine, to 20 m long; leaves 5-25 cm long, heart-shaped; flowers dioecious, small, white, fragrant; *aerial tubers 1-10 cm in diam.; fruit a capsule. Native to Asia. Escapes from cultivation. Wooded disturbed sites. Infreq. all Fla. Spr-sum.

WILD YAM
Dioscorea floridana Bartl.

Perennial vine, to 3 m long; leaves to 15 cm long,
heart-shaped; flowers dioecious, small, white;
aerial tubers none; fruit a capsule. Dry to moist
woods. Infreq. CF, NF, WF. N to N.C. Spr.

(not shown)

SUNDEW FAMILY   *DROSERACEAE*

Carnivorous herbs; leaves alternate, basal or whorled, modified as insect traps by having jaw-like appendages or tentacle-like hairs producing sticky goo; flowers bisexual; sepals 4
or 5; petals 4 or 5; ovary superior; fruit a capsule.

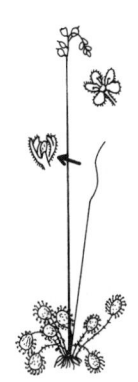

DWARF SUNDEW
Drosera brevifolia Pursh

Perennial; leaves 4-10 mm long; flower stalk 1-8 cm
tall, with glandular hairs; sepals 3-3.5 mm long;
flowers 11-15 mm wide, white, pink or purple;
*capsules approximately same size as sepals. Moist
to wet sites. Infreq. CF, NF, WF. W to La., E to
Tenn. and N.C. Spr.

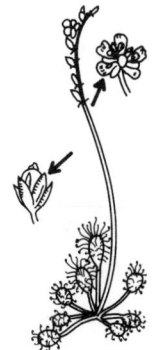

COMMON PINK SUNDEW
Drosera capillaris Poir.
[*D. leucantha* Shinners]

Perennial; leaves to 8 cm long, spoon-shaped,
with tentacle-like hairs, in basal rosettes 2-10 cm
in diam.; flower stems 4-35 cm tall, smooth;
*petals to 7 mm long, pink; capsules equal to
sepals; seeds brown, with 14-16 longitudinal
ridges. Moist to wet acid sites. Freq. all Fla.
W to Tex., N to N.C. Spr-sum.

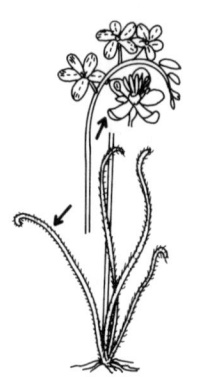

DEW-THREAD
Drosera filiformis Raf.

Perennial; *leaves 8-25 cm long, filiform-shaped,
with purple glandular hairs; flower stems 6-22 cm
long, smooth; *petals to 1.5 cm long, purple; capsule broad, shorter than sepals; seeds black, with
a crater-like surface. Moist sites. Infreq. WF.
N to Mass. Spr.

NARROW-LEAF SUNDEW
Drosera intermedia Hayne

Perennial; leaves to 20 mm long; flower stalk to
20 cm tall, smooth; sepals 5-5.5 mm long; flowers
10-12 mm wide, white to pink; capsules approximately same size as sepals; seeds black, papillose.
Swamps, creeks. Rare CF, NF, WF. W to Tex.,
N to Canada. Spr-fall.

DEW-THREAD
Drosera tracyi Macfarlane

Perennial; leaves to 40 cm long, filiform-shaped,
with green tentacles; flower stems 25-45 cm tall,
with green glandular hairs; petals to 3 cm long,
purple to pink; capsule broad, shorter than sepals;
seeds black, with a crater-like surface. Moist to
wet sites. Freq. WF. W to La., E to Ga. Spr.

(not shown)

EBONY FAMILY   *EBENACEAE*

Trees or shrubs; leaves mostly alternate; flowers mostly unisexual (dioecious); sepals 4-7; petals 3-7; ovary mostly superior; fruit a berry or capsule.

PERSIMMON
Diospyros virginiana L.

Perennial tree, to 35 m tall; leaves 7-15 cm long; male and female flowers on different plants (dioecious); sepals 4-5; petals 4-5, to 11 mm long, green to yellow; berries 3-5 cm in diam., yellow to orange, edible; seeds 3-8. Dry to moist sites. Freq. all Fla. W to Tex., N to Iowa, E to Del. Spr.

CROWBERRY FAMILY   *EMPETRACEAE*

Evergreen shrubs; leaves overlapping; flowers mostly unisexual (dioecious); floral parts 4-6, in 2 whorls; ovary superior; fruit a drupe.

FLORIDA ROSEMARY
Ceratiola ericoides Michx.

Perennial shrub, to 3 m tall, round or moundlike, evergreen; leaves 8-12 mm long; flowers to 1.5 mm long, yellow or red; drupes 2-3 mm wide, yellow, olive or red. Dry woods, scrub. Freq. all Fla. W to Miss., N to S.C. Spr-sum.

HEATH FAMILY	***ERICACEAE***

Acidic soil-dwelling trees, shrubs, vines or herbs; leaves mostly alternate, evergreen; flowers bisexual; sepals 4-7; petals 4-7; ovary mostly superior; fruit a drupe, capsule or berry.

PIPEWOOD
Agarista populifolia (Lam.) Judd
[*Leucothoe populifolia* (Lam.) Dippel]

Perennial, 1-4 m tall; leaves 3-10 cm long, evergreen, finely reticulate-veined; flowers racemose, narrow urceolate; flower stalks 7-10 mm long; *petals 5, to 8 mm long, white; bracts much shorter than flower stalks; *capsules 5-6 mm in diam. Low, swampy sites. Infreq. CF, NF, WF. N to S.C. Spr.

TAR FLOWER
Befaria racemosa Vent.

Perennial shrub, 0.4-2.5 m tall; leaves 2-6 cm long, evergreen, thick; sepals 7; petals 7, 2-3 cm long; *flowers fragrant, white or pinkish, sticky; capsules 6-8 mm in diam. Pinelands. Common all Fla. N to Ga. Wint-sum.

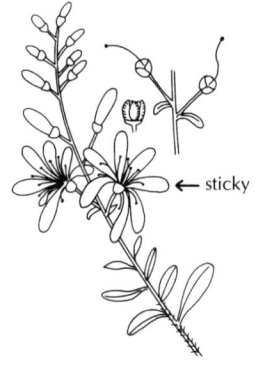

← sticky

DWARF HUCKLEBERRY
Gaylussacia dumosa (Andr.) Torr. & Gray

Perennial rhizomatous shrub, 0.1-1 m tall; leaves 2-6 cm long, deciduous to semievergreen, with amber glandular dots below; *petals 5, 5-9 mm long, white to pink; bracts of raceme leaflike; drupes 6-8 mm in diam., berry-like, black. Scrub, acid swamps to well-drained areas. Freq. all Fla. W to La., N to Canada. Spr.

DOTTED DANGLEBERRY
Gaylussacia frondosa (L.) Torr. & Gray

Perennial rhizomatous shrub, to 3 m tall; leaves
to 7 cm long, deciduous; petals 5, over 3 mm
long, green to white with pink tinge; *drupes to
8 mm in diam., blue. Low, woody sites. Infreq.
CF, NF, WF. W to La., E to N.H. Spr.

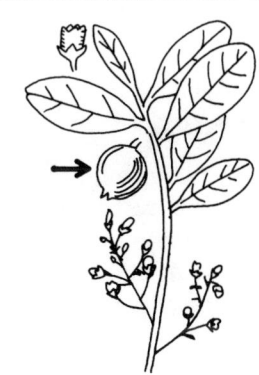

DOG-HOBBLE
Leucothoe axillaris (Lam.) D. Don

(not shown)

Perennial, 1-2 m tall; leaves to 15 cm long,
evergreen; lower surface of leaves minutely hairy
when young, not reticulate-veined; racemes
dense; flower stalks 1-4.5 mm long, stout; petals
5, to 8 mm long, white; bracts half as long as
flower stalks; capsules to 5 mm in diam. Moist,
woody sites. Rare CF, NF, WF. W to La.,
E to Va. Spr.

FETTER BUSH
Leucothoe racemosa (L.) Gray

Perennial shrub, 1-4 m tall; leaves 2-7 cm long;
sepals 5; petals 5, to 8 mm long, white with pink
highlights; capsules to 5 mm long. Low wet sites.
Infreq. CF, NF, WF. W to La., N to Mass. Spr.

RUSTY LYONIA
Lyonia ferruginea (Walt.) Nutt.

Perennial shrub or tree, to 6 m tall; leaves to
7.5 cm long, evergreen, with rusty hairs covering
lower surface; leaf margins smooth, rolled under;
*petals to 4 mm long, white to pink; capsules to
6 mm long. Dry to wet sites, scrub. Freq. CF,
NF, WF. E to Va. Spr.

STAGGER BUSH
Lyonia fruticosa (Michx.) G. S. Torr.

Perennial, 1-3 m tall; leaves 4-6 cm long, leathery, evergreen, with rusty scales below; flowers on current season twigs; petals to 5 mm long, white; capsules to 5 mm long. Scrub, dry sites. Freq. all Fla. N to S.C. Spr.

MALEBERRY
Lyonia ligustrina (L.) DC.

Perennial shrub, 1-4 m tall; twigs round; leaves 2-8 cm long, deciduous, (none on old wood), hairy; leaf margins with fine sharp teeth; sepals 5; *petals 5, to 3.5 mm long, white; capsules 2.5-4 mm wide. Dry woods, wet sites. Freq. all Fla. W to Tex., N to Ohio, E to Maine. Spr-sum.

SHINY LYONIA or FETTER BUSH
Lyonia lucida (Lam.) K. Koch

Perennial shrub or tree, to 4 m tall; leaves 2-8 cm long, evergreen, leathery, with translucent veins along margin; lower leaf surfaces with minute glands; leaf margins smooth, rolled under; *petals to 1 cm long, white or rose; *capsules to 5 mm long. Scrub or swamps. Freq. all Fla. W to La., N to N.C. Spr.

LARGE FLOWERED STAGGERBUSH
Lyonia mariana (L.) D. Don

Perennial, to 2 m tall; leaves 2-10 cm long, membranous, deciduous; flowers on previous season twigs; petals 5, 7-14 mm long, white to pinkish; filaments appendaged; ovary superior; *capsules to 7 mm long. Moist sites. Infreq. CF, NF, WF. W to Tex., E to Pa. and Conn. Spr.

SWAMP HONEYSUCKLE
Rhododendron viscosum (L.) Torr.
[*R. serrulatum* (Small) Millais]

Perennial shrub, to 5 m tall; leaves 1-7 cm long, deciduous; flowers 2-lipped, opening after new leaves; floral tube 15-35 mm long, white; stamens exserted; capsules 1-2 cm long with glandular hairs. Wet, swampy sites. Infreq. CF, NF, WF. W to Miss., N to Ohio, E to Maine. Spr-sum.

SPARKLEBERRY
Vaccinium aboreum Marsh.

Perennial shrub or small tree, to 10 m tall; leaves 2-7 cm long, entire to serrulate, glandular-dentate, leather-like, evergreen, deciduous; *flowers campanulate, closed in bud; petals 5-8 mm long, white to pinkish; stamens included; anthers with dorsal awns; *berries 5-7 mm in diam., black. Dry woody sites. Freq. CF, NF, WF. W to Tex., N to Mo., E to Va. Spr.

DARROW BLUEBERRY
Vaccinium darrowii Camp

Perennial shrub, 20-60 cm tall; leaves 5-15 mm long, not stipitate glandular beneath, glaucous on lower surfaces, evergreen; flowers 6-8 mm long, white or pink; berries 4-6 mm in diam., blue, glaucous. Dry sites. Freq. CF, NF, WF. W to Tex., N to Ga. Spr-early sum.

ELLIOTT'S BLUEBERRY
Vaccinium elliottii Chapm.

Perennial, 1-3 m tall; leaves to 3 cm long, deciduous; petals to 6 mm long, white or pink; berries to 1 cm long, black. Low sites. Infreq. CF, NF, WF. W to Tex., E to Va. Spr.

HIGH BUSH BLUEBERRY
Vaccinium fuscatum Ait.
[*V. corymbosum* L.]

Perennial shrub, 1-3 m tall; leaves 2-8 cm long,
deciduous; flowers 5-11 mm long, white, pink
or red; berries 5-10 mm in diam., black or blue,
glaucous. Dry to wet sites. Freq. CF, NF, WF.
W to Tex., N to Canada. Spr-early sum.

SHINY BLUEBERRY
Vaccinium myrsinites Lam.

Perennial shrub, 20-60 cm tall, rhizomatous;
leaves 8-15 mm long, evergreen, with glandular
hairs beneath; flowers 5-8 mm long, urceolate,
white to pink; berries 5-8 mm in diam., black to
bluish. Dry to moist sites. Common all Fla.
W to Ala., E to S.C. Wint-spr.

DEERBERRY
Vaccinium stamineum L.
[*V. floridanum* (Nutt.) Greene]

Perennial shrub, to 2 m tall; branchlets hairy;
leaves 1.5-10 cm long, entire, thin, deciduous;
flowers open in bud; stamens exserted; petals 5,
5-8 mm long, white; berries to 10 mm in diam.,
pink to purple. Dry, woody sites. Common all
Fla. W to La., N to Canada. Spr.

PIPEWORT FAMILY   ERIOCAULACEAE

Aquatic or terrestrial herbs; leaves basal or on stems; flowers unisexual (monoecious or dioecious); sepals 2 or 3; petals 2 or 3; ovary superior; fruit a capsule.

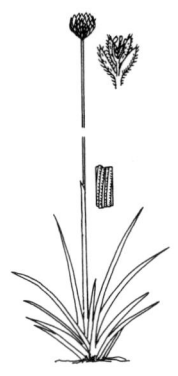

SOFT-HEAD PIPEWORT
Eriocaulon compressum Lam.

Perennial; leaves to 30 cm long; flower stalks to 1 m tall, 10- to 12-ridged, with sheaths longer than leaves; flower heads to 2 cm wide, soft; involucres 3-4 mm high, gray. Wet sites. Common all Fla. W to Tex., E to N.J. Spr-sum.

HARD-HEAD PIPEWORT
Eriocaulon decangulare L.

Perennial; roots septate; leaves to 50 cm long; flower stalks 0.3-1.2 m tall, 8- to 14-ridged, with sheaths shorter than leaves; flower heads to 2 cm wide, hard; involucres 2-4 mm long, straw-colored. Moist sites. Common all Fla. W to Tex., E to N.J. Sum.

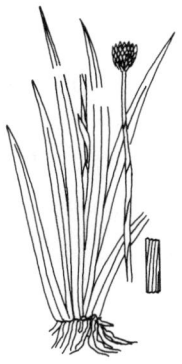

BLACK-HEAD PIPEWORT
Eriocaulon ravenelii Chapm.

Perennial, 4-30 cm tall; stems 5- or 6-ridged; roots septate; leaves 2-3 cm long, with dots; heads 3-5 mm in diam., gray-brown. Low wet sites, swamps. Infreq. all Fla. W to Tex., N to S.C. Sum-fall.

LITTLE WHITE BOG-BUTTON
Lachnocaulon anceps (Walt.) Morong

Perennial; leaves 2-6 cm long; flower stalks 5-
45 cm long; flower heads over 5 mm wide; seeds
to 0.5 mm long, ridged, dull. Moist, wet sites.
Common all Fla. W to Tex., E to N.J. Spr-sum.

LITTLE BOG-BUTTON
Lachnocaulon beyrichianum Sporl. ex Koern.

Perennial; leaves 2.5-4 mm broad; flower stalks
15-23 cm tall; flower heads 3-4.5 mm wide, gray-
whitish; seeds to 0.5 mm long, smooth, lustrous.
Wet sites. Common CF, NF, WF. N to N.C.
Sum-fall.

BROWN BOG-BUTTON
Lachnocaulon engleri Ruhl. in Engler

Perennial, 5-30 cm long; stems smooth, 3- to 5-
ridged, twisted; leaves to 3 cm long; heads 3-8 mm
long, brown. Moist sites. Infreq. CF, NF, WF.
Sum-fall.

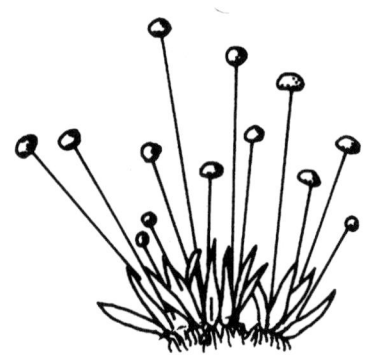

LITTLE BROWN BOG-BUTTON
Lachnocaulon minus (Chapm.) Small

Perennial; leaves to 4 cm long; flower stalks 3-
30 cm tall, hairy; flower heads 3-4 mm wide,
gray, gray-brown; involucres to 1 mm long;
seeds to 0.5 mm long. Moist sites. Infreq. all Fla.
W to Ala., N to N.C. Sum-fall.

SHOE BUTTONS
Syngonanthus flavidulus (Michx.) Ruhl. in Engler

Perennial; roots unbranched, fleshy; leaves to
6 cm long; flower stalks to 30 cm tall, 5-ridged;
flower heads to 10 mm wide, gray; involucres to
3 mm long, yellow. Wet flatwoods, marshes.
Common all Fla. W to Ala., E to N.C. Spr.-sum.

SPURGE FAMILY	***EUPHORBIACEAE***

Herbs, shrubs or trees, mostly succulent with watery or milky
sap; leaves mostly alternate; stipules present; flowers unisexual
(monoecious or dioecious); sepals many to 0; petals many to 0;
ovary superior; fruit mostly a capsule.

THREE-SEEDED MERCURY
Acalypha gracilens Gray

Annual, 10-80 cm tall; stems hairy; leaves 2-6 cm
long, elliptic; petioles to 1 cm long; flowers
green; male and female reproductive parts in separate flowers (unisexual) but on same spike
(monoecious); capsules to 2 mm long, hairy. Dry
sites. Freq. CF, NF, WF. W to Tex., N to Mass.
Spr-fall, all yr S.

HORNBEAM THREE-SEEDED MERCURY
Acalypha ostryifolia Riddell

(not shown)

Annual, 30-80 cm tall; stems with fine hairs;
leaves 3-8 cm long, ovate; petioles to 7 cm long;
flowers green; male and female reproductive
parts in separate flowers (unisexual) in separate
spikes; male spikes axillary; female spikes terminal; capsules to 2.5 mm long, warty. Sandy, disturbed areas. Infreq. all Fla. W to Tex., N.Mex.,
Ariz., Iowa, Okla., N to Pa. and N.J. Sum-fall.

Acalypha rhomboidea Raf.

Annual, to 70 cm tall; stems slightly hairy; leaves
to 10 cm long, rhombic-ovate; petioles over 1 cm
long; flowers green, unisexual, monoecious; capsules to 2 mm long, hairy. Wet sites. Infreq. NF,
WF. W to Tex., N to Canada. Sum.

(not shown)

SAND DUNE SPURGE
Chamaesyce bombensis (Jacq.) Dugand

(not shown)

Annual; stems lying flat; leaves to 1 cm long,
fleshy, entire; cyathia 1 per leaf axil; gland
appendages minute, white; capsules to 2 mm
long, smooth. Coastal dunes. Infreq. SF, CF.
W to Tex., N to Va. All yr S.

ROUND-LEAVED SPURGE
Chamaesyce cordifolia (Ell.) Small

Annual, to 60 cm long; stems lying flat on
ground, matting, smooth; cyathia 1 per leaf axil;
leaves 0.3-1 cm long; gland appendages minute,
white, green, petal-like; capsules to 2 mm long,
smooth. Dry sites. Freq. CF, NF, WF. W to Tex.,
N to N.C. Spr-fall.

GARDEN SPURGE
Chamaesyce hirta (L.) Millsp.

Annual, 20-40 cm tall; stems branching or
ascending, hairy; leaves 1-5 cm long, hairy;
cyathia several in axillary cymes; glands with
minute white appendages; capsules to 1 mm
long, hairy. Disturbed sites. Freq. all Fla.
W to Tex., N to S.C. Sum-fall, all yr S.

ERECT SPURGE
Chamaesyce hypericifolia (L.) Millsp.

Annual, to 60 cm tall; stems smooth, ascending
or erect; leaves 1-3 cm long; leaf margins
toothed; cymes subtended by 2 bracts, rarely
branched; flowers white with red tips; capsules
to 2 mm long, smooth. Disturbed sites. Freq. SF,
CF, NF. Ga. and Tex. All yr.

HYSSOP SPURGE
Chamaesyce hyssopifolia (L.) Small

Annual, to 80 cm tall; stems erect to ascending,
smooth to slightly hairy; leaves 1-4 cm long;
cyathia several, in peduncled cymes; gland
appendages white or pink, petal-like; capsules
1.5-2 mm long, smooth. Dry to wet disturbed
sites. Common all Fla. N to N.C. Spr-fall, all yr S.

SPOTTED SPURGE
Chamaesyce maculata (L.) Small

Annual, to 50 cm long; stems erect to prostrate,
hairy; leaves to 1.5 cm long; cyathia axillary;
gland appendages petal-like, white, pink; capsules to 1.5 mm long, hairy. Dry to wet waste
sites. Freq. all Fla. N to N.C., W to Tex. Spr-fall.

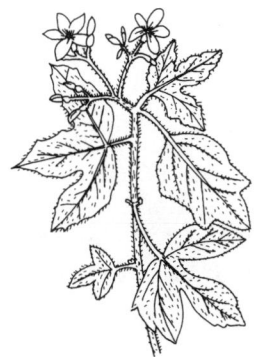

TREAD SOFTLY
Cnidoscolus stimulosus (Michx.) Engelm. & Gray

Perennial, 0.1-1.2 m tall; stems covered with
stinging hairs; leaves 10-30 cm wide; sepals 5,
1-1.5 cm long, petal-like, white; petals 0; capsules to 1 cm long. Dry sites. Common all Fla.
W to La., E to Va. Spr-fall, all yr S.

SILVER LEAF CROTON
Croton argyranthemus Michx.

Perennial, 30-60 cm tall; leaves 3-6 cm long,
oblong, elliptic or linear, widest near middle,
with silvery scales on undersides, often brown-
flecked; petioles 2-10 mm long; male flowers
with white petals; capsules to 5 mm long,
3-lobed. Dry sites. Freq. CF, NF, WF.
W to Tex., N to Ga. Spr-sum.

WOOLLY CROTON
Croton capitatus Michx.

Annual, 0.5-2 m tall; leaves to 15 cm long, oval
to lanceolate, with tan and gray hairs; flowers in
racemes to 3 cm long; *capsules to 9 mm long.
Disturbed sites. Infreq. CF, NF, WF. W to Tex.,
N to N.Y. Spr-fall, all yr S.

ELLIOTT'S CROTON
Croton elliottii Chapm.

Annual, to 90 cm tall; leaves to 5 cm long, linear   (not shown)
or linear-lanceolate, with gray hairs; flowers in
racemes to 2 cm long; capsules to 5 mm long. Low
sites. Infreq. WF. N to Ala., E to Ga. Sum-fall.

TROPIC CROTON
Croton glandulosus L.

Annual, 10-60 cm tall; stems and leaves covered
with star-shaped hairs; leaves 1-9 cm long, with
2 glands at base of blade; flowers monoecious,
white; racemes to 2 cm long; capsules 5-6 mm
long. Dry to moist sites. Freq. all Fla. W to Tex.,
E to Va. Sum-fall, all yr S.

PINELAND CROTON
Croton linearis Jacq.

Perennial, 0.6-2 m tall; leaves to 7 cm long, yellow
to brown, hairy beneath, used for tea; flowers dioecious, minute, white to yellow; capsules to 5 mm
long. Pinelands. Infreq. SF, Keys, CF. All yr.

BEACH-TEA
Croton punctatus Jacq.

Annual, to 1.2 m tall; leaves to 6 cm long, ovate,
widest near base; petioles 1-3 cm long; no petals;
capsules to 7 mm long. Dry coastal sites. Infreq.
all Fla. W to Tex., E to N.C. Sum-fall, all yr S.

RUSHFOIL
Crotonopsis linearis Michx.

Annual, 30-80 cm tall; upper leaves to 4 cm
long, linear; leaf surfaces covered with silvery,
minute, peltate scales; flower spikes 1-3 cm long,
white; *capsules 2.5-3 mm long, 3-lobed, indehiscent. Dry sites. Infreq. all Fla. W to Tex.,
E to S.C. Sum-fall.

RED AGALOMA
Euphorbia exerta (Small) Coker
[*Agaloma gracilior* (Crong.) Ward]

Perennial, 10-30 cm tall; leaves to 5 cm long, 0.4-
2.5 cm wide, linear, ovate, oblong or orbicular;
cyathia 1 per peduncled cyme, red; cyathia
glands without appendages; capsules to 3.5 mm
wide. Dry sandy, rocky sites. Rare CF. N to N.C.
Spr-sum, all yr S.

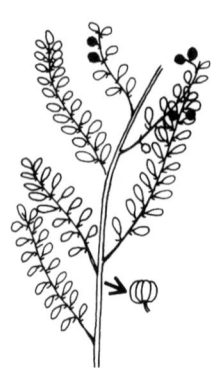

ABNORMAL PHYLLANTHUS
Phyllanthus abnormis Baillon

Perennial, 10-50 cm tall; leaves to 15 mm long; sepals to 3 mm wide, green to purple; *capsules to 3 mm wide, smooth. Coastal areas. Infreq. SF, CF, NF. All yr.

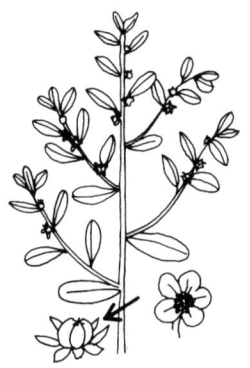

CAROLINA PHYLLANTHUS
Phyllanthus caroliniensis Walt.

Annual, 10-50 cm tall; stems smooth; leaves 5-20 mm long; sepals to 2.5 mm wide, green to red; *capsules to 2 mm wide, smooth. Moist to wet sites. Infreq. all Fla. W to Tex., N to N.C. Spr-fall.

LONG-STALKED PHYLLANTHUS
Phyllanthus tenellus Roxb.

Perennial, to 50 cm tall; leaves to 1 cm long; sepals to 2.5 mm wide; capsules to 2.5 mm wide. Native to Africa. Groves, cultivated and disturbed sites. Freq. all Fla. Spr-sum.

(not shown)

CHAMBER BITTER
Phyllanthus urinaria L.

Annual, 10-50 cm tall; stem leafless; branch leaves to 20 mm long, scattered; sepals to 1 mm wide; capsules to 2.5 mm wide, warty. Moist to wet sites, turf. Infreq. all Fla. W to Tex., N to N.C. Sum-fall.

PAINTED LEAF
Poinsettia cyathophora (Murr.) Kl. & Gke.

Annual, to 1.5 m tall; stems slightly hairy; leaves
5-18 cm long, fiddle-shaped, linear or ovate;
flower bract bases red or green blotched; flowers
monoecious; involucral gland lipped; capsules
4-4.5 mm long. Disturbed sites. Common all
Fla. Sum-fall, all yr S.

WILD POINSETTIA
Poinsettia heterophylla (L.) Kl. & Gke.

Annual, 0.3-1.2 m tall; stems slightly hairy; leaves
5-15 cm long, elliptic, ovate or obovate; floral
bract bases green- or purple-spotted; flowers
monoecious; involucral gland circular; capsules
3.5-4 mm long. Occasionally cultivated.
Disturbed sites. Common all Fla. W to Tex.,
N to Va. Sum-fall, all yr S.

CASTOR-BEAN *Ricinus communis* L.

Annual or perennial shrub, tree or herb, 1-5 m
tall; leaves to 40 cm wide, palmately lobed, with
petiole attached to lower leaf surface; male and
female flowers separate but on same plant; sepals
5; petals 0; male flowers yellow; female flowers
green; capsules 12-20 mm in diam., covered with
soft spines. Native to Africa. Escapes from cultivation. Disturbed sites. Freq. all Fla.
Throughout SE U.S. Sum-fall, all yr S.

CORKWOOD
Stillingia aquatica Chapm.

Perennial, 0.6-3 m tall; wood lighter than cork;
branches reddish or purplish; leaves to 8 cm
long; *flowers green, yellow or red; capsules to
0.5 cm long, to 1 cm wide. Wet sites. Infreq all
Fla. W to Miss., N to S.C. Spr-sum.

QUEEN'S DELIGHT
Stillingia sylvatica L.

Perennial, 0.25-1.2 m tall; leaves 2-15 cm long;
sepals 2-3, yellow; petals 0; stamens 2-3; female
flowers solitary at base of spike; male flowers at
top of spike; capsules 5-15 mm wide. Dry sites.
Common all Fla. W to Tex., E to Va. Spr-fall.

TWINING TRAGIA
Tragia cordata Michx.

Perennial, 0.25-1.3 m long; stems twining; leaves
5-11 cm long; petals 0; *male flower stalks to
2.5 mm long; stamens 3-5; female flower stalks
1-3 mm long; capsules 5-7 mm long. Dry sites.
Rare WF. W to Tex., N to Mo. Sum-fall.

SOUTH FLORIDA TRAGIA
Tragia saxicola Small

Perennial, to 20 cm tall; leaves to 2.5 cm long, to
1.5 cm wide, oval to rounded; leaf stems to 1 cm
long; flowers green to purple; capsules to 7.5 mm
wide, hairy. Pinelands. Endemic SF. All yr.

(not shown)

EASTERN TRAGIA
Tragia urens L.

Perennial, 10-40 cm tall; leaves 2-5 cm long, to
2 cm wide, linear with angled bases, hairy; leaf
stems to 3 mm long; flowers brown to red; no
petals; *capsules to 1 cm wide, hairy. Dry sites.
Infreq. all Fla. W to Tex., N to Va. Spr-fall.

NETTLE-LEAVED TRAGIA
Tragia urticifolia Michx.

Perennial, to 60 cm tall; leaves to 6 cm long, to 4 cm wide, oval to triangular, with angled to lobed bases; leaf stems to 1.5 cm long; flowers green; *capsules to 1 cm wide, hairy. Dry sites. Rare WF. W to Tex., N to Va. Spr-fall.

BEECH or OAK FAMILY   *FAGACEAE*

Trees or shrubs; leaves alternate; flowers unisexual (monoecious); male flowers in long slender catkins (aments); female flowers on short spikes; floral parts bract-like with 4-7 lobes; ovary inferior; fruit a nut set in a cup of hardened bracts (acorn).

CHAPMAN'S OAK
Quercus chapmanii Sarg.

Perennial shrub or tree, to 15 m tall; leaves 5-10 cm long; margins entire; nuts to 25 mm long; acorn cups to 10 mm deep. Scrub or dry sites. Freq. all Fla. N to S.C. Wint-spr.

SAND LIVE OAK
Quercus geminata Small
[*Q. virginiana* Mill. var. *maritima* (Small) Sarg.]

Perennial shrub or tree, to 10 m tall; leaves 3-6 cm long, with raised venation; margins revolute; nuts to 17 mm long; acorn cups to 15 mm deep. Scrub or dry sites. Freq. all Fla. W to Miss., N to N.C. Spr.

DWARF LIVE OAK
Quercus minima (Sarg.) Small

Perennial branching shrub, to 1 m tall; stems
underground; leaves 3-10 cm long, nearly flat;
acorn cups to 15 mm wide, smooth within,
peduncled, maturing first year. Pine flatwoods.
Freq. all Fla. W to Ala., N to Ga. Spr.

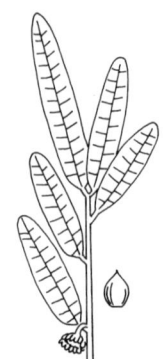

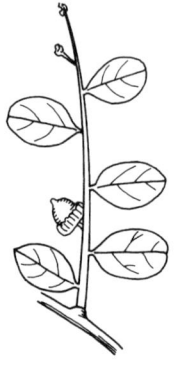

MYRTLE OAK
Quercus myrtifolia Willd.

Perennial shrub or tree, to 6 m tall; leaves 2-8 cm
long; margins revolute; nuts to 14 mm long;
acorn cups to 4 mm deep. Scrub, ridges. Infreq.
all Fla. Spr.

RUNNING OAK
Quercus pumila Walt.

Perennial branching shrub, 30-60 cm tall; stems
underground; leaves to 12 cm long, with revolute
margins; acorn cups 12-15 mm wide, hairy within, mostly sessile. Dry flatwoods. Infreq. all Fla.
W to Miss., E to N.C. Spr.

(not shown)

FUMITORY FAMILY   *FUMARIACEAE*

Herbs or shrubs; leaves alternate or basal; flowers bisexual; sepals 2; petals 4; ovary superior; fruit a capsule.

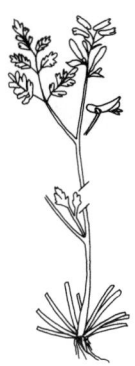

HARLEQUIN
Corydalis micrantha (Engelm.) A. Gray

Annual, 10-40 cm tall; leaves to 15 cm long, finely divided; petals 9-15 mm long, yellow, spurred; capsules 1-2 cm long. Disturbed sites. Freq. CF, NF, WF. W to Mo., N to Minn. Spr.

COMMON FUMITORY
Fumaria officinalis L.

Annual, 0.2-1 cm tall; stems branching, ascending; leaflets to 4 cm long; *flowers to 9 mm long, pink with purple tinge; *capsules to 2.5 mm in diam. Disturbed sites. Infreq. all Fla. W to Tex., N to Canada. Wint-spr.

GENTIAN FAMILY   *GENTIANACEAE*

Herbs or shrubs; leaves opposite, whorled or alternate; flowers bisexual; sepals 4-13; petals 4-13; ovary superior; fruit a capsule or berry.

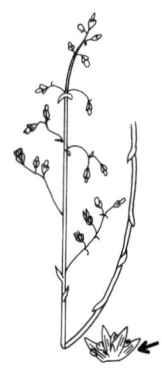

SCREW STEM
Bartonia paniculata (Michx.) Muhl.

Similar to *Bartonia virginica* except: sprawling; stems twisted, purplish; leaves scale-like, opposite and alternate; *petals greenish white, with apex acute to acuminate. Wet flatwoods. Infreq. NF, WF. W to Tex., N to Canada. Sum-fall.

WHITE BARTONIA
Bartonia verna (Michx.) Muhl.

Annual or biennial, 3-20 cm tall; leaves 1-3 mm long; sepals 4, to 3.5 mm long; petals 4, 0.5-1 cm long, white; capsules 4-6 mm long. Low, moist sites. Infreq. all Fla. W to Tex., N to N.C. Wint-spr.

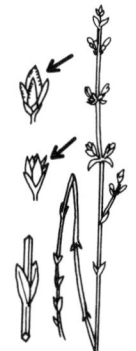

YELLOW BARTONIA
Bartonia virginica (L.) BSP.

Annual, 10-45 cm tall; stems erect, greenish yellow above, purple below; leaves 1-4 mm long, scale-like, opposite; sepals 4; *petals 4, greenish yellow to white, with apex rounded to mucronate; stamens 4; *capsules 3-5 mm long. Low, wet sites. Rare CF, WF. W to La., N to Canada. Sum-fall.

CATCHFLY GENTIAN
Eustoma exaltatum (L.) G. Don

Annual, 10-90 cm tall; leaves 2-15 cm long, sessile; sepals 5-6, persistent; petals 5-6, 1.8-2 cm long, rose-purple or white; capsules 2-3 cm long. Dry to wet coastal sites. Infreq. SF, CF, NF. All yr.

CATESBY GENTIAN
Gentiana catesbaei Walt.

Perennial, 20-70 cm tall; stems hairy; leaves to 7 cm long; *petals to 5 cm long, dark blue, violet or light blue; capsules to 2.5 cm long. Wet sites. Infreq. NF, WF. E to Va. Fall.

SOAPWORT GENTIAN
Gentiana saponaria L.

Perennial, 30-80 cm tall; stems smooth; leaves to 8 cm long; *petals to 5 cm long, blue, violet or white; capsules to 1.5 cm long. Moist to wet sites. Infreq. WF. E to Va., N to N.Y. Fall.

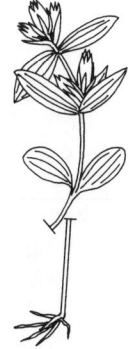

WHITE GENTIAN
Gentiana villosa L.

Perennial, 10-60 cm tall; leaves to 8 cm long; petals 4-5 cm long, greenish white or purplish; capsules 2-3 cm long. Disturbed sites. Rare WF. W to La., N to N.J. Fall.

ROSE PINK SABATIA
Sabatia angularis (L.) Pursh

Annual or biennial, 20-90 cm tall; stems 4-sided,
winged; leaves 1.5-3 cm long, clasping; sepals 5;
petals 5, 15-20 mm long, pink; *capsules to 5 mm
long. Moist sites. Rare NF, WF. W to Tex.,
N to Mich., E to Conn. Sum.

(not shown)

TEN-PETAL SABATIA
Sabatia bartramii Wilbur

Perennial; leaves to 10 cm long, succulent; upper
leaves narrower than stem; basal leaves rosulate;
petals 2.2-3.2 cm long. Pineland ponds. Infreq.
NF, WF. N to Ga., W to Miss. Sum-fall.

NARROW-LEAVED SABATIA
Sabatia brevifolia Raf.
[*S. elliottii* Steud.]

Annual, to 60 cm tall; leaves 0.3-2 cm long;
sepals tubular, with tubes to 7 mm long; petals
tubular, with tubes to 13 mm long, white; capsules 4-5 mm long. Moist, low sites. Common
all Fla. W to Ala., E to S.C. Spr-fall.

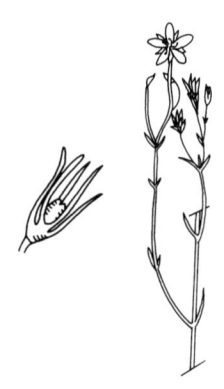

COASTAL PLAIN SABATIA
Sabatia calycina (Lam.) Heller

Perennial, 10-40 cm tall; leaves 2.5-6 cm long,
thin; basal leaves absent; petals 5-7, 7-15 mm
long, red to pink, rarely white, with yellow spot
on floral tube; capsules 7-8 mm long. Low, wet
sites. Freq. all Fla. W to Tex., E to Va. Sum.

BOG MARSH-PINK
Sabatia campanulata (L.) Torr.

Perennial, 10-60 cm tall, tufted with several stems; leaves 1-2 cm long, linear; basal leaves absent; petals pink with yellow throat. Wet flat-woods. Freq. NF, WF. W to La., N to Mass. Sum-fall.

No basal leaves

WHITE SABATIA
Sabatia difformis (L.) Druce

(not shown)

Perennial, 10-80 cm tall; leaves 1-4 cm long; petals to 2 cm long, white; capsules to 6 mm long. Flatwoods. Freq. CF, NF, WF. E to Ga., N to N.J. Spr-sum.

TEN-PETAL MARSH-PINK
Sabatia dodecandra (L.) BSP.

Perennial, 0.3-1 m tall; leaves 1.5-6 cm long, thin; upper leaves wider than stem; basal leaves often absent; petals 5-13, 1-2 cm long, red to purple or red to pink, rarely white, with yellow spot on floral tube; capsules to 8 mm long. Wet sites. Infreq. CF, NF, WF. W to Tex., E to S.C. Sum-fall.

LARGE-FLOWERED SABATIA
Sabatia grandiflora (Gray) Small

Annual, 0.4-1.2 m tall; leaves 3-10 cm long, succulent; upper leaves narrower than stem; petals 5, 1.5-2.5 cm long, red to pink, with yellow spot on floral tube; capsules 8-10 mm long. Low sites. Freq. all Fla. Endemic. W to Ala. All yr.

STAR SABATIA
Sabatia stellaris Pursh

Annual, 15-50 cm tall; leaves 1-4 cm long, thin;
upper leaves wider than or equal to stem; basal
leaves often absent; petals 5, 1-1.5 cm long, rose
to pink, rarely white, with yellow spot on floral
tube; capsules 5-10 mm long. Bogs. Freq. all
Fla. W to La., N to Mass. Sum-fall.

GERANIUM FAMILY   *GERANIACEAE*

Herbs or shrubs; leaves opposite or alternate; flowers bisexual;
sepals mostly 5; petals mostly 5; ovary superior; fruit a capsule
or schizocarp; upon maturation, fruits splitting apart into 5
sections by peeling backwards while remaining attached to
apex of capsule.

CRANESBILL
Geranium carolinianum L.

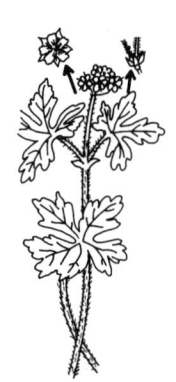

Annual, 20-60 cm tall; stems covered with short
hairs; leaves to 5 cm long, 2-8 cm wide, palmately divided; *flowers paired, pink-violet; *capsules
2-5 cm long. Disturbed sites. Common all Fla.
W to Calif., N to Canada. Spr-sum.

GOODENIA FAMILY   *GOODENIACEAE*

Herbs or shrubs; leaves mostly alternate; flowers bisexual; sepals 5; petals 5; ovary mostly inferior; fruit a drupe, berry, nut or capsule.

BEACHBERRY
Scaevola plumieri (L.) Vahl

Perennial shrub or herb, 0.3-1.8 m tall; leaves 3-6 cm long; flowers 2-3 cm long, white to pinkish, villous within; *berries 1-1.5 cm long, black, juicy. Coastal sites. Freq. SF, CF. All yr.

GRASS FAMILY   *GRAMINEAE*

Herbaceous or woody plants; leaves parallel-veined; sheaths open; flowers bisexual or unisexual (monoecious or dioecious); flowers in spikelets with glumes, lemna and palea; ovary superior; fruit a grain or caryopsis.

WOODS GRASS
Oplismenus setarius (Lam.) Roem. & Schult.

Perennial, 15-35 cm tall; stems creeping; spikelets to 3 mm long. Moist shady woods. Common all Fla. W to Tex., E to N.C. Sum-fall, all yr S.

SEA OATS
Uniola paniculata L.

Perennial, 1-2 m tall; leaves to 60 cm long; flowers in panicles 20-30 cm long; spikelets with up to 20 flowers. Coastal dunes. Common all Fla. W to Tex., E to Va. Sum-wint.

ST. JOHN'S WORT FAMILY   *GUTTIFERAE*

Herbs, shrubs or trees; leaves opposite, with punctate glands; flowers bisexual or unisexual (dioecious); sepals 4 or 5; petals 4 or 5; ovary superior; fruit a capsule or berry.

SHORT-LEAVED SANDWEED
Hypericum brachyphyllum (Spach) Steud.

Perennial shrub, 0.5-1.5 m tall; leaves 3-9 mm long, to 1 mm wide; petals 5, to 8 mm long, yellow; *capsules 4-5 mm long. Low, wet sites. Freq. all Fla. W to Miss., E to Ga. Sum-fall.

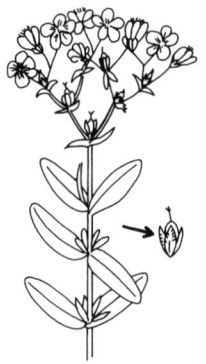

CLUSTER-LEAF ST. JOHN'S-WORT
Hypericum cistifolium Lam.

Perennial shrub, 0.3-1 m tall; leaves 1.5-4 cm long, with branchlets clustered in axils; petals 5, 4-8 mm long, yellow; *capsules to 5 mm long, 1-celled. Moist to wet sites. Common all Fla. W to Tex., N to N.C. Sum-fall.

ST. PETER'S-WORT
Hypericum crux-andreae (L.) Crantz
[*H. stans* (Michx.) Adams & Robson]

Perennial shrub, 0.3-1 m tall; leaves 1-4 cm long;
petals 4, to 1.8 cm long, yellow; capsules to 9 mm
long. Dry to moist sites. Infreq. all Fla. W to Tex.,
N to Ky. Sum-fall.

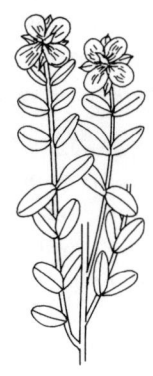

SANDWEED
Hypericum fasciculatum Lam.

Perennial shrub, 1-2 m tall; larger leaves 13-
17 mm long; petals 5, 6-9 mm long, yellow;
capsules 3-5 mm long. Low, wet sites. Common
all Fla. W to Tex., N to N.C. Spr-fall.

BEDSTRAW ST. JOHN'S-WORT
Hypericum galioides Lam.

Perennial shrub, 0.3-1.7 m tall, bushy; leaves
0.5-3 cm long; *petals 5, to 7 mm long, yellow;
*capsules to 5 mm long. Low, wet sites. Infreq.
CF, NF, WF. W to Tex., N to Ky. Sum-fall.

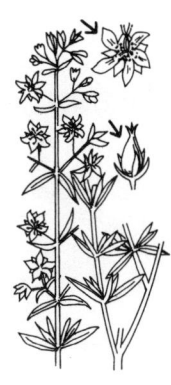

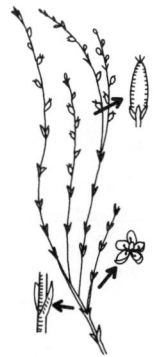

PINEWEED
Hypericum gentianoides (L.) BSP.

Annual, 10-50 cm tall; *leaves 1-5 mm long;
*petals 5, 1.5-4 mm long, yellow to orange; *cap-
sules 4-5 mm long. Dry to moist disturbed sites.
Freq. all Fla. W to Tex., N to Canada. Sum-fall.

ST. ANDREW'S-CROSS
Hypericum hypericoides (L.) Crantz
[*Ascyrum hypericoides* L.]

Perennial shrub, 0.3-1.5 m tall; stems erect to diffuse; leaves 0.8-3.5 cm long; sepals 4; outer sepals 5-12 mm long; inner sepals minute; petals 4, 6-10 mm long, yellow; stamens many; capsules 7-8 mm long; styles 2. Wet to dry sites. Common all Fla. W to Tex., N to Pa., E to Va. Sum-fall, all yr S.

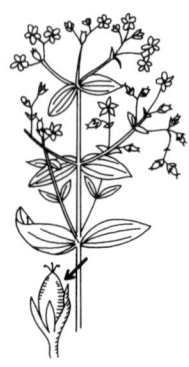

DWARF ST. JOHN'S-WORT
Hypericum mutilum L.

Perennial herb, 10-80 cm tall; leaves 1-5 cm long; petals 5, 2-3.5 mm long, yellow; *capsules 3-4 mm long. Wet sites. Common all Fla. W to Tex., N to Canada. Sum-fall.

MYRTLE-LEAVED ST. JOHN'S-WORT
Hypericum myrtifolium Lam.

Perennial shrub, 0.3-1 m tall, evergreen; leaves 1-3 cm long; petals 5, 10-15 mm long, yellow; capsules 5-8 mm long. Low, wet sites. Freq. all Fla. W to Miss., N to Ga. Spr-fall.

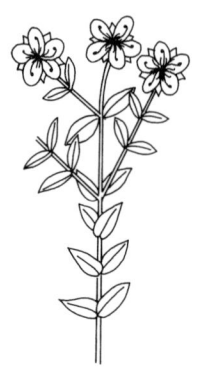

MATTED SANDWEED
Hypericum reductum P. Adams

Perennial shrub, 10-50 cm tall; stems lying flat, often matted; young stems 6-angled, with auriculate nodes; leaves to 5 mm long; petals 5, to 7 mm long, yellow; *capsules 6-10 mm long. Dry woods to moist sites. Freq. all Fla. W to Ala., N to N.C. Spr-sum.

HAIRY ST. JOHN'S-WORT
Hypericum setosum L.

Annual or biennial herb, 30-75 cm tall; stems densely hairy; leaves 3-15 mm long, densely hairy; petals 5, 4-7 mm long, orange-yellow; capsules 4-5 mm long. Moist sites. Rare CF, NF, WF. W to Tex., N to Va. Sum.

(not shown)

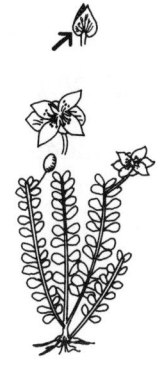

LITTLE ST. ANDREW'S-CROSS
Hypericum suffruticosum P. Adams & Robson

Perennial shrub, 7-15 cm tall; leaves 3-9 mm long; petals 4, to 7 mm long, yellow; *capsules to 5 mm long. Pinelands. Infreq. CF, NF, WF. W to Miss., E to Ga. Spr-sum.

HEART-LEAVED ST. PETER'S-WORT
Hypericum tetrapetalum Lam.
[*Ascyrum tetrapetalum* (Lam.) Vail]

Perennial shrub, 0.1-1 m tall; leaves 1-2.5 cm long; sepals 4; outer sepals leaflike; petals 4, to 21 mm long, yellow; capsules to 6 mm long, with 3 persistent styles. Low, wet sites. Common all Fla. N to Ga. Spr-fall, all yr S.

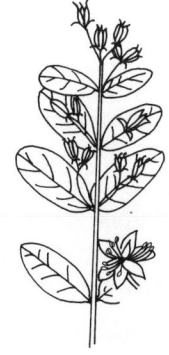

MARSH ST. JOHN'S-WORT
Triadenum virginicum (L.) Raf.

Perennial, 20-60 cm tall; leaves 2-7 cm long, sessile; petals 4-7 mm long, pink with green or purple highlights; capsules 0.8-1 cm long. Marshy, swampy sites. Infreq. CF, NF, WF. W to Tex., N to Canada. Sum.

WALTER'S MARSH ST. JOHN'S-WORT
Triadenum walteri (Gmel.) Gl.

Perennial, 0.3-1 m tall; leaves to 15 cm long,
with the lower petiolate; petals to 6.5 mm long,
pink with green or purple highlights; capsules
to 1 cm long. Swamps. Infreq. CF, NF, WF.
W to Tex., N to N.J. Sum.

(not shown)

BLOOD ROOT FAMILY　　　　　*HAEMODORACEAE*

Fibrous-rooted, rhizomatous or tuberous herbs; underground
portion often red in color; leaves mostly basal; flowers bisexual; floral parts 6, in 1 to several whorls; ovary inferior or
superior; fruit a capsule.

RED ROOT
Lachnanthes caroliniana (Lam.) Dandy

Perennial, 0.3-1.2 m tall; roots and rhizomes red;
basal leaves to 30 cm long, 5-15 mm wide, overlapping; flowers and stems covered with white
hairs; flowers yellow; petals persisting; stamens 3,
exserted; capsules 5-6 mm in diam. Low, wet
sites. Common all Fla. W to La., N to Canada.
Spr-fall.

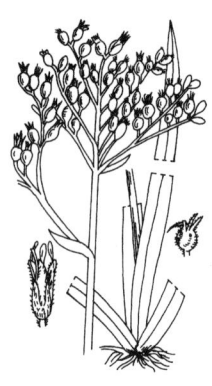

GOLDEN-CREST
Lophiola americana (Pursh) Wood

Perennial, 50-80 cm tall; roots and rhizomes
white to brown; leaves to 30 cm long, 3-10 mm
wide, overlapping; *flowers and stems covered
with white hairs; flowers yellow; stamens 6,
included; *capsules to 2 mm long; style persistent. Moist to wet sites. Freq. WF. W to Miss.,
N to Canada. Spr-sum.

WATER-MILFOIL FAMILY   *HALORAGACEAE*

Aquatic or moist land-dwelling herbs; leaves alternate, whorled or opposite; flowers bisexual or unisexual; sepals 2-4 or 0; petals 2-4 or 0; ovary inferior; fruit 1 to several nutlets.

PARROT'S-FEATHER
Myriophyllum aquaticum (Vell.) Verdc.
[*M. brasiliense* Camb.]

Perennial, to many cm long, forming mats; leaves 2-5 cm long, divided, feathery, whorled, mostly emersed; flowers rare, axillary, female only; nutlets to 2 mm long. Sluggish water. Freq. all Fla. W to Tex., N to Va. Spr-fall.

WATER MILFOIL
Myriophyllum heterophyllum Michx.

Perennial; leaves whorled, mostly submersed; emersed leaves divided into 12-28 sections; petals to 2.5 mm long; *fruits to 2.5 mm long. Ponds, lakes, slow water. Infreq. all Fla. W to N.Mex., N to Canada. Spr-fall.

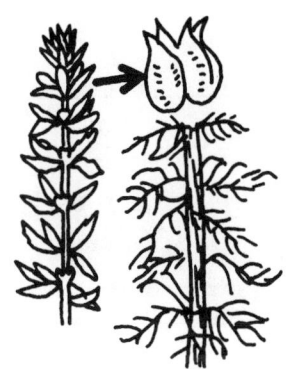

(not shown)

GREEN PARROT'S-FEATHER
Myriophyllum pinnatum (Walt.) BSP.

Perennial; leaves whorled, mostly submersed; emersed leaves pinnately divided into 8-10 segments; petals to 1.5 mm long; fruits to 1.5 mm long. Ditches, ponds. Infreq. all Fla. W to Tex., N to N.Y. Spr-fall.

MARSH MERMAID WEED
Proserpinaca palustris L.

Perennial, to 60 cm long; stems ascending to erect; leaves to 6 cm long, toothed when emersed, deeply divided otherwise; flowers minute, perfect; sepals 3, green or white; no petals; *fruits 4-5 mm long. Wet sites. Freq. all Fla. W to Tex., N to Canada. Spr-fall.

CUT-LEAF MERMAID WEED
Proserpinaca pectinata Lam.

Perennial, to 50 cm tall; stems prostrate, rooting, ascending to erect; all leaves 1-2.5 cm long, divided like a comb (pectinate); flowers minute, perfect; sepals 3, green or white; no petals; *fruits 3-4 mm long. Wet sites. Freq. all Fla. W to Tex., N to Canada. Spr-fall.

BUCKEYE FAMILY *HIPPOCASTANACEAE*

Trees or shrubs; leaves opposite; flowers bisexual or unisexual; sepals 5; petals 4 or 5; ovary superior; fruit a capsule.

RED BUCKEYE
Aesculus pavia L.

Perennial, 1-12 m tall; leaflets 6-17 cm long; *petals to 4 cm long, red; *capsules to 5 cm in diam. Woods. Infreq. CF, NF, WF. W to La., E to Va. Spr.

FROG'S-BIT FAMILY   *HYDROCHARITACEAE*

Aquatic herbs; leaves whorled or spiraled; flowers bisexual or unisexual (dioecious); sepals 3; petals 3; ovary inferior; fruit a capsule or berry.

BRAZILIAN ELODEA
Egeria densa Planch.
[*Elodea densa* (Planch.) Caspary]

Perennial, submersed, rooted or with free-floating pieces; upper leaves 4-6, 1.5-3 cm long, whorled; petals 7-11 mm long, white. Native to South America. Escapes from cultivation. Fresh water. Freq. CF, NF, WF. W to Tex., E to Va. Spr-fall.

HYDRILLA
Hydrilla verticillata (L. f.) Royle

Perennial; male and female flowers on separate plants (dioecious), submersed, rooting to several m below surface; stems branching below and forming tubers that germinate, also branching at surface; leaves mostly 4-8 but occasionally 2-3 in a whorl (verticel), 6-15 mm long, sharply toothed on margin, having 1 to few teeth on underside midvein making leaves rough to touch; flowers solitary, white, axillary from a spathe; male flowers short-stalked; female flowers 8-12, to 4 mm long; fruits 2- to 3-seeded. Shallow fresh or brackish water. Freq. all Fla. N to Ga. Spr-sum.

(not shown)

TAPE-GRASS
Valisneria americana Michx. [*V. neotropicalis* Marie-Victorin]

Perennial, forming new plants from runners; male and
female flowers on separate plants (dioecious), submersed;
leaves to 60 cm long, 3-10 mm wide, net-veined; flowers
white, minute; male flowers 200, on spadix enclosed in
spathe; female flower 1 per leaf axil, on stalks 0.5-1 m
long; capsules 7-15 cm long, 300- to 500-seeded. Still,
flowing water. Infreq. all Fla. W to Tex., N to Canada.
Spr-fall, all yr S.

WATERLEAF FAMILY *HYDROPHYLLACEAE*

Herbs or shrubs; leaves alternate or opposite; flowers bisexual;
sepals 5; petals 5; ovary superior; fruit a capsule.

TALL HYDROLEA
Hydrolea corymbosa Macbr. ex Ell.

Perennial, 20-70 cm tall; leaves 2-6 cm long;
*petals to 10 mm long; styles 2; *capsules oval.
Wet sites. Infreq. all Fla. N to Ga. Sum.

HYDROLEA
Hydrolea ovata Nutt.

(not shown)

Similar to *Hydrolea corymbosa* except: leaves
ovate or ovate-lanceolate, with conspicuous
spines; petals 10-14 mm long. Wet sites. Ga.,
W to Tex. Sum-fall.

HAIRY HYDROLEA
Hydrolea quadrivalvis Walt.
[*Nama quadrivalve* (Walt.) Kuntze]

Perennial, 0.2-1 m tall; stems with long hairs; leaves 4-12 cm long; petals to 5 mm long, blue; styles as long as ovary; *capsules to 6 mm long. Wet sites. Rare CF, NF, WF. W to La., E to Va. Sum.

IRIS FAMILY   *IRIDACEAE*

Herbs arising from rhizomes, corms or bulbs; leaves overlapping at base; flowers bisexual; sepals 3, petal-like; petals 3; ovary mostly inferior; fruit a capsule.

ANGLEPOD BLUE-FLAG
Iris hexagona Walt.
var. *savannarum* (Small) Foster

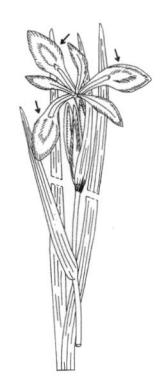

Perennial, 0.5-1.2 m tall; leaves to 1 m long, may exceed stem in length; flowers terminal, axial, purple, violet, blue or white; capsules to 8 cm long. Wet sites. Common all Fla. W to Tex., E to S.C. Spr-sum.

SLENDER BLUE-FLAG
Iris prismatica Pursh ex Ker.

(not shown)

Perennial, 30-60 cm tall; leaves to 40 cm long, 2-5 mm wide; sepals 4-5 cm long; petals 3-5 cm long, purple; capsules to 4 cm long. Low sites. Ga., N to Canada. Spr-sum.

BAY BLUE-FLAG
Iris tridentata Pursh

Perennial, 20-40 cm tall; leaves 15-40 cm long,
to 1 cm wide; sepals 6-9 cm long; petals to 2 cm
long, violet; capsules to 3 cm long. Low sites.
Infreq. NF, WF. N to N.C. Spr-sum.

DWARF IRIS
Iris verna L.

Perennial, 15-50 cm tall; leaves 10-15 cm long,
to 1.3 cm wide; petals 2-7 cm long, violet;
*capsules to 2.5 cm long. Dry woods. Rare WF.
W to Miss., N to Pa. Spr.

SOUTHERN BLUE-FLAG
Iris virginica L.

Perennial, 0.3-1 m tall; leaves 0.4-1.1 m long, to
3 cm wide; petals 7-10 cm long, blue to purple;
*capsules to 6 cm long. Low, wet sites. Infreq.
CF, NF, WF. W to Tex., N to Del. Spr.

FALL-FLOWERING IXIA
Nemastylis floridana Small

Perennial, 0.4-1.5 m tall; bulbs to 2 cm in diam.;
leaves 7-15 cm long; flowers 25-50 cm wide,
open 4-6 p.m.; capsules to 1.2 cm long. Low,
wet sites. Infreq. to rare eastern CF, eastern NF.
Endemic. Fall.

BARTRAM'S IXIA
Salpingostylis coelestina (Bartr.) Small
[*Nemastylis coelestina* (Bartr.) Nutt.;
Sphenostigma coelestinum (Bartr.) Foster]

Similar to *Nemastylis floridana* except: 20-50 cm
tall; bulbs to 1.2 cm in diam.; flowers 4.5-6 cm
wide; style long, not surrounded by filament
tube. Flatwoods. Rare eastern NF. Spr.

WIDE-WINGED BLUE-EYED-GRASS
Sisyrinchium angustifolium Mill.

Perennial; leaves to 30 cm long, to 4 mm wide;
flowers stems to 60 cm tall, broad-winged;
spathes to 2 cm long, 4-5 mm wide; *flowers
blue-violet; *capsules 4-6 mm long. Wet sites.
Infreq. all Fla. Spr-sum.

NARROW-WINGED BLUE-EYED-GRASS
Sisyrinchium atlanticum Bickn.

Perennial, 10-70 cm tall; leaves to 70 cm long;
flower stems to 40 cm tall, winged; spathes 2
together; flowers blue or occasionally white;
capsules 3-5 mm long. Disturbed open sites.
Freq. all Fla. N to Canada. Spr-sum.

YELLOW BLUE-EYED-GRASS
Sisyrinchium exile Bickn.

Annual, 6-15 cm tall; leaves to 2.5 cm long,
1-3 mm wide; petals to 6 mm long, yellow or
blue-white, with brown or red patch; capsules to
3 mm in diam. Low, moist sites. Infreq. CF, NF,
WF. W to Tex., N to Ga. Spr.

SANDHILL BLUE-EYED-GRASS
Sisyrinchium nashii Bickn.
[*S. arenicola* Bickn.]

Perennial; leaves to 30 cm long, to 4 mm wide;
flower stems l5-50 cm tall, narrowly winged and
margined; spathes 10-20 mm long; *flowers blue;
capsules 2-4 mm long. Flatwoods, sandhills.
Common all Fla. W to Miss., E to Va. Spr-sum.

ANNUAL BLUE-EYED-GRASS
Sisyrinchium rosulatum Bickn.

Annual, to 30 cm tall; stems ascending; leaves
2-8 cm long; flowers white, lavender, rose or
blue; petals with yellow eye and rose-purple eye
ring; capsules to 3 mm long. Moist to wet sites.
Freq. all Fla. W to Tex., N to N.C. Spr.

SCRUB BLUE-EYED-GRASS
Sisyrinchium solstitiale Bickn.
[*S. xerophyllum* Greene]

Perennial; leaves to 30 cm long, to 6 mm wide;
flower stems 20-65 cm tall, flat and winged,
fibrous at base; spathes 2-2.5 cm long; flowers
blue; capsules 5-8 mm long. Dry woods, scrub.
Freq. all Fla. Sum-fall.

KRAMERIA FAMILY　　　　　*KRAMERIACEAE*

Herbs or shrubs; leaves alternate; flowers bisexual; sepals 4 or 5; petals 5; ovary superior; fruit round, spiny.

SAND BUR
Krameria lanceolata Torr.

Perennial, to 30 cm long; stems prostrate to ascending, covered with silky hairs; leaves to 2.5 cm long, hairy; *flowers 2.5 cm wide, purple or red; pods to 12 mm in diam., spiny. Dry sites. Infreq. CF, NF, WF. N to Ga., W to N.Mex. and Kans. Spr-sum.

MINT FAMILY　　　　　　*LABIATAE*

Herbs or shrubs; stems square; glands or hairs present, usually releasing fragrant aroma; leaves opposite; flowers bisexual; sepals 5 or 2; petals 5; ovary superior; fruit 4 nutlets.

LAVENDER BASIL
Calamintha ashei (Weatherby) Shinners
[*Clinopodium ashei* (Weatherby) Shinners]

(not shown)

Perennial shrub, 10-50 cm tall; leaves to 1 cm long, linear; leaf margins revolute; petals 1.2-1.5 cm long, pink to purple. Scrub, dry sites. Infreq. CF, NF. Spr.

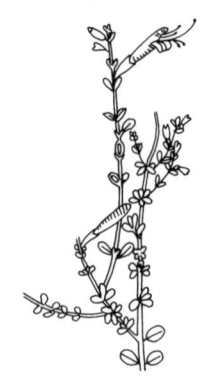

RED BASIL
Calamintha coccinea (Nutt.) Benth.
[*Clinopodium coccineum* (Nutt.) Kuntze]

Perennial shrub, 30-90 cm tall; leaves 0.5-2 cm
long, obovate, hairy on lower surface; leaf margins smooth; petals 3-5 cm long, usually scarlet,
rarely orange or yellow. Scrub, dry sites. Infreq.
CF, NF, WF. W to Miss., E to Ga. Spr-fall.

HORSE-BALM
Collinsonia canadensis L.

Perennial, 30-90 cm tall; leaves 2-25 cm long;
petals to 1.4 cm long, yellow; nutlets to 2.5 mm
long. Moist woods. Rare WF. W to Ark.,
N to Canada. Sum-fall.

PENINSULA DICERANDRA
Dicerandra densiflora Benth. ex DC.

Annual, 10-40 cm tall; leaves to 4 cm long; flowers many; petals to 1.2 cm long, pink to purple.
Dry sites. Infreq. CF, NF. E to Ga. Sum-fall.

(not shown)

DICERANDRA
Dicerandra linearifolia (Ell.) Benth.

Annual, 0.2-1 m tall; leaves 1-5 cm long; flowers
few; petals to 1.5 cm long, white, pink or purple.
Dry sites. Infreq. CF, NF, WF. N to Ala., E to Ga.
Sum-fall.

MUSKY MINT
Hyptis alata (Raf.) Shinners

Perennial, 0.4-2.5 m tall; stems hairy; leaves 4-12 cm long; flowers in heads; sepals to 8 mm long; petals to 5 mm long, white with violet spots; nutlets to 1.3 mm long. Low, moist sites. Freq. all Fla. W to Tex., N to N.C. Spr-fall, all yr S.

COMMON BITTER MINT
Hyptis mutabilis (A. Rich.) Briq.

Perennial, to 2 m tall; stems hairy; leaves 2-7 cm long; flowers in spreading spikes; *sepals to 7 mm long; petals 3-6 mm long, bluish violet; nutlets to 1.5 mm long. Native to tropical America. Disturbed sites. Common all Fla. W to La., E to Ga. Spr-fall.

MINTWEED
Hyptis pectinata (L.) Poir.

Perennial, to 1.5 m tall; stems slightly hairy; leaves to 8 cm long; flowers in clustered spikes; sepals to 5 mm long; petals to 4 mm long, blue; nutlets to 1 mm long. Disturbed sites. Freq. SF, CF. All yr.

(not shown)

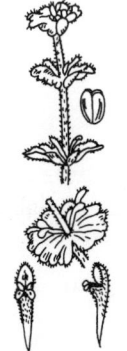

HENBIT
Lamium amplexicaule L.

Annual, to 35 cm tall; stems lying flat, ascending, hairy; leaves 0.6-3.5 cm long, sessile; petals 1.2-1.8 cm long, violet; nutlets to 2 mm long. Native to Europe. Open sites. Infreq. CF, NF, WF. W to Calif., N to Canada. Wint-spr.

LION'S EARS
Leonotis nepetifolia (L.) R. Br.

Annual, 0.3-2 m tall; stems hairy; leaf blades 4-
12 cm long; flowers in globose heads, 2-2.5 cm
long, yellow to scarlet; *nutlets to 3 mm long.
Open, disturbed sites, often fence rows and pastures. Infreq. all Fla. W to La., N to N.C. Sum-fall.

SIBERIAN MOTHERWORT
Leonurus sibiricus L.

Biennial, 0.9-1.2 m tall; leaves to 10 cm long;
*petals to 1.2 cm long, purple; *nutlets to 2 mm
long. Disturbed sites. Rare NF, WF. W to La.,
N to Pa. Spr-fall.

AMERICAN BUGLEWEED
Lycopus americanus Muhl. ex Bart.

Perennial, 30-70 cm tall; leaves to 12 cm long;
lower or occasionally all leaves pinnatified;
upper leaves toothed; sepal lobes subulate apically, exceeding nutlets; petals to 1.5 mm long,
white; stamens exserted; nutlets 1-1.5 mm long.
Moist to wet sites. Infreq. WF. W to Calif.,
N to Canada. Sum-fall.

(not shown)

SESSILE-LEAVED BUGLEWEED
Lycopus amplectens Raf.

Perennial, 0.2-1 m tall; leaves 3-6 cm long;
*flowers to 4 mm long, white; *nutlets to 1.5 mm
long. Low wet sites. Infreq. NF, WF. W to Miss.,
N to Mass. Sum-fall.

VARIABLE WATER HOARHOUND
Lycopus rubellus Moench

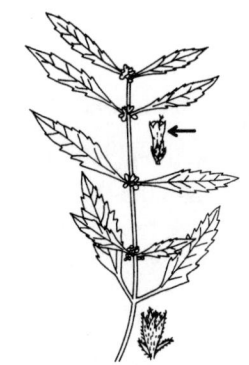

Perennial, 0.4-1.2 m tall; leaves 3-15 cm long,
variable, green to purple, elliptic to lanceolate to
ovate-elliptic; *flowers in dense axillary cymes;
sepal lobes acuminate apically, larger than nutlets; petals 3.5-4 mm long, white; stamens 2,
exserted; nutlets 1-1.6 mm long, included. Moist
to wet sites. Freq. all Fla. W to Tex., N to Ill.,
E to Maine. Sum-fall.

BUGLEWEED
Lycopus virginicus L.

(not shown)

Perennial, 0.1-0.8 m tall; leaves to 14 cm long,
elliptic to ovate; sepal lobes obtuse to acute apically, shorter than or equal to nutlets; petals 2.5-
3 mm long, white; stamens not exserted. Moist
to wet sites. Infreq. NF, WF. W to Tex., N to Pa.,
E to Mass. Sum-fall.

PEPPERMINT
Mentha piperita L.

Perennial, 20-70 cm tall; leaves 1.5-5 cm long;
flowers in clusters 1-2 cm in diam.; *sepals to
4 mm long; *petals to 4 mm long, violet. Native
to Europe. Disturbed sites. Rare NF, WF.
W to Tex., N to Canada. Sum-fall.

SPEARMINT
Mentha spicata L.

Perennial, 20-80 cm tall; stems smooth; leaves
2-7 cm long; sepals to 2 mm long; *petals to
2.5 mm long, pink or violet; nutlets to 2 mm
long. Native to Europe. Moist to wet sites.
Rare WF. W to Tex., N to Canada. Sum-fall.

FALSE PENNYROYAL
Micromeria brownei (Sw.) Benth.

Perennial, 5-40 cm tall; stems creeping, ascending
or erect; leaves 0.5-2 cm long; flowers solitary,
7-10 mm long, pink to white; stamens 4; nutlets
minute. Low, wet sites. Freq. all Fla. W to Tex.,
N to Ga. Spr-sum.

HORSE MINT
Monarda punctata L.

Perennial, 0.3-1 m tall; stems hairy; leaves 2-8 cm
long; petals to 2 cm long, yellow or white, with
purple spots; nutlets to 1.5 mm long. Disturbed,
usually dry sites. Freq. all Fla. W to Tex., N to
Minn. Sum-fall.

SMALL'S FALSE DRAGON-HEAD
Physostegia leptophylla Small

Perennial, 0.3-1.5 m tall; basal leaf blades 4-8 cm
long; upper leaves not much smaller; exposed
portion of petals 2-2.5 cm long, violet, purple or
pink; nutlets 2-3.5 mm long. Wooded river
swamps, fresh and brackish marshes. Infreq. CF,
NF, WF. E to Va. Spr-sum.

(not shown)

(not shown)

PURPLE FALSE DRAGON-HEAD
Physostegia purpurea (Walt.) Blake
[*P. denticulata* (Ait.) Britt.]

Perennial, 0.3-1 m tall; lower leaves 2-14 cm
long, thick; upper leaves much smaller than
lower; petals 2-2.5 cm long, pink to purple; nutlets to 2 mm long. Moist open pinelands. Freq.
all Fla. W to Miss., E to Va. Spr-sum.

OBEDIENT PLANT
Physostegia virginiana (L.) Benth.

Perennial, 0.3-1.5 m tall; leaves 3-12 cm long;
petals to 3 cm long, rose or purple; nutlets to 4 mm
long. Low woody sites, shores. Infreq. CF, NF,
WF. N to Canada. Sum.

PENNYROYAL
Piloblephis rigida (Bartr. ex Benth.) Raf.

Perennial, 10-70 cm tall; stems lying flat and
diffuse, hairy; leaves to 1 cm long; flowers in
panicles, light purple; nutlets minute. Dry sites.
Common and endemic SF, CF, NF. All yr.

CARPENTER WEED
Prunella vulgaris L.

Perennial, 5-40 cm tall; stems erect or diffuse;
leaves 2-7.5 cm long; sepals to 11 mm long;
*petals to 1.8 cm long, purple, violet or white;
nutlets to 2 mm long. Moist disturbed areas.
Infreq. WF. W to La., N to Canada. Spr-fall.

BLUE SAGE
Salvia azurea Lam.

Perennial, 0.3-1.2 m tall; leaves 3-10 cm long,
oblong, lanceolate or linear; sepals to 9 mm long;
petals 13-15 mm long, blue or white; nutlets to
3 mm long. Dry sites. Infreq. CF, NF, WF.
W to Tex., N to N.C. Spr-fall.

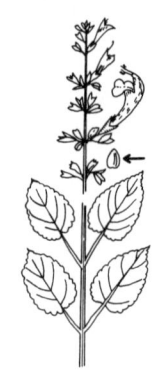

TROPICAL SAGE
Salvia coccinea Buchoz ex Epling

Perennial, to 70 cm tall; stems branching; leaves
1-8 cm long; flowers in panicles, red; *nutlets to
2.5 mm long. Disturbed sites, hammocks. Freq.
all Fla. W to Tex., N to S.C. Spr-fall.

LYRE-LEAVED SAGE
Salvia lyrata L.

Annual, 10-60 cm tall; stems hairy; leaves
5-20 cm long, lyre-shaped; sepals to 11 mm long;
petals to 2.5 cm long, blue to purple; nutlets 2-
2.3 mm long. Disturbed sites. Common all Fla.
W to Tex., N to Conn. Spr-fall.

WEST INDIAN SAGE
Salvia occidentalis Sw.

Annual, to 2 m tall; stems hairy; leaves 2-6 cm
long, elliptic to ovate; sepals to 3.5 mm long;
petals to 5 mm long, blue; nutlets to 2 mm long.
Disturbed sites. Infreq. SF, CF. Spr-fall.

SOUTHERN SAGE
Salvia riparia HBK.
[*S. privoides* Benth.]
[*S. setosa* Fern.]

Annual, to 1.8 cm tall; stems hairy; leaves 1-3 cm
long, ovate; margins toothed; sepals to 6 mm
long; petals 5-6 mm long, blue or white; nutlets
to 2 mm long. Disturbed, shady, dry sites. Freq.
SF, CF, NF. Spr-fall.

SMALL WHITE SAGE
Salvia serotina L.

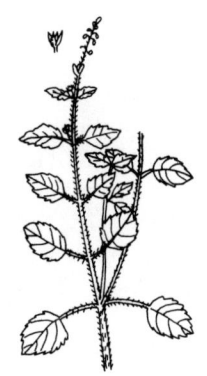

Annual or perennial, 0.1-0.7 m tall; stems hairy; leaves 1-4 cm long, ovate, with bluntly toothed margins; flowers in panicles, 5-10 cm long; sepals 5-8 mm long, with glandular hairs; sepal lobes sharp-tipped; petals 6-10 mm long, white or blue; stamens 2; nutlets to 2 mm long, included. Low, moist, shady sites. Infreq. SF, CF. Spr-fall.

CENTRAL FLORIDA SKULLCAP
Scutellaria arenicola Small

Perennial, 20-60 cm tall; stems hairy; leaves to 6 cm long, with toothed margins; flowers 2-2.5 cm long, blue, pink-purple or white; nutlets 1-1.5 mm long. Dry sites. Freq. CF, NF. Endemic to central Fla. Sum.

COMMON LARGE SKULLCAP
Scutellaria integrifolia L.

Perennial, 20-70 cm tall; stems hairy; leaves to 4 cm long; upper leaf margins smooth or blunt-toothed; *flowers 1.5-3 cm long, blue or pink; nutlets to 1 mm long. Dry to wet sites. Freq. CF, NF, WF. W to Tex., E to Mass. Spr-sum.

ROUGH SKULLCAP
Scutellaria integrifolia L.
var. *hispida* Benth

Similar to *Scutellaria integrifolia* except: stems with spreading hairs. Wet sites. Freq. CF, NF, WF. W to Tex., N to Ala. Spr-fall.

MARSH SKULLCAP
Scutellaria lateriflora L.

Perennial, to 1 m tall; stems smooth or hairy;
leaves to 6 cm long, with toothed margins; *sepals
with glandular hairs; *flowers 6-7 mm long, blue,
pink or white; *nutlets to 1 mm long. Low, wet
sites. Infreq. WF. W to Calif., N to Canada.
Sum-fall.

ROLLED-LEAF SKULLCAP
Scutellaria multiglandulosa (Kearney) Small

Perennial, 10-35 cm tall; stems with glandular
white hairs; leaves 1-4 cm long, with revolute margins; flowers 2-2.5 cm long, blue or white, with
lower lip longer than upper; nutlets to 1.5 mm
long. Dry sites. Freq. NF, WF. N to Ga. Spr.

HAIRY SKULLCAP
Scutellaria ovalifolia Pers.
[*S. pilosa* Michx.]

Perennial, 10-60 cm tall; stems hairy; leaves 1.5-
8 cm long; leaf margins crenate; flowers 11-14 mm
long, blue or white; nutlets to 1 mm long. Dry
woods. Rare NF, WF. W to Tex., E to N.Y.
Spr-sum.

HEART-LEAF SKULLCAP
Scutellaria ovata Hill

Perennial, 0.1-1 m tall; stems and leaves hairy;
leaves 3-12 cm long; upper leaves cordate; leaf
margins crenate; flowers 15-21 mm long, blue-
violet; nutlets to 1.5 mm long. Dry to wet sites.
Rare WF. W to Tex., E to Pa. Spr-sum.

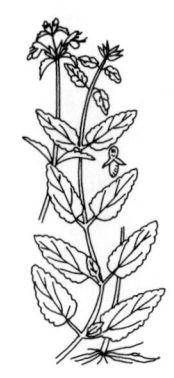

FLORIDA BETONY
Stachys floridana Shuttlew.

Perennial, 10-40 cm tall; stems hairy; leaves 1-6 cm
long; flowers in panicles; petals 10-15 cm long,
pink to white; nutlets to 1.5 mm long. Moist to wet,
open sites. Freq. CF, NF, WF. W to Tex., N to Va.
Spr-fall.

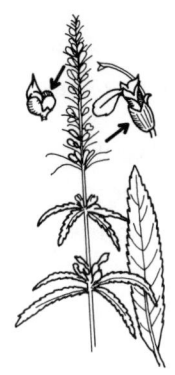

WOOD SAGE
Teucrium canadense L.

Perennial, 0.3-1.2 m tall; stems with gray hairs;
leaves 5-14 cm long, hairy beneath; margins
toothed; *petals 10-20 mm long, purple to pink;
stamens 4; *nutlets to 2 mm long. Low, wet sites.
Freq. all Fla. W to Tex., E to Maine. Sum.

BLUE CURLS
Trichostema dichotomum L.

Annual, 0.1-2 m tall; leaves 2-7 cm long, to 2 cm
wide, with glandular hairs; sepals to 6 mm long;
petals to 5 mm long, blue to white; nutlets to 2 mm
long. Dry sites. Freq. all Fla. W to Tex., N to
Maine. Sum-fall.

NARROW-LEAVED BLUE CURLS
Trichostema setaceum Houtt.

Annual, to 50 cm tall; leaves to 4 cm long, to 6 mm
wide; *sepals to 5.5 mm long; petals to 4 mm long,
blue; nutlets to 2 mm long. Dry sites. Rare CF;
freq. NF, WF. W to Tex., N to Conn. Sum-fall.

BLUE CURLS
Trichostema suffrutescens Kearney

Perennial or biennial, 20-40 cm tall; leaves 1-2 cm
long, to 5 mm wide; sepals to 6 mm long; petals to
8 mm long, blue; nutlets to 1.5 mm long. Scrub.
Infreq. SF, CF, NF. Endemic to Fla. Spr.

(not shown)

LAUREL FAMILY   *LAURACEAE*

Trees, shrubs or vines; leaves mostly alternate, with oil-pro-
ducing glands emitting aromatic smell when crushed; flowers
bisexual or unisexual; floral parts 6, in several whorls; ovary
mostly superior; fruit a drupe or berry.

LOVE VINE
Cassytha filiformis L.

Perennial, parasitic vine; plant fragrant; flowers
*3-6 per spike; sepals 6, unequal in length, white
or green; stamens 9; *drupes 5-7 mm in diam.
Disturbed areas, hammocks. Freq. SF, CF. All yr.

PEA FAMILY   *LEGUMINOSAE*

Herbs, shrubs or trees; leaves alternate, mostly divided into several leaflets (pinnately compound); flowers bisexual or unisexual; sepals 5; petals 5, 1 or 0; petals separate and distinct, but mostly butterfly-like (papilionaceous) with lowest 2 petals joined to form the keel, lateral 2 petals the wings and uppermost petal the standard; fruit opening along 2 sides (legume or pod) or between seeds (loment).

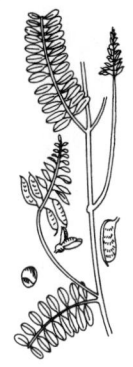

ROSARY PEA
Abrus precatorius L.

Perennial vines climbing to 3 m long; leaflets numerous, 9-18 mm long; flowers to 1.5 cm long; petals pink-purple; legumes to 3 cm long, rectangular, in clusters; seeds bright red with black base, poisonous. Dry thickets or any disturbed area. Freq. SF, CF, NF. Sum.

SWEET ACACIA
Acacia farnesiana (L.) Willd.

Perennial, to 4 m tall; stems armed with pairs of spines 1-3 cm long; leaflets 10-20 pairs, 4.5-6.5 mm long; flower spikes to 1 cm in diam., round, yellow-orange; legumes 3.5-8 cm long. Disturbed, coastal sites. Infreq. all Fla. W to Tex. Spr, all yr S.

KEY ACACIA
Acacia pinetorum (Small) Herm.

Perennial, to 4 m tall; stems to 3 cm long, with smooth spines; leaves bipinnate; leaflets 1.5-4 mm long, smooth; flowers to 4 mm long; *spikes to 9 mm in diam.; legumes 2.5-7 cm long. Pinelands, coastal sites. Rare SF, CF. Spr, or infrequently all yr.

SHY-LEAF
Aeschynomene americana L.

Annual, to 2 m tall; stem erect to reclining, hairy;
leaflets 15-60, 3-12 mm long, 2- to 3-veined;
petals 6-10 mm long, yellow with purple lines;
*legumes 2-4 cm long; joints 3-9. Disturbed
sites. Common all Fla. N to Ga. Spr-sum.

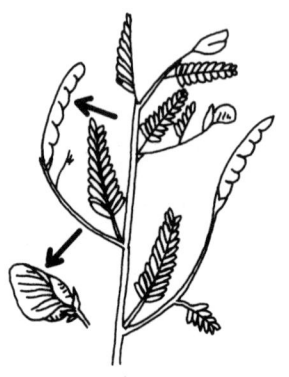

INDIAN JOINT-VETCH
Aeschynomene indica L.
[*A. virginica* (L.) BSP.]

Annual, 0.2-1.5 m tall; leaflets 50-70, to 1 cm
long; *flowers 1-1.5 cm long, yellow; *legumes
to 5 cm long, jointed. Low wet sites. Infreq. all
Fla. W to Tex., E to N.J. Sum-fall.

EVERGLADES SHY-LEAF
Aeschenomene pratensis Small

Annual, to 2 m tall; stems smooth; leaflets 15-25,   (not shown)
to 6 mm long; sepals to 4.5 mm long; petals 1-
1.3 cm long; legumes to 4 cm long. Wet sites.
Infreq. SF. All yr.

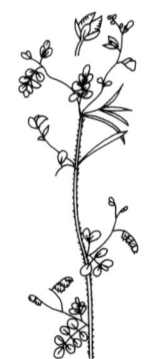

PROSTRATE JOINT-VETCH
Aeschynomene viscidula Michx.

Annual, 0.1-1.2 m long; stems prostrate, with
sticky glandular hairs; leaflets 3-9, 3-12 mm long;
petals 4-6 mm long, yellow with pink spots;
legumes to 2 cm long; joints 1-4, hairy. Dry
sites. Freq. all Fla. W to Miss., E to S.C. Spr-fall.

MIMOSA
Albizia julibrissin Durazz.

Perennial shrub or tree, to 10 m tall, spreading; leaflets 10-40, 10-17 mm long; pinnae 10-30; flowers perfect; petals 5, 6-8 mm long, distinct, pink; stamens numerous, 2-3 cm long; legumes 8-15 cm long. Native to Asia. Disturbed sites. Infreq. CF, NF, WF. W to La., E to Md. Spr-sum.

ALYCE-CLOVER
Alysicarpus vaginalis (L.) DC.

Annual, to 1 m long; stems diffuse; leaves 1-5 cm long, simple, round or linear; petals purple or pink; standard to 6 mm long; legumes to 2 cm long. Native to Old World. Low, disturbed sites. Infreq. all Fla. Sum-fall.

BASTARD INDIGO
Amorpha fruticosa L.

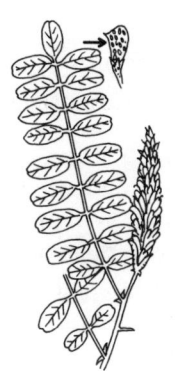

Perennial shrub, 1.5-4 m tall; new growth smooth to hairy; leaflets 9-35, 1-5 cm long, with abrupt sharp tip; *flowers to 6 mm long, with numerous flowers per raceme, red, purple or blue; legumes 5-8 mm long, curved. Moist to wet sites. Freq. all Fla. W to Calif., N to Canada. Spr-sum.

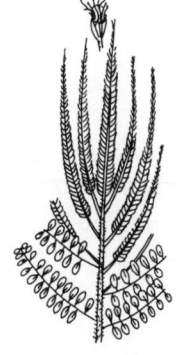

INDIGO BUSH
Amorpha herbacea Walt.

Perennial, 0.3-1.5 m tall; leaflets numerous, with midrib exserted into a swollen tip; sepals 5; petal 1, 4-5 mm long, white to blue to purple; stamens 10; legumes (pods) 4-5 mm long, covered with resinous glands, hairy. Dry, disturbed sites. Infreq. all Fla. W to Okla., N to N.D., E to N.C. Spr-sum.

GROUND NUT
Apios americana Medicus

Perennial twining vine, 1-3 m long; leaflets 5-7 or
rarely 3, each to 7 cm long; sepals to 5 mm long;
petals to 11 mm long, chocolate or purplish;
legumes 5-12 cm long. Thickets, low sites.
Infreq. all Fla. W to Tex., N to Canada. Sum-fall.
Roots used for food by Indians.

PEANUT
Arachis hypogaea L.

Annual, 20-70 cm long; stems lying flat or erect;
leaflets 4, 1.5-5 cm long; *stipules tapering to
point; flowers to 1.5 cm long, yellow; legumes to
8 cm long, underground, used as food. Native to
South America. Escapes from cultivation.
Disturbed sites and often old cultivated areas.
Rare CF, NF, WF. W to Miss., N to N.C.
Abundant locally. Sum-fall.

HAIRY MILK VETCH
Astragalus villosus Michx.

Perennial, 10-20 cm long; stems diffuse; stems,
leaves and fruits with long soft hairs (villous);
leaflets 9-15, 3-10 mm long; flowers 0.9-1 cm
long, yellow; *legumes to 2 cm long. Dry and
disturbed sites. Infreq. CF, NF, WF. W to Ala.,
N to S.C. Spr-sum.

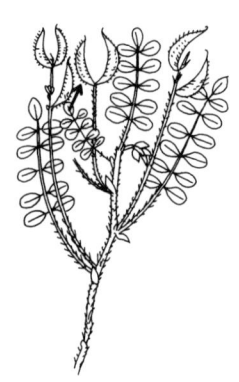

(not shown)

PINELAND WILD INDIGO
Baptisia lanceolata (Walt.) Ell.

Perennial, to 1 m tall; stems curving upwards;
leaflets 3, 4-12 cm long; flowers to 2 cm long,
yellow; legumes 1.5-2.5 cm long. Dry sites.
Infreq. CF, NF, WF. E to Ga. Spr.

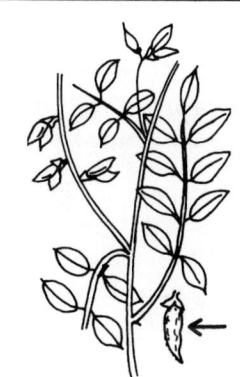

LECONTE'S WILD INDIGO
Baptisia lecontei Torr. & Gray

Perennial, to 1 m tall, bushy; stems covered with
hairs; leaflets 3, 1-4 cm long; flowers 1-1.5 cm
long, yellow or white; *legumes to 1 cm long.
Dry sites. Freq. CF, NF, WF. N to Ga. Spr-sum.

(not shown)

CAT-BELLS
Baptisia perfoliata (L.) R. Br.

Perennial, to 1 m tall; stems arching; leaves 5-
10 cm long, simple, sessile, orbicular to ovate;
flowers yellow, solitary, axillary; legumes 1-1.5 cm
long. Dry sites. Rare CF. N to N.C. Spr.

POWDER PUFF
Calliandra haematocephala Hassk.

Perennial shrub or small tree, to 5 m tall; leaves
to 7 cm long; flowers clustered; *stamens to 4 cm
long, red. Native to Bolivia. Disturbed sites.
Rare CF, WF. Fall-spr.

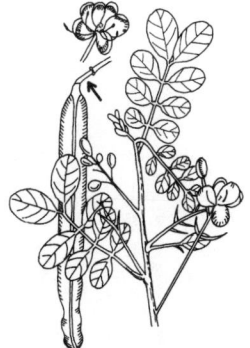

CHRISTMAS SENNA
Cassia bicapsularis L.

Perennial, to 1.5 m tall; leaflets 6-8, 1.5-3 cm
long, punctate on lower side, having gland
between lowermost pair; *capsules 6-15 cm long,
stipitate. Native to tropical America. Cultivated.
Canal banks, mangroves. Common all Fla. All yr.

FLORIDA BUTTERFLY-PEA
Centrosema arenicola (Small) Hermann
[*C. floridanum* (Britt.) Lakela]

Perennial, trailing or climbing vine, to 1.5 m
long; leaves 3, 2-8 cm long; lateral lobes of lower
sepal lip shorter than central one; petals to 2 cm
long, blue-purple; legumes to 14 cm long. Open
sites. Rare CF, NF. Spr-fall.

CLIMBING BUTTERFLY-PEA
Centrosema virginianum (L.) Benth.

Perennial, climbing or twining vine, to 1.5 m
long; leaves 3, 3-11 cm long, with distinct marginal veins; lateral lobes of lower sepal lip equal
central one; petals to 4 cm long, blue-purple;
*legumes to 12 cm long, very narrow. Disturbed
and open areas, margins of woods. Freq. all Fla.
W to Tex., N to N.J. Spr-fall.

PARTRIDGE-PEA
Chamaecrista fasciculata (Michx.) Greene
[*Cassia fasciculata* Michx.;
Cassia chamaecrista L.]

Annual, 0.1-2.4 m tall; stems erect to diffuse; leaf
petioles with round glands; leaflets 12-50, 9-20 mm
long; petals 8-21 mm long, yellow; legumes 4-5 cm
long. Dry sites. Common all Fla. W to Tex., E to
Mass. Spr-fall, all yr S.

WILD SENSITIVE PLANT
Chamaecrista nictitans (L.) Moench
[*Cassia nictitans* L.]

(not shown)

Annual, 10-40 cm tall; stems smooth or with
hairs incurved; leaf petioles with glands wider
than stalk; leaflets 18-36, 7-15 mm long; petals
3-8 mm long, yellow; legumes 2.5-4.5 cm long.
Dry and disturbed sites. Infreq. all Fla. W to
Tex., E to Vt. Sum-fall.

FLATWOODS PARTRIDGE-PEA
Chamaecrista nictitans (L.) Moench
var. *aspera* (Muhl. ex Ell.) Irwin & Barneby
[*Cassia aspera* Muhl. ex Ell.]

Annual, 30-70 cm tall; stems diffuse or prostrate,
hairy; leaf petioles with glands barely wider than
stalk; leaflets 30-62, 5-15 mm long; petals 3-7 mm
long, yellow; legumes 1.5-2.5 cm long, hairy. Dry
sites. Common all Fla. N to S.C. Sum-fall.

HAIRY PARTRIDGE-PEA
Chamaecrista pilosa (L.) Greene
[*Cassia pilosa* (Chapm.) L.]

Annual, to 1.5 m long; stems lying flat or trailing,
hairy; leaflets 8-10, to 2 cm long; stipules heart-
shaped; flowers to 1 cm long, yellow; legumes to
2.5 cm long. Dry, disturbed sites, along railroad
tracks. Infreq. SF, CF. All yr.

TWO-LEAF SENNA
Chamaecrista rotundifolia (Pers.) Greene
[*Cassia rotundifolia* Pers.]

Perennial, to 2 m across; stems prostrate, branching; petioles glandless; leaflets 2, to 1.5 cm long;
*stipules heart-shaped; flowers to 5 mm long, yel-
low; legumes to 2.5 cm long. Disturbed sites,
along railroad tracks. Infreq. CF. All yr.

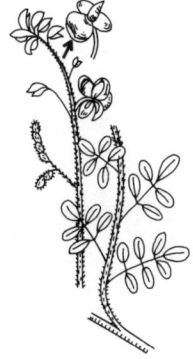

ALICIA
Chapmannia floridana Torr. & Gray

Perennial, 20-90 cm tall; stems with sticky hairs;
leaflets 5-7, 5-20 mm long; *standard 10-14 mm
long, yellow; legumes 1-3 cm long; joints 1-4.
Dry sites. Freq. SF, CF, NF. Spr-sum.

SANDHILL BUTTERFLY-PEA
Clitoria fragrans Small

Perennial, 10-50 cm tall; stems covered with a
white film; leaflets 2-5 cm long; sepals to 1.6 cm
long; *petals to 3.5 cm long, purple; standard to
4.7 cm long; *legumes to 7 cm long. Dry sites,
scrub. Rare and endemic CF. Spr-sum.

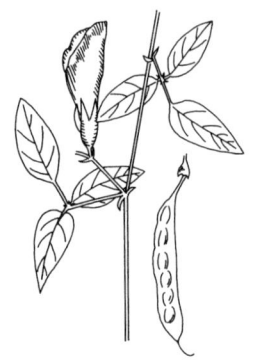

BUTTERFLY-PEA
Clitoria mariana L.

Perennial, 0.3-1.2 m long; stems trailing; leaflets
3, 2-7 cm long; sepal tube longer than lobes;
flowers 4-6 cm long, with 1-3 flowers per
raceme, blue to white; stamens joined together
by their stalks into one bundle; legumes 4-8 cm
long; fruit stalk 1-2 cm long. Pinelands. Infreq.
all Fla. W to Tex., N to N.Y. Spr-fall.

SLENDERLEAF CROTALARIA
Crotalaria brevidens Benth.
[*C. intermedia* Kotschy]

Annual, to 1.5 m tall; leaflets 3, to 12 cm long;
flowers 1.5-2 cm long, yellow; legumes to 5 cm
long. Disturbed sites. Infreq. CF, NF, WF.
N to N.C. Sum-fall.

(not shown)

LANCELEAF CROTALARIA
Crotalaria lanceolata E. Meyer

Annual, to 1 m tall; leaflets 3, 2-9 cm long;
*flowers 0.6-0.9 cm long, yellow; legumes 3-4 cm
long. Disturbed sites. Infreq. CF, NF, WF. N to
S.C. Sum-fall.

SMOOTH CROTALARIA
Crotalaria pallida Ait.
[*C. mucronata* Desv.]

Annual, 1-1.5 m tall; stems hairy; leaflets 3, 2-7 cm
long; petals 1.2-1.4 cm long, yellow; legumes 3-
4.5 cm long. Disturbed sites. Freq. all Fla. N to
N.C. Spr-fall.

SMALL RATTLE BOX
Crotalaria pumila Ortega

Annual or perennial, to 1.2 m long; stems lying
flat; leaflets 3, 0.5-1.2 cm long; flowers to 1 cm
long, yellow; legumes 1-1.5 cm long. Hammocks,
pinelands, dunes. Infreq. SF, CF, NF. Spr-fall,
all yr S.

RATTLE BOX
Crotalaria purshii DC.

Perennial, 10-50 cm tall; leaf 1, to 8 cm long;
flowers 0.9-1 cm long, yellow; legumes 2-3 cm
long. Dry pinewoods, disturbed areas. Infreq. all
Fla. W to La., E to Va. Sum-fall.

RABBIT-BELLS
Crotalaria rotundifolia (Walt.) Gmel.
[*C. maritima* Chapm.]

Perennial, to 30 cm long; stems lying flat, hairy;
leaf 1, 1-3 cm long; petals 5-11 mm long, yellow;
legumes 2-3 cm long. Disturbed sites. Common
all Fla. W to La., N to N.C. Spr-fall.

RATTLE BOX
Crotalaria sagittalis L.

Annual or perennial, 10-50 cm tall; stems hairy;
leaves 3-8 cm long; stipules fused to stem; petals
to 5 mm long, yellow; legumes to 4 cm long.
Dry sites. Infreq. NF. W to Tex., N to Wis.,
E to Ga. and Vt. Spr-fall.

SHOWY CROTALARIA
Crotalaria spectabilis Roth

Weedy annual, 0.5-2 m tall; stems finely hairy;
leaf 1, to 20 cm long, simple; stipules more than
5 mm long; petals 1.5-2.5 cm long, yellow;
legumes to 5 cm long. Disturbed sites. Common
all Fla. W to Miss., E to Va. Sum-fall.

THREE CORNER PRAIRIE CLOVER
Dalea carnea (Michx.) Poir.
[*Petalostemon carneum* Michx.]

Perennial, 0.5-1 m tall; stems erect; leaflets 5-9,
to 1 cm long, punctate glandular below; flowers
spikes 2-4 cm long, oblong; petals pink-purple;
legumes to 3 mm long, straight. Dry flatwoods.
Freq. all Fla. W to Ala., N to Ga. Spr-fall.

THREE CORNER PRAIRIE CLOVER
Dalea carnea (Michx.) Poir.
var. *gracilis* (Nutt.) Barneby

(not shown)

Perennial, to 1 m long; stems sprawling; leaflets
5-9, to 1.5 cm long; flower spikes 1-2 cm long,
oblong; petals white; legumes to 2 mm long.
Moist to wet sites. Infreq. WF. W to Miss.,
E to Ga. Fall.

LEGUMINOSAE 211

GLOBE-HEADED PRAIRIE CLOVER
Dalea feayi (Chapm.) Barneby
[*Petalostemon feayi* Chapm.]

Perennial, 20-50 cm tall; stems erect; leaflets 3-11,
to 2 cm long; flower spikes globose; petals pink or
purple; legumes to 3 mm long, straight. Dry sites.
Common CF, NF, WF. N to Ga. Spr-sum.

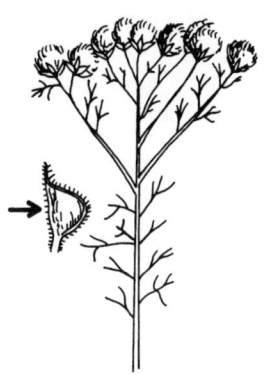

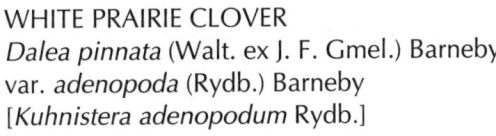

SUMMER FAREWELL
Dalea pinnata (Walt. ex J. F. Gmel.) Barneby
[*Petalostemon pinnatum* (Walt. ex J. F. Gmel.)
Blake]

Perennial, 0.3-1.2 m tall; leaflets 3-7, 5-12 mm
long, glandular-dotted; flowers white; flower spikes
8-12 mm long, subtended by an involucre of
bracts; standard 6-8 mm long; *legumes to 2 mm
long; seeds 1-2. Dry sites. Freq. all Fla. W to
Miss., N to N.C. Sum-fall.

WHITE PRAIRIE CLOVER
Dalea pinnata (Walt. ex J. F. Gmel.) Barneby
var. *adenopoda* (Rydb.) Barneby
[*Kuhnistera adenopodum* Rydb.]

Perennial, 0.5-1 m tall; leaflets 3-7, 5-9 mm long;
*flowers small, spicate, white; involucre 9-11 mm
long; legumes to 2.5 mm long. Dry sites. Freq.
all Fla. N to N.C. Sum-fall.

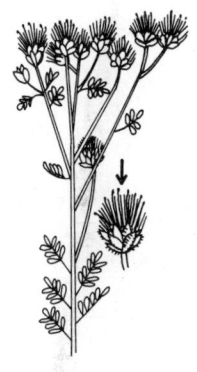

HAIRY BEGGARWEED
Desmodium canescens (L.) DC.

Perennial, 0.6-1.5 cm tall; stems erect to ascending, hairy; leaflets 3-10 cm long; petals to 1 cm
long, pink, purple or white; *loments to 5 cm
long, with 2-6 joints. Fields, woods. Rare CF,
NF, WF. W to Tex., N to Canada. Sum-fall.

SMALL-LEAVED TICK-TREFOIL
Desmodium ciliare (Muhl. ex Willd.) DC.

Perennial, 0.4-1 m tall; stems hairy; leaflets 1-
2.5 cm long; petals to 5 mm long, pink to purple;
*loments to 1.5 cm long, hairy, with 1-3 joints.
Dry sites. Infreq. CF, NF, WF. W to Tex., N to
Canada. Sum-fall.

RHOMB-LEAVED TICK-TREFOIL
Desmodium floridanum Chapm.

Perennial, to 50 cm tall; stems erect, hairy;
leaflets to 9 cm long, hairy on lower surface;
petals 6-7.5 mm long, purple; loments to 3 cm
long, with 3-5 joints. Flatwoods, woods. Infreq.
all Fla. W to La., N to Va. Sum-fall.

CREEPING BEGGARWEED
Desmodium incanum DC.
[*D. canum* (J. F. Gmel.) Schinz & Thellung]

Perennial, to 1 m tall; stems hairy; leaflets 2-6 cm
long; petals to 6 mm long, purple; loments to 3 cm
long, hairy, with 3-7 joints. Hammocks, woods.
Common all Fla. All yr.

SMOOTH TICK-TREFOIL
Desmodium laevigatum (Nutt.) DC.

Perennial, to 1.2 m tall; stems erect, smooth;
leaflets 4-8 cm long; petals 0.8-1 cm long, pink
to purple; loments to 3 cm long, with *2-5 joints.
Woods. Infreq. NF, WF. W to Tex., N to N.J.
Sum-fall.

SANDHILL ROUND-LEAF BEGGARWEED
Desmodium lineatum DC.

Perennial, 50-70 cm long; stems prostrate;
leaflets 1-5, 0.7-3 cm long; petals 4-6 mm long,
purple; loments to 1.5 cm long, hairy, with 2-3
joints. Flatwoods, sandhills, moist woods.
Infreq. CF. W to La., N to Md. Sum-fall.

MARYLAND TICK-TREFOIL
Desmodium marilandicum (L.) DC.

Perennial, 0.6-1.5 m tall; stems smooth; leaflets
1-2.5 cm long; petals 4-6 mm long, purplish;
*loments to 1 cm long, with 1-3 joints. Disturbed
woods. Infreq. all Fla. W to La., N to Mass.
Sum-fall.

TWO-STEM BEGGARWEED
Desmodium nudiflorum (L.) DC.

Perennial, 5-30 cm tall; flower stems 0.3-1 m tall;
leaflets to 10 cm long; petals to 8 mm long, pink,
purple or white; loments to 3.5 cm long, with 1-4
joints. Woods. Infreq. NF, WF. W to Tex., N to
Canada. Sum-fall.

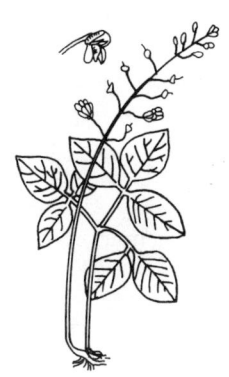

LANCE-LEAVED BEGGARWEED
Desmodium paniculatum (L.) DC.

Perennial, 0.5-1.2 m tall; stems hairy; leaflets 0.2-
8 cm long; petals 5-8 mm long, purple; loments
to 4 cm long, with 3-6 joints. Sandhills, dry
woods, hammocks. Freq. all Fla. W to Tex.,
N to Canada. Sum-fall.

WHITE TICK-TREFOIL
Desmodium pauciflorum (Nutt.) DC.

Perennial, 10-70 cm tall; stems lying flat or
ascending, erect; leaflets 3, 2.5-8 cm long; petals
4-7 mm long, white; fruit segments 1-3; each segment 9-14 mm long. Rich woods. Infreq. WF.
W to Tex., E to N.Y. Sum-fall.

STIFF TICK-TREFOIL
Desmodium strictum (Pursh) DC.

Perennial, to 1.2 m tall; stems erect, smooth;
leaflets to 5 cm long; petals 3-5 mm long, purple;
loments to 1.5 cm long, with *1-3 joints. Dry sites.
Infreq. all Fla. W to Tex., N to N.J. Sum-fall.

NARROWLEAF TICK-TREFOIL
Desmodium tenuifolium Torr. & Gray

Perennial, to 1.5 m tall; stems erect; leaflets to
8 cm long; petals 4-5 mm long, pink-purple;
loments to 1 cm long, with 1-3 joints. Flatwoods,
sandhills. Infreq. CF, NF, WF. W to La., N to
N.C. Sum-fall.

(not shown)

FLORIDA BEGGARWEED
Desmodium tortuosum (Sw.) DC.

Annual, 0.5-3.5 m tall; stems erect, hairy; leaflets
2-14 cm long; petals 4-6 mm long, purple;
*loments to 2.5 cm long, with 2-7 joints.
Disturbed sites. Common all Fla. W to Tex.,
N to Ga. Sum-fall.

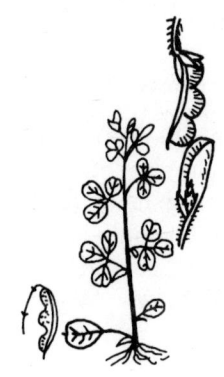

SAGOTIA BEGGARWEED
Desmodium triflorum (L.) DC.

Perennial, to 60 cm long; stems creeping, trailing, hairy; leaflets 0.3-1.2 cm long; petals to 5 mm long, pink to purple; loments to 1.5 cm long, with 3-5 joints. Native to Old World. Disturbed sites. Freq. all Fla. Sum-fall.

VELVET-LEAVED TICK-TREFOIL
Desmodium viridiflorum (L.) DC.

Perennial, to 2 m tall; stems hairy; leaflets to 11 cm long, hairy beneath; petals to 9 mm long, pink; loments to 4 cm long, with *2-6 joints. Dry sites. Infreq. CF, NF, WF. W to Tex., N to N.Y. Sum-fall.

CHEROKEE BEAN
Erythrina herbacea L.

Perennial herb, 0.6-1.2 m tall; stems spiny; leaflets 3-10 cm long, round to 3-lobed; petals red; standard 3.5-5.5 cm long; legumes to 21 cm long. Dry sites to moist woods. Freq. all Fla. W to Tex., N to N.C. Spr-fall. Becoming shrub or tree 3-8 m tall in CF and SF.

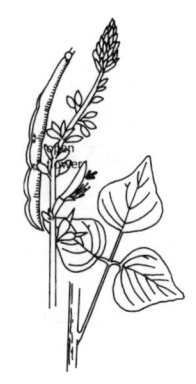

WHITE MILK-PEA
Galactia elliottii Nutt.

Perennial, climbing or prostrate vine, 0.5-1.5 m long; stems hairy; leaflets 5-9, 2-4 cm long; petals white; standard 11-15 mm long; legumes 3-5 cm long. Dry to moist sites. Common CF, NF, WF. N to N.C. Spr-fall.

Galactia erecta (Walt.) Vail

Perennial, 20-40 cm tall; stems erect, smooth to
lightly hairy; leaflets 1.4-4 cm long; sepals 4.5-
5.5 mm long, hairy; petals purple or white;
standard 6-8.5 mm long; keel to 6.5 mm long;
legumes to 2 cm long, less than 5 mm wide. Wet
sites. Infreq. WF. W to Miss., N to N.C. Spr-fall.

(not shown)

FLORIDA MILK-PEA
Galactia floridana Torr. & Gray

Perennial, to 1.5 m long; stems lying flat or trailing,
with gray hairs; leaflets to 4.5 cm long; sepals 4, 5-
6 mm long, hairy; petals purple; standard to 1.4 cm
long; keel to 1 cm long; legumes to 5 cm long, usually 5.5 mm or less wide. Dry sites. Freq. all Fla.
W to Ala. Sum-fall.

Galactia mollis Michx.

Perennial, to 4 m long; stems trailing, climbing,
hairy; leaflets to 5 cm long, hairy on both surfaces; sepals to 4.5 mm long, hairy; petals red
or purple; standard to 8 mm long; keel 5-6 mm
long; legumes to 4 cm long, usually 5.5 mm or
less wide, tomentose. Dry to wet sites. Infreq.
CF, NF, WF. N to N.C. Spr-fall.

(not shown)

PROSTRATE MILK-PEA
Galactia regularis (L.) BSP.

Perennial, to 1.2 m long; stems prostrate, hairy,
with climbing tips; leaflets 2-4.5 cm long; sepals
4, to 6 mm long, smooth; petals purple; standard
to 1.5 cm long; keel 1-1.4 cm long; legumes to
4.5 cm long, usually 5.5 mm or less wide, with
short hairs. Dry sites. All Fla. W to La., E to Va.
Spr-sum.

BIG CLIMBING MILK-PEA
Galactia regularis (L.) BSP. 'Baltzellii'

Similar to *Galactia regularis* except: stems twining; leaflets to 7 cm long; sepals 9-10 mm long; standard to 1.9 cm long. Dry sites. Infreq. CF around Lake Griffin, Lake Co. Sum-fall.

TWINING MILK-PEA
Galactia volubilis (L.) Britt.

Perennial; stems twining, climbing, hairy; leaflets 2-5 cm long; sepals 4, 5-6 mm long, hairy; petals pink; standard to 1.4 cm long; keel 9-11 mm long; legumes to 5 cm long, usually 5.5 mm or less wide. Dry pinelands, dry open woods, dry disturbed areas. Common all Fla. W to Tex., N to N.Y. Sum.

WILD INDIGO
Indigofera caroliniana Mill.

Perennial, 0.5-2 m tall; stems erect, smooth; leaves 5-10 cm long; leaflets 9-15, 1-2.5 cm long; flowers 5-6.5 mm long, yellow-brown; legumes to 1 cm long, straight. Dry sites, scrub. Freq. all Fla. W to La., N to N.C. Spr-fall.

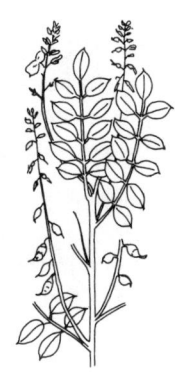

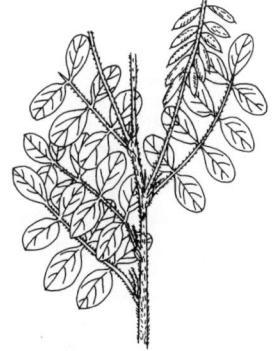

HAIRY INDIGO
Indigofera hirsuta Harv.

Annual, 30-90 cm tall; stems covered with long red-brown hairs; leaflets 5-9, to 3 cm long; flowers to 4.5 mm long; legumes to 2 cm long, 4-angled, covered with long red-brown hairs. Native to Africa. A cultivated cover crop. Disturbed dry sites. Common all Fla. Fall.

Indigofera miniata Ortega

Perennial, to 30 cm long; stems prostrate; leaflets 5-9, to 1 cm long; sepals to 5 mm long; petals to 10 mm long, red; legumes 1-2 cm long. Dry sites. Rare CF, SF. W to Tex. Sum-fall.

(not shown)

BLANKET INDIGO
Indigofera pilosa Poir.

Annual or biennial, sprawling; leaflets 3, with the terminal larger; petals rose-red; keel petals spurred; anthers with appendage; legumes 12-30 mm long. Escapes from cultivation. Disturbed sites. Rare CF, NF. Sum-fall.

CREEPING INDIGO
Indigofera spicata Forsk.
[*I. hendecaphylla* Jacq. or
I. endecaphylla of various authors]

Annual or biennial, to 90 cm long; stems prostrate; leaflets 7-9, to 1 cm long; sepals to 3 mm long; petals 7-8 mm long, red to purple; legumes to 1.8 cm long. Disturbed sites. Infreq. SF, CF, NF. Sum-fall.

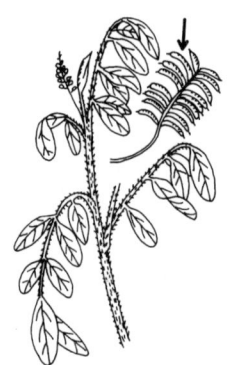

ANIL INDIGO
Indigofera suffruticosa L.

Perennial shrub, 0.5-2 m tall; leaflets 7-15, hairy on lower side; flowers 4-5 mm long, orange; *legumes 1.5-3 cm long. Various dry to moist sites. Infreq. SF, CF, NF. W to Tex., N to N.C. Sum-fall, all yr S.

JAPANESE CLOVER
Kummerowia striata (Thunb.) Schindler
[*Lespedeza striata* (Thunb.) Hook. & Arn.]

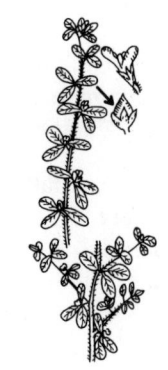

Annual, 10-40 cm tall; stems erect to diffuse,
hairy; leaflets 0.8-1.5 cm long, elliptic, obovate
or oblong; stipules ovate; petals to 6 mm long,
pink; *legumes 3-4 mm long. Native to Asia.
Disturbed sites. Infreq. all Fla. W to Tex., N to
N.J. Sum-fall.

NARROW LEAF BUSH-CLOVER
Lespedeza angustifolia (Pursh) Ell.

Perennial, to 1 m tall; stems with fine hairs;
leaflets 3, 10-35 mm long, linear; flowers white;
*legumes to 4 mm long. Dry to moist sites.
Common CF, NF, WF. N to Mass. Sum-fall.

ROUND HEADED BUSH-CLOVER
Lespedeza capitata Michx.

Perennial, 0.5-1.6 m tall; stems hairy; leaves
2.5-3.5 cm long, with silver hairs; petals to 1 cm
long, yellowish white; legumes 4-6 mm long.
Dry to moist sites. Rare WF. W to La., N to
Minn. Sum-fall.

HAIRY BUSH-CLOVER
Lespedeza hirta (L.) Hornem.

Perennial, 0.5-1.5 m tall; stems hairy; leaflets 3,
12-40 mm long, elliptic to obovate; *flowers in
heads, yellow; legumes 7-8 mm long, with 1
large joint. Dry sites. Infreq. CF, NF, WF.
W to Tex., N to Mich., E to Maine. Sum-fall.

WAND-LIKE BUSH-CLOVER
Lespedeza intermedia (S. Wats.) Britt.
[*L. frutescens* (L.) Britt.]

Perennial, 0.3-1 m tall; stems erect, hairy or
smooth; leaflets to 3.5 cm long, oblong; stipules
subulate; petals to 8 mm long, purple; *legumes
to 7 mm long, hairy. Dry sites. Infreq. WF.
W to Miss., N to Canada. Sum-fall.

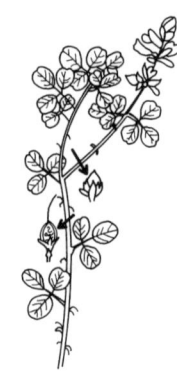

TRAILING BUSH-CLOVER
Lespedeza procumbens Michx.

Perennial, to 1.2 m long; stems lying flat, hairy;
leaflets to 2.5 cm long, ovate; stipules awl-
shaped; petals to 8 mm long, pink to purple;
*legumes to 6 mm long. Dry sites. Rare NF, WF.
W to Tex., N to N.C. Sum-fall.

CREEPING BUSH-CLOVER
Lespedeza repens (L.). Bart.

Perennial, 15-70 cm long; stems lying flat,
branching; leaflets 3, 6-16 mm long, ovate;
flowers few, white, blue to purple; *legumes 3.5-
4.5 mm long, hairy. Dry sites. Common NF.
W to Tex., N to Mo., E to Pa. Sum-fall.

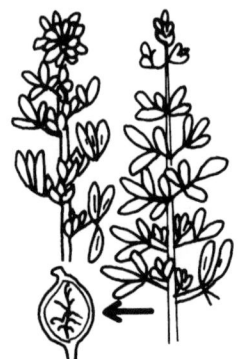

SLENDER BUSH-CLOVER
Lespedeza virginica (L.) Britt.

Perennial, to 1.1 m tall; stems erect to recurving,
slightly hairy; leaflets to 3.5 cm long, linear; stipules linear; petals to 7 mm long, rose to purple;
*legumes to 6 mm long. Dry sites. Infreq. WF.
W to Tex., N to Canada. Sum-fall.

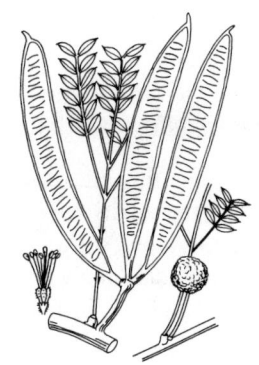

LEAD TREE
Leucaena leucocephala (Lam.) de Wit

Perennial shrub, to 10 m tall; leaves twice
pinnate; leaflets 6-12 mm long; petals to 6 mm
long, white; pods 10-20 cm long. Cultivated
and escapes from cultivation. Disturbed areas,
pastures. Infreq. SF, CF. W to Tex. All yr.

SKY-BLUE LUPINE
Lupinus diffusus Nutt.
[*L. cumulicola* Small]

Perennial, to 1.5 m tall; stems lying flat; leaves
3-15 cm long, evergreen; petals 12-15 mm long,
blue; standard having white spot; stamens 10,
united; legumes to 4 cm long, hairy. Dry sites,
scrub. Freq. all Fla. W to Miss., E to N.C.
Wint-sum.

SUNDIAL LUPINE
Lupinus perennis L.

Similar to *Lupinus diffusus* except: to 70 cm tall;
stems creeping; leaves palmately compound,
with 7-11 leaflets; petals blue, white or pink.
Dry sites. Freq. NF, WF. W to La., E to Maine.
Sum-fall.

(not shown)

LADY LUPINE
Lupinus villosus Willd.

(not shown)

Similar to *Lupinus diffusus* except: stems, petioles and leaves villous; petals pink to purple;
standard with reddish purple spot. Sandhills,
other dry sites. Freq. CF, NF, WF. W to Miss.,
E to N.C. Spr.

PHASEY BEAN
Macroptilium lathyroides (L.) Urban
[*Phaseolus lathyroides* L.]

Perennial, to 1 m tall; stems branching; leaflets 3,
3.5-6 cm long; sepals 4-6 mm long; petals to 1.5 cm
long, red or purple, showy; legumes to 10 cm long.
Native to tropical America. Disturbed sites, pastures. Freq. all Fla. Spr-fall.

SPOTTED MEDICK
Medicago arabica (L.) Huds.

(not shown)

Annual, 10-60 cm long; stems lying flat to erect;
leaflets 3, 1-3 cm long, as wide as long, reddish
purple-spotted; flowers 4-5 mm long, yellow;
legumes 5-6 mm wide, coiled, covered with
strongly curved spines. Disturbed sites. Rare NF,
WF. W to Tex., N to Canada. Spr-fall.

BLACK MEDICK
Medicago lupulina L.

Annual, 10-60 cm long; stems prostrate to creeping; leaflets 3, 0.5-2 cm long, as wide as long;
stipules entire; flowers 1.5-2 mm long, yellow;
legumes 2-3 mm wide, curved, not spiny.
Disturbed sites, turf. Common all Fla. All U.S.
Spr-sum.

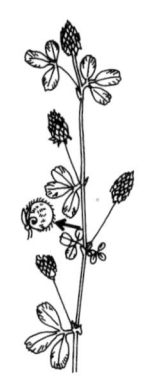

CALIFORNIA BUR-CLOVER
Medicago polymorpha L.

Annual, 20-50 cm long; stems prostrate, spreading;
leaflets 3, 6-15 cm long, longer than wide; *stipules lobed like teeth of comb; flowers 2.5-4 mm
long, yellow; *legumes 4-8 mm wide, coiled, covered with nearly straight spines. Disturbed sites.
Infreq. CF, NF, WF. W to Tex., N to Canada.
Wint-spr.

WHITE SWEET-CLOVER
Melilotus alba Desr.

Biennial, 0.6-2.5 m tall; leaflets 3, 1-3 cm long;
flowers white; standard 4-5 mm long; legumes
2.5-4 mm long; seeds 1-4. Disturbed sites. Freq.
all Fla. N to Canada. Spr-fall.

SOUR SWEET-CLOVER
Melilotus indica (L.) All.

Annual, to 50 cm tall; stems erect; leaflets 3, to
2.5 cm long; *flowers to 2 mm long, yellow;
legumes to 2.5 mm long. Disturbed sites. Infreq.
all Fla. W to Tex., N to Canada. Spr-sum.

BABY'S BATH BRUSH
Mimosa strigillosa Torr. & Gray

Perennial, to 2 m long; stems prostrate, covered
with long hairs; leaflets in 5-8 pairs, to 1 cm long;
flowers spikes to 2.5 cm long, oblong; petals
pink; legumes 2-3 cm long, hairy. Dry open sites.
Infreq. all Fla. W to Tex., E to Ga. Spr-sum.

SMALL-HEAD YELLOW PUFF
Neptunia pubescens Benth.
[*N. floridana* Small]

Perennial, to 2 m long; stems ascending to prostrate, matting; leaves 2-4 mm long, twice compound, numerous; *stipules ovate; *flower spikes
long-peduncled; flowers to 1 cm long, yellow;
legumes 2.5-3.5 cm long. Pinelands, coastal
sites. Infreq. all Fla. W to La. Spr-fall, all yr S.

JERUSALEM THORN
Parkinsonia aculeata L.

Perennial shrub or tree; leaves 20-40 cm long;
leaf stalks (rachis) winged; leaflets several; petals
0.9-1.7 cm long, yellow; *legumes 5-15 cm long.
Native to southwestern U.S. Escapes from cultivation. Disturbed sites. Infreq. all Fla. W to Tex.
and Calif., N to S.C. Spr-sum.

TRAILING WILD BEAN
Phaseolus sinuatus Nutt.

Perennial, 1-4 m long; stems trailing, climbing,
smooth; leaflets 2-4 cm long, round or 2- to 3-
lobed; petals 7-9 mm long, lavender; legumes to
4 cm long. Dry sites. Infreq. all Fla. N to N.C.
Sum-fall.

BUCKROOT
Psoralea canescens Michx.

Perennial, 30-90 cm tall, bushy; stems and leaves
covered with gray hairs (canescent); leaflets 1 or
3, 1.5-6 cm long, glandular-punctate; *flowers
8-15 mm long, with few flowers per raceme, blue
to violet; *legumes to 2 cm long. Dry sites. Freq.
CF, NF, WF. W to Ala., N to Va. Spr-sum.

FINE-LEAF PSORALEA
Psoralea lupinellus Michx.

Perennial, 20-60 cm tall; leaves divided into 5-7
leaflets each to 8 cm long; *petals to 6 mm long,
blue or violet; *legumes to 1 cm long. Dry sites.
Infreq. CF, NF, WF. N to N.C. Spr-sum.

SAMSON'S SNAKEROOT
Psoralea psoralioides (Walt.) Cory
[*Orbexilum pedunculata* (Mill.) Rydb.]

Perennial, 30-80 cm tall; stems smooth to hairy;
leaflets 3, 3-7 cm long, elliptic to oval; petals to
7 mm long, purple; *legumes to 4 mm long.
Dry sites. Infreq. NF, WF. W to Tex., N to Va.
Spr-sum.

SCURF-PEA
Psoralea virgata Nutt.

(not shown)

Perennial, to 70 cm tall; stems hairy; leaflet mostly 1, to 8 cm long, linear; petals to 8 mm long,
purple; legumes to 4 mm long. Flat pinelands.
Infreq. NF. N to Ga. Spr-sum.

KUDZU VINE
Pueraria lobata (Willd.) Ohwi.

Perennial, to 30 m long, trailing or climbing;
young stems hairy; older stems woody; leaflets 3,
7-20 cm long; *petals to 2 cm long, violet-purple
to reddish; *legumes to 5 cm long, hairy.
Roadsides, woods. Infreq. CF, NF, WF.
W to Tex., E to Va. Sum-fall.

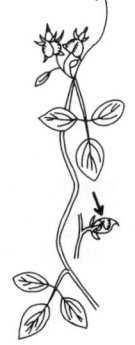

ASHY RHYNCHOSIA
Rhynchosia cinerea Nash

Perennial; stems prostrate, covered with stiff gray
hairs; leaflets 3, 1-3 cm long; sepals and petals
6-8 mm long, yellow; *legumes to 15 mm long.
Dry sites. Freq. SF, CF, NF. Spr-sum, all yr S.

TWINING RHYNCHOSIA
Rhynchosia difformis (Ell.) DC.

Perennial, climbing or prostrate vine, 0.5-1 m long; stems with downward-directed hairs; leaflets 1 or 3, but mostly 3, 1.5-4 mm long; the 1 leaflet less long than wide; the 3 leaflets ovate; sepals and petals 7-10 mm long, yellow; *legumes 16-20 mm long, covered with long hairs. Dry woods and clearings. Infreq. CF, NF, WF. W to Tex., N to Va. Sum-fall.

ONE-LEAF RHYNCHOSIA
Rhynchosia michauxii Vail

Perennial; stems prostrate, covered with gray hairs; leaflets 1, 2.5-5 cm long; *sepals and petals 12-17 mm long, yellow; *legumes 11-16 mm long. Pinelands. Freq. all Fla. Spr-fall, all yr S.

SMALL RHYNCHOSIA
Rhynchosia minima (L.) DC.

Perennial; stems branching, clambering, twining, covered with downward-directed hairs; leaflets 3, 0.1-4 cm long; sepals to 3.5 mm long; petals to 5.5 mm long, yellow; *legumes 10-15 mm long. Dry to moist sites. Freq. all Fla. W to Tex., E to Ga. Spr-fall, all yr S.

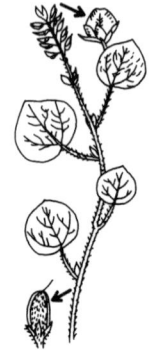

DOLLAR WEED
Rhynchosia reniformis DC.

Perennial, 5-25 cm tall; stems erect, with short hairs; leaflets 1 or 3, but mostly 1; all leaflets less long than wide; *sepals and petals 7-10 cm long, yellow; *legumes 10-18 mm long, covered with short hairs. Pinelands. Freq. all Fla. W to Tex., N to Va. Spr-fall.

TALL RHYNCHOSIA
Rhynchosia tomentosa (L.) Hook. & Arn.

Perennial, 15-90 cm tall; stems erect, with dense
hairs; leaflets 3, 2.5-7 cm long, oblong, elliptic
or obovate, with woolly hairs covering lower
surface; *sepals and petals 4-9 mm long, yellow;
legumes 12-21 mm long, covered with short
hairs. Dry sites. Infreq. CF, NF, WF. W to La.,
N to Del. Spr-fall.

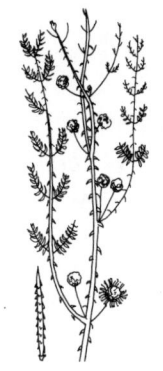

SMOOTH-LEAF SENSITIVE BRIER
Schrankia microphylla (Dry. ex Sm.) Macbr.

Perennial, to 2 m long; stems prostrate to spreading, angled, prickly; leaves bipinnate; leaflets 20-
32, 3-8 mm long, sensitive to touch and light;
flower spikes to 2 cm in diam., usually round,
pink; legumes 3.5-15 cm long, 4-sided.
Sandhills, pinelands, disturbed sites. Freq. all
Fla. W to Tex., E to Ky. and Va. Spr-fall, all yr S.

SENSITIVE BRIER
Schrankia uncinata Willd.

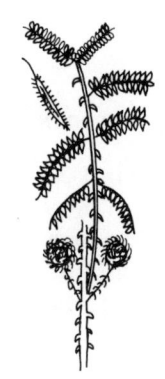

Perennial, 0.6-1.2 m long; stems spreading or
prostrate, covered with spines; leaflets in 8-15
pairs, 4-8 mm long; lower leaflet surfaces with
lateral veins raised; flower spikes to 1 cm long,
rounded; petals pink-purple; legumes 5-15 cm
long, covered with spines. Dry sites. Freq. SF,
CF, NF. N to Ala., W to Tex. Spr-sum.

PRIVET SENNA
Senna ligustrina (L.) Irwin & Barneby
[*Cassia ligustrina* L.]

Perennial, 0.6-2 m tall; stems partly woody,
smooth; petiole glands elongate; leaflets 12-16,
2-6 cm long; stipules linear; flowers to 10 mm
long, yellow; legumes to 12 cm long, *stipitate.
Hammocks, disturbed sites. Freq. SF, CF. All yr.

SICKLE POD
Senna obtusifolia (L.) Irwin & Barneby
[*Cassia obtusifolia* L.]

Annual, 0.4-1.5 m tall; stems smooth; leaf petioles with glands; leaflets 4-6, 1.5-8 cm long, obovate; petals 8-17 mm long, yellow; legumes 10-20 cm long, curved. Disturbed sites. Freq. all Fla. W to Tex., E to Va. Sum-fall.

COFFEE WEED
Senna occidentalis (L.) Link
[*Cassia occidentalis* L.]

Annual, 0.5-1.5 m tall; stems smooth; leaf petioles with globular glands; leaflets 8-12, 3.5-8 cm long, ovate-lanceolate; petals 15-19 mm long, yellow; legumes 8-13 cm long, with light margins. Disturbed sites. Freq. all Fla. W to Tex., E to Va. Sum-fall.

RATTLE BUSH
Sesbania drummondii (Rydb.) Cory

Perennial shrub, 1-3 m tall; leaflets 12-40; stipules linear; flowers 1.5-2.5 cm long, yellow; legumes 5-9 cm long, 4-winged, many-seeded. Moist to wet disturbed areas, sometimes along coasts. Rare WF. W to Tex. Spr-fall.

(not shown)

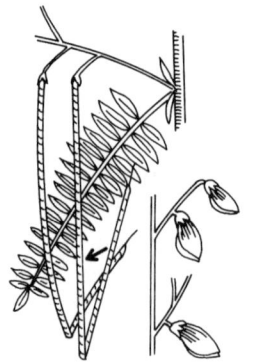

HEMP SESBANIA
Sesbania macrocarpa Muhl.
[*S. emerus* (Aubl.) Urban]

Annual, 1-4 m tall; stems smooth; leaflets 20-70, 1-3 cm long; flowers 13-20 mm long; standard yellow on upper surfaces, purple or red on lower surfaces; *legumes 10-20 cm long. Low, disturbed sites. Freq. all Fla. W to Tex., N to N.Y. Sum-fall.

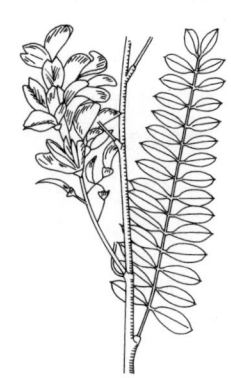

DAUBENTONIA or SPANISH GOLD
Sesbania punicea (Cav.) Benth.

Perennial shrub, 1-3 m tall; leaflets 12-40, 1-3 cm
long; flowers in racemes; petals 1.5-2.5 cm long,
orange-red; legumes 5-8 cm long, 4-winged. Wet
disturbed sites. Freq. CF, NF, WF. W to Tex.,
N to N.C. Spr-fall.

BLADDER POD
Sesbania vesicaria (Jacq.) Ell.

Annual, 1-4 m tall; leaflets 20-52, 1.5-3 cm long;
stipules linear; flowers 6-9 mm long, yellow with
red markings; *legumes 2-8 cm long; seeds 2.
Moist to wet disturbed areas. Freq. all Fla.
W to Tex., N to N.C. Sum-fall.

NECKLACE-POD
Sophora tomentosa L.
var. *truncata* Torr. & Gray

Perennial, to 3 m tall; leaflets 13-21, 2.5-6 cm long,
hairy beneath; petals 18-25 mm long; stamens 10,
distinct; legumes 6-15 cm long, stipitate. Coastal
sites. Infreq. SF, CF, NF. W to Tex. All yr.

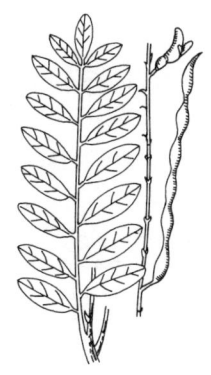

TRAILING WILD BEAN
Strophostyles helvola (L.) Ell.

(not shown)

Annual, to 10 m long; stems trailing; leaflets to
4 cm long; flowers purple; standard to 1.5 cm
long; legumes to 10 cm long. Moist sites. Infreq.
NF, WF. W to Tex., N to Canada. Sum-fall.

COMMON WILD BEAN
Strophostyles umbellata (Muhl.) Britt.

Perennial, to 1.5 m long; stems trailing; leaflets
1-5 cm long; flowers pink or purple; standard to
1.5 cm long; legumes to 5 cm long. Dry sites.
Infreq. CF, NF, WF. W to Tex., N to N.Y. Sum-fall.

PENCIL FLOWER
Stylosanthes biflora (L.) BSP.

Perennial, 15-60 cm long; stems lying flat to
erect; leaflets 3, 1-4 cm long, with lower surfaces
hairy; flower stems with long hairs; *flowers 5-
9 mm long, orange-yellow; *legumes 4-5 mm
long, with short beak. Dry sites. Infreq. CF, NF,
WF. W to Tex., N to N.J. Spr-fall

SOUTHERN PENCIL FLOWER
Stylosanthes hamata (L.) Taub.

Perennial, to 60 cm long; stems trailing, hairy;
leaflets 3, 1-4 cm long, hairy; *petals to 4 mm
long, yellow; *legumes to 2 mm long, 2-jointed,
covered with white hairs. Disturbed sites. Infreq.
SF, CF, NF. Endemic. All yr.

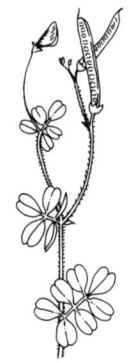

GOLDEN HOARY-PEA
Tephrosia chrysophylla Pursh

Perennial; stems prostrate, ascending, hairy;
petioles to 3 mm long; leaflets 5-7, with parallel
veins running from midrib to margin; petals white
or purple; standard 17-20 mm long; legumes 5-
6 mm wide. Dry sites. Freq. all Fla. W to Miss.,
N to Ga. Spr-fall.

LONG-STALKED HOARY-PEA
Tephrosia florida (Dietr.) Wood

Perennial, to 60 cm long; stems prostrate to erect, hairy; leaflets 7-17, 1-5 cm long; sepals to 5 mm long; *petals to 1.8 cm long, purple or white; flower stems flattened; legumes 3-5 cm long. Pinelands. Freq. all Fla. N to N.C. Spr-fall.

RUSTY HOARY-PEA
Tephrosia hispidula (Michx.) Pers.

Perennial, to 60 cm long; stems lying flat to erect; leaflets 9-23, 0.7-2.2 cm long; sepals to 6 mm long; petals to 1.8 cm long, white or pink; flower stems rounded; legumes to 5 cm long. Wet flatwoods. Infreq. NF, WF. Spr-fall.

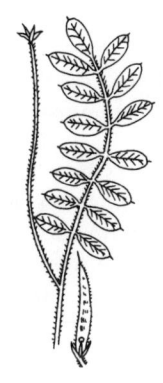

HOARY-PEA
Tephrosia rugelii Shuttlew.

Perennial, to 70 cm long; stems diffuse, monopodial, hairy; leaflets 3-11, 8-17 mm long; petals white to purple; standard 1.5-2 cm long; inflorescence nearly equal to, to shorter than, leaves; pods 3.5-4 cm long. Pinelands. Infreq. all Fla. Endemic. Spr-fall.

SAND PEA
Tephrosia spicata (Walt.) Torr. & Gray

(not shown)

Perennial, 30-60 cm long; stems lying flat to ascending, hairy; petioles over 5 mm long; leaflets 5 or 7-13, elliptic, linear; petals white, pink, red or purple; standard 16-19 mm long; legumes 3-5 cm long. Dry sites. Freq. all Fla. W to La., E to Va. Spr-fall.

GOAT'S RUE
Tephrosia virginiana (L.) Pers.

Perennial, 20-70 cm tall; stems and lower leaf
surfaces covered with gray hairs; leaflets 7-37,
9-30 mm long; flower stems and sepals covered
with gray hairs; petals 14-21 mm long, with yellow or white standard and rose-colored wings;
legumes 2.5-5.5 cm long. Dry sites. Infreq. CF,
NF, WF. W to Tex., N to Maine. Spr-fall.

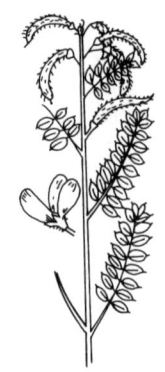

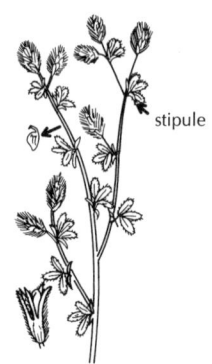

stipule

RABBIT-FOOT CLOVER
Trifolium arvense L.

Annual, 10-40 cm tall; stems hairy; leaflets 1-
2.5 cm long, hairy; flower heads to 2 cm long,
oblong, white or pink; *legumes to 1 cm long.
Native to Eurasia. Dry waste sites. Infreq. CF,
NF, WF. N to Canada. Spr-sum.

LOW HOP CLOVER
Trifolium campestre Schreb.

Annual, 5-40 cm long; stems prostrate; leaflets 3,
6-15 mm long; heads 8-12 mm in diam., 20- to
40-flowered; flowers 2.5-5.5 mm long, yellow;
legumes to 3 mm long. Disturbed areas. Infreq.
CF, NF, WF. W to Miss., N to N.C. Spr-fall.

(not shown)

SMALL HOP CLOVER
Trifolium dubium Sibth.

Annual, 5-30 cm long; stems ascending; leaflets
3, 6-12 mm long; *heads 5-8 mm in diam., 3- to
15-flowered; *flowers 2.5-3 mm long, yellow;
legumes to 2.5 mm long. Native to Europe.
Disturbed sites. Infreq. NF, WF. W to Tex.,
N to Canada. Spr-fall.

WHITE CLOVER
Trifolium repens L.

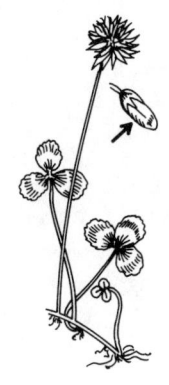

Perennial, 10-40 cm long; stems branching or
creeping, smooth; leaflets 3, 1-3 cm long; *flowers pediceled; pedicels 6-12 mm long; heads
1-3 cm in diam., white or pink; legumes 4-5 mm
long. Disturbed sites. Freq. all Fla. Common
throughout most of U.S. Sum, all yr S.

SAND VETCH
Vicia acutifolia Ell.

Perennial, 0.5-1.2 m long; stems sprawling or
climbing; leaflets 2-6, 1.5-4 cm long; flowers
to 7.5 mm long, with 8-10 flowers per raceme,
white to blue; *legumes 2.5-3 cm long, with
short stalk. Moist to wet sites. Common all Fla.
W to Ala., N to S.C. Spr-sum, all yr S.

FLORIDA VETCH
Vicia floridana S. Wats.

Perennial, to 80 cm long; stems sprawling or
reclining; leaflets 2-6, 0.5-1.5 cm long; flowers to
6 mm long, with 2-8 flowers per raceme, white to
blue; legumes to 1.5 cm long. Moist to wet sites.
Freq. CF, NF, WF. Spr-sum.

OCALA VETCH
Vicia ocalensis Godfrey & Kral

(not shown)

Perennial, to 1.2 m tall; stems reclining to climbing; leaflets 4-6, 3-5 cm long; flowers 15-18,
10-12 mm long, bluish white; standard to 10 mm
long; legumes 4-4.5 cm long; seeds 8-12. Wet
sites. Rare CF, NF. Endemic. Spr-fall.

NARROW-LEAF VETCH
Vicia sativa L.
subsp. *nigra* (L.) Erh.
[*V. angustifolia* L.]

Annual, to 1 m long; stems ascending or clambering; leaflets 8-14, 1-4 cm long; *flowers 1-1.8 cm
long, axillary, subsessile, purple-white; standard
to 15 mm long; legumes 4-8 cm long. Disturbed
sites. Infreq. NF, WF. W to Tex., N to Canada.
Spr-sum.

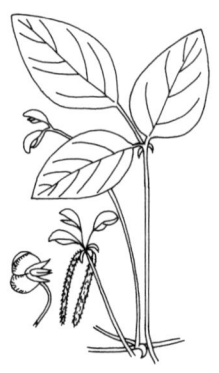

YELLOW VIGNA
Vigna luteola (Jacq.) Benth.

Perennial, to 3 m long; stems trailing or climbing,
hairy; leaflets 3, 2-8 cm long; petals to 1.9 cm
long, yellow; legumes to 6 cm long. Low, disturbed sites. Freq. all Fla. W to Tex., N to N.C.
All yr.

ZORNIA
Zornia bracteata Walt. ex Gmel.

Annual or perennial, 10-80 cm long; stems trailing to erect, hairy; leaflets 2 or 4, 1-3 cm long;
petals 0.9-1.4 cm long, yellow; legumes to 1.2
cm long. Various disturbed sites. Infreq. all Fla.
W to Tex., E to Va. Spr-sum.

DUCKWEED FAMILY   *LEMNACEAE*

Aquatic, small floating herbs consisting of frond-like structures; stems and leaves 0; flowers unisexual (monoecious); sepals 0; petals 0; fruit a utricle.

PIMPLED DUCKWEED
Lemna perpusilla Torr.

Annual; plant to 3 mm long, to 2.5 mm wide, 1-5 in a group, light green. Stagnant water. Infreq. NF. W to Kans., N to Mass. Sum.

GIANT DUCKWEED
Spirodela polyrhiza (L.) Schleid.

Annual, 3-10 mm long; plant body rounded; roots 5-20, often purple beneath. Quiet fresh water. Infreq. all Fla. W to Calif., N to Canada. Sum.

FEW-ROOT DUCKWEED
Spirodela punctata (Meyer) Thomps.
[*S. oligorhiza* (Kurz) Hegelm.]

Annual, 2-5 mm long; plant body rounded, 2- to 5-connected, with 2-5 roots each. Quiet fresh water. Common all Fla. W to Calif., N to N.C. All yr S.

BLADDERWORT FAMILY   *LENTIBULARIACEAE*

Aquatic or moist terrestrial herbs; carnivorous (gaining nutrients from small insects); carnivory accomplished by specialized trapping devices of leaves with sticky hairs or of bladders; leaves alternate or in basal rosettes; flowers bisexual; sepals 2-5; petals 5; ovary superior; fruit a capsule.

BLUE BUTTERWORT
Pinguicula caerulea Walt.

Perennial; rosettes 5-10 cm in diam.; leaves to 6 cm long; flower stems to 20 cm tall; flowers to 3 cm wide, dark purple to light purple; capsules to 1 cm wide. Low wet sites. Freq. all Fla. N to N.C. Spr.

GODFREY'S BUTTERWORT
Pinguicula ionantha Godfrey

Similar to *Pinguicula primuliflora* except: flowers to 2 cm wide, with violet throat. Wet sites. Infreq. WF. Spr.

(not shown)

YELLOW BUTTERWORT
Pinguicula lutea Walt.

Perennial; stems with glandular hairs; leaves 1-6 cm long, with sticky hairs; flower stems 10-30 cm tall; petals greater than 2 cm long, 2-3.5 cm wide, yellow; capsules 4-8 mm in diam. Low moist to wet sites. Infreq. all Fla. W to La., N to N.C. Spr.

CHAPMAN'S BUTTERWORT
Pinguicula planifolia Chapm.

Perennial; rosettes to 18 cm in diam.; leaves to
8 cm long; flower stems to 35 cm long; flowers to
3 cm wide, purplish red; capsules to 5 mm wide.
Wet sites. Freq. WF. W to Miss. Spr.

(not shown)

PROLIFEROUS BUTTERWORT
Pinguicula primuliflora Wood & Godfrey

(not shown)

Perennial; rosettes to 15 cm in diam.; leaves to
8 cm long; flower stems to 15 cm long; flowers
to 3 cm wide, light violet with yellow throat; capsules to 5 mm wide. Wet sites. Infreq. WF.
W to Miss., E to Ga. Spr.

SMALL BUTTERWORT
Pinguicula pumila Michx.

Perennial; leaves 1-2.5 cm long, basal, with
sticky hairs covering upper surface; flower stems
to 10 cm tall; petals 1-1.5 cm wide, white to
violet or rose, rarely yellow; spur to 3 mm long;
*capsules 4-5 mm wide. Moist, wet, acid sites.
Freq. all Fla. W to Tex., N to N.C. Wint-spr.

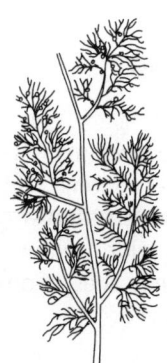

BLADDERWORT
Utricularia sp.

Perennial, forming green masses in quiet water.
Freq. all Fla. W to La., N to Canada. All yr.

HORNED BLADDERWORT
Utricularia cornuta Michx.

Perennial, terrestrial; flower stalks 10-40 cm tall
with 1 bract and 2 bractlets; petals to 2.5 cm
long, yellow to orange; spur to 1.2 cm long;
*capsules to 3.5 mm in diam. Wet sites. Freq.
all Fla. W to Tex., N to Canada. Sum-fall.

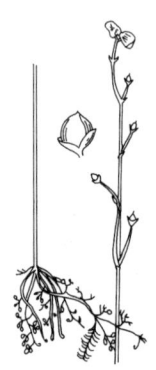

SLENDER STEMMED BLADDERWORT
Utricularia fibrosa Walt.

Perennial, aquatic; plants singular or mat-forming;
submerged stems 10-20 cm long, with several
leaf-like branches and bladders; flower stems 10-
40 cm tall; petals to 9 mm long, yellow; capsules
to 4 mm wide. Wet sites. Infreq. all Fla. W to
Tex., N to Mass. Spr-fall.

FLORIDA BLADDERWORT
Utricularia floridana Nash

Perennial, aquatic; flower stems to 60 cm long,
to 10 cm tall, with 1 bract; leafy branches 2-5 cm
wide; petals to 1 cm long, yellow; spur shorter
than lower lip; capsules to 8 mm in diam. Wet
sites. Infreq. CF, NF, WF. N to S.C. Spr-sum.

FLAT STEM BLADDERWORT
Utricularia foliosa L.

Perennial, aquatic, free-floating; main axes strap-
like; submersed stems 0.9-3 m long, with several
leaf-like branches and bladders; flower stems 10-
30 cm tall; *petals 1.7 cm wide, yellow; *capsules to 7 mm wide. Aquatic sites. Common all
Fla. W to La. Spr-fall, all yr S.

TANGLED BLADDERWORT
Utricularia gibba L.
[*U. biflora* Lam.]

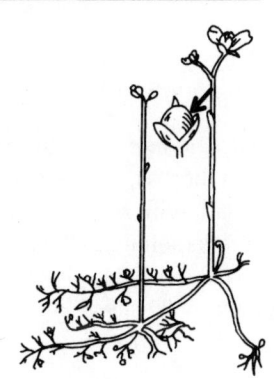

Perennial, aquatic; stems creeping, forming mats; flower stalks 2-15 cm tall with 1 bract; petals to 8 mm long, yellow; spur shorter or longer than lower lip; *capsules to 4 mm in diam. Aquatic sites, wet margins. Common all Fla. W to Tex., N to Maine. Spr-fall.

FLOATING BLADDERWORT
Utricularia inflata Walt.

Perennial, aquatic; flower stems to 40 cm long, above water to 20 cm tall, with 1 bract and several inflated side branches keeping plant and flowers floating; branchlets to 4 cm long, to 2 cm wide; petals to 2 cm long, yellow; spur half length of lower lip; *capsules to 6 mm in diam. Quiet water. Common all Fla. W to Tex., N to Del. Wint-spr.

RUSH BLADDERWORT
Utricularia juncea Vahl

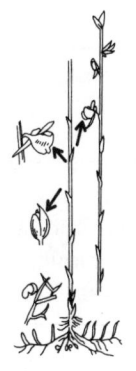

Perennial, terrestrial; below ground stems leaf-like; flower stems 10-40 cm tall; *petals 6-15 mm long, yellow; *capsules to 2.8 mm wide. Muddy to wet sites. Freq. all Fla. W to Tex., N to N.Y. Sum-fall.

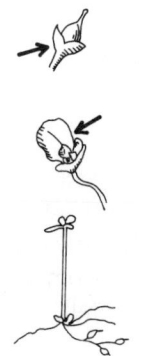

LITTLE BLADDERWORT
Utricularia olivacea Wright ex Griseb.

Perennial, aquatic; plants singular or mat-forming; stems short, not leaf-like, but bearing bladders; *petals 1-2 mm long, white; *capsules to 1 mm long. Ponds, lakes. Infreq. CF, WF. N to N.J. Fall.

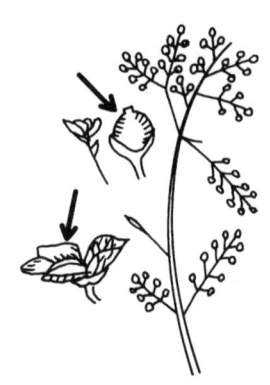

PURPLE BLADDERWORT
Utricularia purpurea Walt.

Perennial, aquatic; stems submerged, unattached;
branches whorled, bladder-bearing; flower stems
5-15 cm tall, with bracts attached to stems by
their centers (peltate); *flowers 1-5 per raceme;
petals to 1.2 cm long, pink to purple; *capsules to
4 mm long. Wet sites. Freq. all Fla. W to Tex.,
N to Canada. Sum-fall.

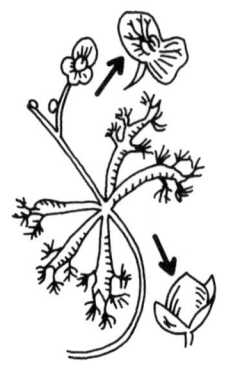

FEW-FLOWERED FLOATING BLADDERWORT
Utricularia radiata Small

Perennial, aquatic; flower stems to 15 cm long,
above water to 6 cm tall, with 1 bract and several
inflated side branches keeping plant and flowers
floating; branches to 3 cm long, to 5 mm wide;
*petals to 1.4 cm long, yellow; spur length of
lower lip; *capsules to 5 mm in diam. Wet sites.
Infreq. all Fla. W to Tex., N to Canada. Spr-sum.

BLADDERWORT
Utricularia simulans Pilger
[*U. fimbriata* HBK.]

Perennial, terrestrial; leaves absent; only bladders
present; flower stalks to 10 cm tall with 1 bract
and 2 fringed (fimbriate) bractlets; sepals fringed;
petals to 8 mm long, yellow; spur about as long
as lower lip; capsules to 4 mm long. Low wet
sites. Common SF, CF. Sum-fall.

(not shown)

WET SAND BLADDERWORT
Utricularia subulata L.

Perennial, terrestrial; flower stalks 3-20 cm tall
with single bract 1-2 mm long attached to stem
by its center (peltate); *petals 0.5-1.2 cm long,
yellow; spur about as long as lower lip; *capsules
to 2 mm in diam. Wet sites. Common all Fla.
W to Tex., N to Canada. Spr-fall.

LILY FAMILY   *LILIACEAE*

Herbs mostly arising from rhizomes, bulbs, corms or fleshy roots; leaves either basal and grass-like or succulent, or on stems and oblong to elliptic in shape or needle-like; flowers bisexual; sepals 3; petals 3; floral parts collectively termed tepals; ovary mostly superior; fruit a capsule or berry.

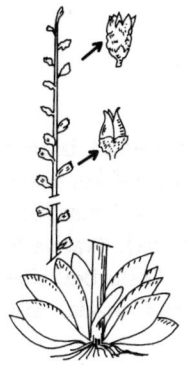

LATE FLOWERING COLICROOT
Aletris aurea Walt.

Perennial, 30-80 cm tall; leaves 3-8 cm long, in basal rosettes; *sepals and petals to 7 mm long, yellow; *capsules to 5 mm long. Moist sites. Infreq. NF, WF. W to Tex., N to Md. Spr-fall.

SOUTHERN COLICROOT
Aletris farinosa L.

Perennial; leaves 4-15 cm long; flower stalks 0.3-1.2 m tall; flowers 7-10 mm long, cylindric, white; capsules to 4 mm long. Dry to wet sites. Infreq. SF, CF. W to Tex., N to Minn., E to Maine. Spr-sum.

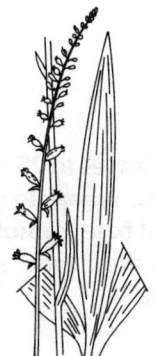

YELLOW COLICROOT
Aletris lutea Small

Perennial; leaves to 15 cm long, to 2 cm wide; flower stems 0.3-1.1 m tall; flowers 6-9 mm long, yellow; capsules to 4 mm long, beaked. Low, wet sites. Common all Fla. W to La., E to Ga. Spr-sum.

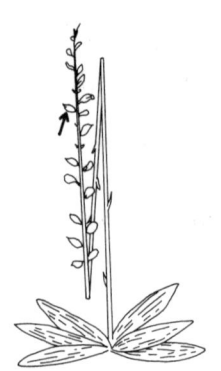

WHITE COLICROOT
Aletris obovata Nash

Perennial; leaves 2-8 cm long, with narrow trans-
parent margins; flower stalks to 80 cm tall; *flowers
5-6 mm long, obovoid, white. Moist sites. Infreq.
CF, NF, WF. W to Miss., N to S.C. Spr-sum.

COMMON ASPARAGUS
Asparagus officinalis L.

Perennial, 1-2 m tall; stems erect, branching
above, glaucous; leaves scale-like; branchlets 6-
16 mm long, terete; *tepals 4-7 mm long, green;
*berries 6-8 mm in diam., red. Native to Europe.
Escapes from cultivation. Disturbed sites. Rare
SF, NF, CF. W to La., N to Canada. Spr-sum.

ASPARAGUS FERN
Asparagus setaceus (Kunth.) Jessop

Perennial, woody, climbing; branchlets axillary **(not shown)**
to spines, forming fern-like ornamental sprays;
flowers white; fruits purple. Disturbed sites.
Infreq. SF, CF. All yr.

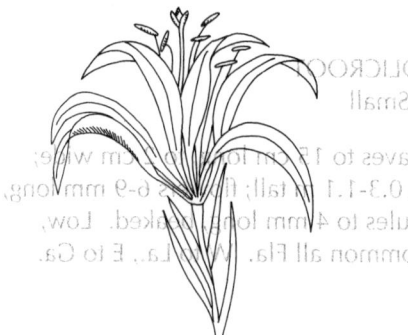

CATESBY'S LILY
Lilium catesbaei Walt.

Perennial, 30-80 cm tall; leaves to 25 cm long,
linear; tepals 7-11 cm long, orange to red, with
yellow and purple spots at base; capsules 2-5 cm
long. Low, moist, woody sites. Freq. all Fla.
W to La., N to Va. Sum-fall.

INDIAN CUCUMBER ROOT
Medeola virginiana L.

Perennial, 20-90 cm tall; leaves 5-15 cm long,
whorled; *sepals and petals to 11 mm long, green
to yellow; berries to 1.5 cm in diam. Moist
woods. Rare WF. W to La., N to Canada. Spr.

FALSE-GARLIC
Nothoscordum bivalve (L.) Britt.

Perennial, 10-45 cm tall, from bulb; leaves 10-
40 cm long, 2-4 mm wide; flowers 8-15 mm
long, 1-2 cm wide, white; *capsules to 5 mm
long. Disturbed sites. Infreq. all Fla. W to Tex.,
N to Nebr., E to Va. Spr-fall.

AFRICAN FALSE-GARLIC
Nothoscordum inodorum (Ait.) Nichols

Perennial, to 80 cm tall; leaves 30-45 cm long,
8-12 mm wide; flowers 9-12 mm long, white;
capsules 8-10 mm long, stipitate. Native to
Africa. Escapes from cultivation. Disturbed sites.
Rare CF, WF. W to La., E to S.C. Spr.

SOLOMON'S-SEAL
Polygonatum biflorum (Walt.) Ell.

Perennial, 20-90 cm tall or to 1.5 m long; leaves
4-20 cm long; *flowers 10-20 mm long, umbel-
late, green, white or yellow; *berries 8-13 mm in
diam., bluish black. Moist shady sites. Infreq.
CF, NF, WF. W to Tex., N to Canada. Spr-sum.

FEATHER-SHANK
Schoenocaulon dubium (Michx.) Small

Perennial; leaves to 50 cm long, 2-4 mm wide, channeled; flower stems to 90 cm tall; *sepals and petals to 2 mm long, yellow or green; *capsules 7-9 mm long. Dry sites. Infreq. CF, NF. N to Ga. Spr-sum.

WILD-BAMBOO
Smilax auriculata Walt.

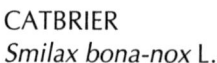

Perennial; stems prickly all over; leaves 2-12 cm long, evergreen, leathery, fiddle-shaped, 3-lobed or linear; *flowers 4-5 mm long; male and female flowers on separate plants (dioecious); berries to 13 mm in diam., black. Sandy open woods, disturbed areas. Common all Fla. W to La., N to N.C. Spr-sum.

CATBRIER
Smilax bona-nox L.

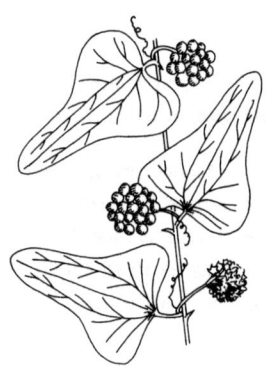

Perennial; stems prickly all over; leaves to 14 cm long, mostly evergreen, fiddle-shaped or 3-lobed; leaf margins with teeth; berries 6-8 mm in diam., black. Woody sites. Freq. all Fla. W to Tex., N to Md. Spr.

BAMBOO VINE
Smilax laurifolia L.

Perennial; stems prickly on lower portion only; leaves 5-15 cm long, evergreen, oblong or linear; flowers to 5 mm long; berries 5-8 mm in diam., black. Swampy sites. Freq. all Fla. W to Tex., N to N.J. Spr-sum.

LILIACEAE 245

SARSAPARILLA VINE
Smilax pumila Walt.

Perennial; stems and leaves woolly, having no
prickles; leaves to 12 cm long, evergreen, oval;
flowers to 3 mm long; berries to 8 mm in diam.,
red. Dry sites. Infreq. CF, NF, WF. W to Tex.,
N to S.C. Fall.

(not shown)

(not shown)

ROUNDLEAF GREENBRIER
Smilax rotundifolia L.

Perennial; stems prickly everywhere except on
nodes; leaves to 10 cm long, mostly evergreen,
oval to rounded; flowers to 6 mm long; berries
to 5 mm in diam., glaucous, black. Moist sites.
Infreq. NF, WF. W to Tex., N to Canada. Spr.

LANCELEAF GREENBRIER
Smilax smallii Morong

Perennial; stems prickly all over; leaves to 10 cm
long, evergreen, lanceolate to elliptic, with green
petioles; flowers 4-5 mm long; berries 5-8 mm
in diam., red. Moist sites. Infreq. all Fla.
N to Va. Spr.

(not shown)

(not shown)

BRISTLY GREENBRIER
Smilax tamnoides L.

Perennial; stems prickly all over; leaves to 12 cm
long, mostly evergreen, fiddle-shaped, 3-lobed or
oval; flowers 4-5 mm long; berries to 5 mm in
diam., black. Hammocks, woods. Infreq. all Fla.
N to Canada. Spr.

CORAL GREENBRIER
Smilax walteri Pursh

Perennial; stems prickly on lower portion only;
leaves 5-14 cm long, deciduous, ovate or ovate-
lanceolate, with red petioles; male and female
flowers 4-7 mm long, yellow-brown, on separate
plants; berries 6-8 mm in diam., red. Moist to
wet woody sites. Infreq. CF, NF, WF. W to La.,
N to N.J. Spr.

(not shown)

WHITE FEATHERLING
Tofieldia glabra Nutt.

Similar to *Tofieldia racemosa* except: flower
stems smooth; petals to 3 mm long. Moist, low
sites. Ga., N to N.C. Sum-fall.

ASPHODEL
Tofieldia racemosa (Walt.) BSP.

Perennial; leaves to 40 cm long; flower stems 30-
70 cm tall, glandular, scabrous; petals 4-5 mm
long, white; capsules to 3 mm long. Low wet
sites. Infreq. WF. W to Tex., E to N.J. Spr-fall.

BELLWORT
Uvularia perfoliata L.

Perennial, 10-80 cm tall; leaves 3-9 cm long,
alternate, sessile; *sepals and petals 1-3 cm long,
yellow; *capsules to 1 cm long. Rich woods.
Rare WF. W to Ark., N to Canada. Spr.

CROW-POISON
Zigadenus densus (Desr.) Fern.

Perennial, 0.5-1 m tall; leaves 1-3, to 1 cm wide, enclosed in sheaths to 10 cm long; *petals to 5 mm long, white to pink or purple; *capsules to 1.2 cm long, to 4 mm in diam. Moist, low sites. Infreq. CF, NF, WF. W to Tex., E to Va. Spr.

CREEPING CAMUS
Zigadenus glaberrimus Michx.

Perennial, 0.5-1.4 m tall, glaucous; leaves to 40 cm long, linear, mostly basal; *tepals 11-15 mm long, white, having 2 glands near base; capsules to 1 cm long. Moist to wet sites. Infreq. WF. W to Tex., E to Va. Sum.

FLY-POISON
Zigadenus muscaetoxicus (Walt.) Zimmerman

Similar to *Zigadenus densus* except: leaves 4, to 2 cm wide, not enclosed in sheaths; capsules broader than long. Leaves and roots toxic. Thickets, woods, dry sites. Infreq. NF, WF. W to Ark., N to N.Y. Sum.

(not shown)

FLAX FAMILY   *LINACEAE*

Herbs or shrubs; leaves alternate; flowers bisexual; sepals 5 or 4; petals 5 or 4; ovary superior; fruit a capsule.

CARTER'S FLAX
Linum carteri Small

(not shown)

Annual, to 60 cm tall; leaves to 3 cm long, alternate; petals to 1.7 cm long, orange to yellow; capsules to 4 mm in diam., ovoid. Disturbed sites. Rare SF. Spr.

FLORIDA FLAX
Linum floridanum (Planch.) Trel.

Perennial, to 1.2 m tall; leaves opposite and alternate; petals to 1 cm long, yellow; capsules 2.5-3.5 mm long, ovoid. Moist to wet sites. Freq. CF, NF, WF. N to Va. Sum.

(not shown)

YELLOW FLAX
Linum medium (Planch.) Britt.
var. *texanum* (Planch.) Fern.

Perennial, 30-70 cm tall; leaves 0.8-2.5 cm long, opposite and alternate; petals 4-8 mm long, yellow; *capsules 2-2.5 mm wide. Flatwoods, open areas. Freq. all Fla. W to Tex., N to Maine. Spr-fall.

BLUE FLAX
Linum usitatissimum L.

Annual, 20-80 cm tall; leaves 1.5-4 cm long;
petals 1.5-2 cm long, blue; *capsules to 9 mm
long. Native to Europe. Disturbed sites. Rare
all Fla. N to N.J. Spr-sum.

LOASA FAMILY   *LOASACEAE*

Herbs, shrubs or trees, with hooked or stinging hairs; leaves
alternate or opposite; flowers bisexual; sepals 4-7; petals
mostly 4 or 5; ovary inferior; fruit a capsule.

POOR MAN'S PATCH
Mentzelia floridana Nutt.

Perennial, to 1 m long; stems brittle, sprawling;
leaves 2-10 cm long, some 3-lobed; leaf hairs
barbed; leaf blades easily attaching to clothing;
petals 1.5-2 cm long, yellow; *capsules to 1.5 cm
long, with hooked hairs. Coastal hammocks and
dunes. Common SF, CF, NF. Spr-fall.

LOGANIA FAMILY · *LOGANIACEAE*

Herbs, vines, shrubs or trees; leaves opposite; stipules present; flowers bisexual; sepals 4 or 5; petals 4 or 5; ovary superior; fruit a capsule or berry.

ODORLESS JESSAMINE
Gelsemium rankinii Small

(not shown)

Perennial, woody high-climbing vine; leaves to 9 cm long; petals to 3 cm long, dull yellow, odorless; capsules to 1.6 cm long; seeds 3-4 mm long, wingless. Wet, wooded sites. Freq. WF. W to La., E to Ga. Wint-spr. All parts poisonous.

YELLOW JESSAMINE or CAROLINA JASMINE
Gelsemium sempervirens (L.) Ait. f.

Perennial, woody high-climbing vine; leaves 4-7 cm long; petals 2-4 cm long, bright yellow, fragrant; capsules to 2.5 cm long; seeds to 1.5 mm long, winged. Wooded sites. Common CF, NF, WF. W to Tex., N to Va. Wint-spr. All parts poisonous.

STALKED MITERWORT
Mitreola petiolata (J. F. Gmel.) Torr. & Gray

Annual, 10-80 cm tall; leaves to 8 cm long, with stalks 3-15 mm long; flowers on one side of branch; sepals 5; petals 5, white; capsules 3-4 mm long, 2-lobed. Moist to wet places. Freq. all Fla. W to Tex., E to Va. Sum-fall.

LOGANIACEAE ▰▰▰▰▰▰▰▰▰▰▰ 251

MITERWORT
Mitreola sessilifolia (J. F. Gmel.) G. Don.

Annual, 10-60 cm tall; leaves 1.5-2 cm long,
lacking stalks; flowers on one side of stem; sepals
5; petals 5, 1-3 mm long, white; stamens 5; capsules to 4 mm long. Moist to wet places. Freq.
all Fla. W to Tex., E to Va. Sum-fall.

RUST WEED
Polypremum procumbens L.

Perennial, 10-30 cm long; stems creeping to
ascending; leaves 1-2.5 cm long; sepals 4, to 3 mm
long; petals 4, to 3 mm long, white; *capsules to
2 mm long. Dry to moist sites. Common all Fla.
W to Tex., N to Pa. and N.J. Spr-fall.

PANHANDLE PINK-ROOT
Spigelia gentianoides Chapm.

Perennial, to 30 cm tall; leaves 1.5-5 cm long;
petals 1.5-2.5 cm long, pink; anthers not exerted;
capsules to 0.5 cm wide. Dry sites. Infreq. NF,
WF. Spr.

(not shown)

INDIAN PINK
Spigelia marilandica L.

Perennial, 15-70 cm tall; leaves 3-12 cm long;
petals 3-4.5 cm long, red, yellow inside; anthers
exerted; *capsules to 1 cm wide. Rich sites.
Infreq. WF. W to Tex., N to Md. Spr.

MISTLETOE FAMILY   *LORANTHACEAE*

Parasitic shrubs; leaves opposite; flowers bisexual or unisexual (dioecious); floral parts 3-5; ovary inferior; fruit a berry or drupe.

MISTLETOE
Phoradendron serotinum (Raf.) M. C. Johnst.

Perennial, branching; leaves 2-16 cm long; flowers in spikes, yellow; berries to 4 mm in diam., white. Apparent in treetops as round masses. Freq. all Fla. W to Tex., N to N.J. Fall-wint.

CLUB-MOSS FAMILY   *LYCOPODIACEAE*

Moss-like, low-growing plants; leaves numerous, evergreen, minute; fruit a sporangia shedding many yellow spores.

FOXTAIL CLUB-MOSS
Lycopodium alopecuroides L.

Perennial, to 60 cm long; stems creeping and mostly arching over vegetation, rooting at tips; leaves to 8 mm long; *sporangia in erect leafy structure resembling and about twice as wide as its stem. Wet open sites. Freq. CF, NF, WF. W to Tex., N to Mass. Sum-fall, all yr S.

SOUTHERN CLUB-MOSS
Lycopodium appressum (Chapm.) Lloyd & Underw.

Perennial, to 50 cm long; stems creeping; leaves
to 8 mm long; *sporangia in erect leafy structure
resembling and only slightly thicker than its stem.
Wet open sites. Infreq. all Fla. W to Tex., N to
Canada. Sum-fall, all yr S.

NODDING CLUB-MOSS
Lycopodium cernuum L.

Perennial, to 80 cm long; stems erect or reclining;
leaves to 13 mm long; sporangia grouped into
cone-like structures (strobili); strobili 6-10 mm
long, nodding. Wet sites. Infreq. all Fla. Ga. to La.
Sum-fall, all yr S.

LOOSESTRIFE FAMILY   *LYTHRACEAE*

Herbs, shrubs or trees; leaves alternate or opposite; flowers
bisexual; sepals 4-6; petals 4, 6 or 0; ovary superior; fruit a
capsule.

WEST FLORIDA WAXWEED
Cuphea aspera Chapm.

(not shown)

Perennial, to 40 cm tall; leaves 1-2.5 cm long,
whorled; sepals to 4 mm long; petals 6, to 7 mm
long, lavender; capsules to 1 cm long. Wet sites.
Infreq. WF. W to La., N to N.C. Sum.

COMMON WAXWEED
Cuphea carthagenensis (Jacq.) Macbr.

Annual, 10-80 cm tall; stems erect, hairy; leaves
2-6 cm long, opposite; sepals 5-7 mm long;
petals 6, 2-3 mm long, green to purple; *capsules
to 5 mm long. Low, wet sites. Freq. all Fla.
W to Tex., N to N.C. Sum-fall.

SWAMP LOOSESTRIFE
Decodon verticillatus (L.) Ell.

Perennial, 0.6-2.5 m long; stems recurved; bark
peeling off in brown strips; leaves to 20 cm long,
opposite or whorled; upper leaf surfaces smooth;
lower leaf surfaces hairy; sepals and petals 5-7,
magenta; stamens 8-10; capsules to 5 mm in
diam. Wet, swampy sites. Infreq. CF, NF, WF.
W to Tex., N to Canada. Sum-fall.

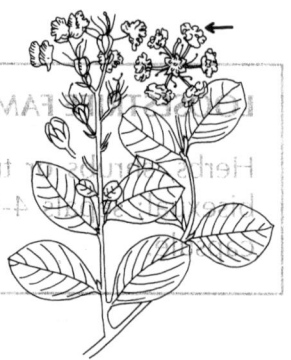

CREPE-MYRTLE
Lagerstroemia indica L.

Perennial shrub or tree, to 7 m tall; leaves 2-7 cm
long; sepals 5-7; *petals 5-7, 10-20 mm long,
purple, pink or white; stamens many; capsules
8-14 mm long. Native to Asia and Australia.
Cultivated, persists and escapes. Disturbed sites.
Infreq. CF, NF, WF. W to Tex., E to Va. Sum-fall.

LANCE-LEAVED LOOSESTRIFE
Lythrum alatum Pursh
var. *lanceolatum* (Ell.) Torr. & Gray
ex Rothrock in G. M. Wheeler
[*L. lanceolatum* Ell.]

Perennial, to 1.3 m tall; stems erect; leaves to
7 cm long, to 1.4 cm wide, opposite and alternate;
petals to 6 mm long, purple; capsules to 5 mm
long. Wet sites. Infreq. all Fla. W to Tex.,
N to N.C. Sum-fall.

CREEPING LOOSESTRIFE
Lythrum flagellare Shuttlw. ex Chapm.

Perennial, sprawling, creeping; leaves 5-10 mm
long, mostly opposite; petals 4-5 mm long, lavender to purple; capsules over 3 mm long. Low,
open sites. Infreq. SF, CF. Spr, all yr S.

SALTMARSH LOOSESTRIFE
Lythrum lineare L.

Perennial, 0.3-1.5 m tall; stems erect; leaves 1-4 cm
long, 1-4 mm wide, opposite; petals 4-5 mm long,
purple or white; *capsules to 4 mm long. Coastal
sites, hammocks. Infreq. all Fla. W to Tex., N to
Canada. Sum-fall.

TOOTHCUP
Rotala ramosior (L.) Koehne

Annual, 10-40 cm tall; stems erect to lying flat;
leaves 1-3 cm long, mostly opposite; petals minute
or 0, white to pink; *capsules to 3.5 mm in diam.
Wet, open sites. Freq. all Fla. W to Tex., N to
Mass. Sum-fall.

MALLOW FAMILY   *MALVACEAE*

Herbs, shrubs or trees; leaves alternate; stipules present; flowers bisexual; sepals 5; petals 5; ovary superior; fruit a capsule.

YELLOW INDIAN MALLOW
Abutilon hirtum (Lam.) Sweet

Perennial, 1-3 m tall; stems and leaves covered with thick soft hairs; leaves 2-20 cm long; petals to 20 mm long, orange-yellow with purple base; *carpels 11 or more, 8-10 mm long. Disturbed and coastal sites. Rare CF, NF. All yr.

RED INDIAN MALLOW
Abutilon hulseanum (Torr. & Gray) Torr. ex Chapm.

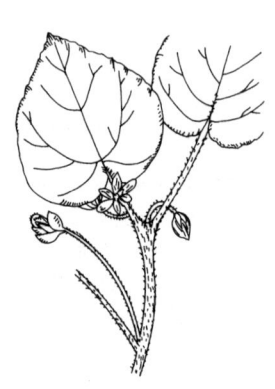

Perennial shrub, to 1.8 m tall; stems covered with shaggy hairs; leaves 5-13 cm long; both leaf surfaces covered with velvety hairs; flowers solitary, to 10 cm long, to 2.5 cm wide, pink to dark red; sepals and petals to 1.3 cm long; capsules to 2 cm long, hairy. Dry disturbed sites. Infreq. SF, CF. All yr.

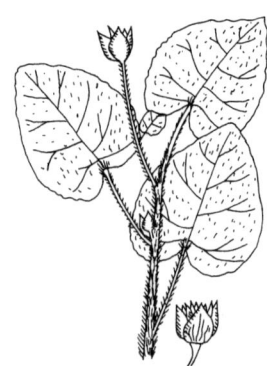

INDIAN MALLOW
Abutilon permolle (Willd.) Sw.

Perennial, 0.5-1.6 m tall; stems with velvety hairs; leaves 2.5-25 cm long, oval to heart-shaped, with velvety hairs; petals to 1.5 cm long, yellow; carpels 7-10, 1.1-1.2 cm long, with short hairs. Disturbed sites. Infreq. SF, CF. All yr.

VELVET LEAF
Abutilon theophrasti Medicus

Annual; stems with velvety hairs; leaves 9-30 cm
long, oval to heart-shaped, with velvety hairs;
petals to 1.5 cm long, orange to yellow; carpels
12-17, 1.5-2 cm long, with long hairs. Native to
Asia. Row crops, disturbed areas. Infreq. SF, CF,
NF. W to Tex., N to Maine. Spr-fall.

(not shown)

CRIMSON-EYED ROSE MALLOW
Hibiscus aculeatus Walt.

Perennial, 0.9-1.2 m tall; stems several, with
rough hairs; leaves to 12 cm long, 5-lobed; petals
to 6 cm long, cream to yellow, with crimson
bases; capsules to 13 mm long. Moist sites.
Freq. NF, WF. W to La., N to S.C. Spr-fall.

RED HIBISCUS
Hibiscus coccineus Walt.

Perennial, to 3 m tall; stems smooth; leaves to
15 cm long, divided like a hand into 5-7 lobes
each 4-25 cm long; flowers solitary; sepals 4-5 cm
long; petals to 10 cm long, red or pink; capsules
7-10 cm long, hairy. Wet, marshy sites. Infreq.
CF, NF, WF. N to Ala., E to Ga. Sum-fall.

(not shown)

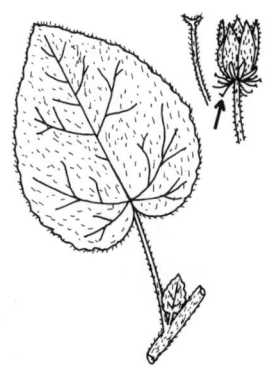

INDIAN RIVER HIBISCUS
Hibiscus furcellatus Desv.

Perennial, 0.9-2.5 m tall; stems hairy; leaves
5- 20 cm long, hairy; petals 6-10 cm long, pink
or purple; *bracts 1-2 cm long, to 1 mm wide;
*capsules to 2.5 cm long, 5-valved. Dry sites.
Infreq. SF, CF, NF. All yr.

SWAMP HIBISCUS
Hibiscus grandiflorus Michx.

Perennial, to 3 m tall; stems covered with star-
shaped hairs; leaves to 18 cm long, 3-lobed; both
leaf surfaces covered with velvety smooth star-
shaped hairs; flowers to 15 cm long; sepals 3-4 cm
long, covered with soft hairs; petals 12-14 cm long,
pink or white, with red base; capsules to 3 cm long,
covered with star-shaped hairs. Swamps. Freq. all
Fla. W to Miss., E to Ga. Sum.

(not shown)

HALBERD-LEAVED MARSH MALLOW
Hibiscus militaris Cav.

(not shown)

Perennial, to 2.5 m tall; stems smooth; leaves to
16 cm long, unlobed or 2-lobed at base of leaves;
flowers solitary; sepals 3-4 cm long; petals 6-8 cm
long, white or pink, with purple base; capsules
3.5-4 cm long. Wet sites. Infreq. WF. W to Tex.,
N to Pa. Sum-fall.

MARSH ROSE MALLOW
Hibiscus moscheutos L.

L.Perennial, 0.8-2 m tall; stems smooth to slightly
hairy; leaves to 20 cm long; both or only lower
leaf surfaces covered with gray felty hairs; flowers
solitary, 10-12 cm long; sepals to 3 cm long;
petals to 10 cm long, white, cream or pink, with
red to purple base; capsules to 3 cm long, smooth,
with short beak. Marshes. Infreq. NF, WF.
W to Tex., N to Md. Sum-fall.

STINGING MALLOW
Kosteletzkya pentaspermum (Bert. ex DC.) Griseb.

(not shown)

Perennial, to 1 m tall; stems and leaves with stinging hairs; leaves 3-7 cm long; flowers solitary;
sepals to 4 mm long; petals to 1 cm long, white;
capsules to 1 cm in diam. Coastal hammocks.
Infreq. SF. All yr.

SEASHORE MALLOW or SALTMARSH MALLOW
Kosteletzkya virginica (L.) Presl ex Gray

Perennial, 0.5-1.5 m tall; entire plant covered with star-shaped hairs; leaves 3-15 cm long; flowers solitary; sepals to 1.2 cm long; petals 2-4 cm long, white or pink; capsules to 1.2 cm in diam. Wet inland and coastal sites. Freq. all Fla. W to Tex., N to N.Y. Spr-fall, all yr S.

TURK'S CAP
Malvaviscus arboreus Cav.

Perennial, to 2 m tall, evergreen; leaves 3-21 cm long; flowers to 7 cm long, red, closed; stamens protruding; fruits 1-2 cm wide, fleshy, red. Disturbed sites. Freq. all Fla. Spr-fall.

SOUTHERN SIDA
Sida acuta Burm.

Perennial, 0.3-1 m tall; leaves 2-10 cm long; petals 1-1.5 cm long, yellow or white; *carpels 7-9, 2-toothed. Disturbed sites. Common all Fla. Spr-fall, all yr S.

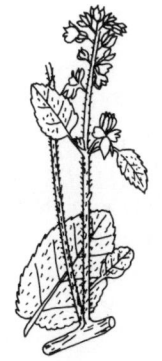

HEARTLEAF SIDA
Sida cordifolia L.

Perennial, to 1.5 m tall; stems covered with velvety hairs; leaves to 14 cm long, heart-shaped, covered with velvety hairs; petals to 7 mm long, yellow; carpels 10-12, to 3 mm long. Native to tropical America. Disturbed sites. Freq. all Fla. Spr-fall.

NARROWLEAF SIDA
Sida elliottii Torr. & Gray

Perennial, 30-50 cm tall; stems hairy; leaves 2-
7 cm long, linear; petals to 1.5 cm long, yellow;
carpels 7-12, each 4-5 mm long. Open sites.
Infreq. all Fla. W to Miss., N to Va. Sum-fall.

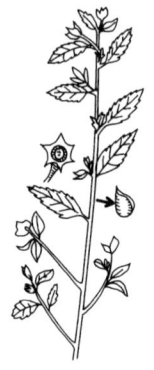

ARROWLEAF SIDA
Sida rhombifolia L.

Annual, 0.5-1.2 m tall; stems hairy; leaves 1.5-
8 cm long, elliptic to rhombic; receptacle thick-
nerved; petals to 1 cm long, yellow; *carpels 10-
12, each to 5 mm long, 1-spined. Disturbed
sites. Common all Fla. W to Tex., N to Va.
Spr-fall, all yr S.

PRICKLY SIDA
Sida spinosa L.

Annual, 10-60 cm tall; stems hairy; leaves 1-5 cm
long, linear to ovate; petals 6-8 mm long, yellow
to orange; *carpels usually 5, each to 4 mm long.
Disturbed sites, row crops. Infreq. all Fla.
W to Tex., N to Mass. Sum-fall.

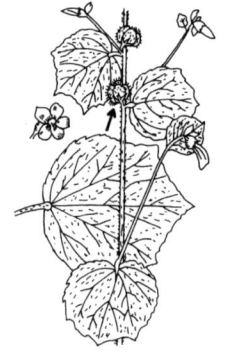

CAESAR WEED
Urena lobata L.

Perennial, 1-3 m tall; entire plant covered with
star-shaped hairs; leaves 5-10 cm long; lower
leaf surfaces covered with woolly hairs; petals
to 1.5 cm long, pink; *capsules to 1 cm wide,
bristly. Disturbed sites. Common SF, CF, NF.
All yr.

ARROWROOT FAMILY   *MARANTACEAE*

Herbs with underground tubers or rhizomes; leaves mostly alternate; petiole bases swollen into pulvinus; flowers bisexual; sepals 3; petals 3; ovary inferior; fruit a capsule.

POWDERY THALIA
Thalia dealbata Roscoe

(not shown)

Perennial, 1-2 m tall; leaves to 50 cm long, to 20 cm wide; flowers crowded on stem; petals to 8 mm long, purple; utricles 10-13 mm in diam., round. Wet sites. S.C., W to Tex. Spr-fall.

SWAMP LILY or ALLIGATOR FLAG
Thalia geniculata L.

Perennial, 2-3.5 m tall; leaves to 80 cm long, to 20 cm wide; *flowers not crowded on stem; petals to 9 mm long, purple; utricles 6-7 mm in diam., oval. Wet sites. Freq. all Fla. N to S.C. Spr-fall.

BOG MOSS FAMILY — *MAYACACEAE*

Aquatic herbs; leaves alternate; flowers bisexual; sepals 3; petals 3; ovary superior; fruit a capsule.

BOG MOSS
Mayaca fluviatilis Aubl.

Perennial, 2-40 cm long, creeping, floating; leaves 0.3-1.4 cm long; *petals to 6 mm long, white, pink or violet; *capsules to 5 mm long. Stream margins. Infreq. CF, NF, WF. W to Tex., E to Ga. Spr-sum.

MEADOW BEAUTY FAMILY — *MELASTOMATACEAE*

Herbs, vines, shrubs or trees; leaves opposite, mostly with 3 prominent parallel veins; flowers bisexual; sepals 4-5; petals 4-5; ovary superior or inferior; fruit a capsule or berry.

TALL MEADOW BEAUTY
Rhexia alifanus Walt.

Perennial, to 1 m tall; stems smooth; leaves to 8 cm long, with glandular hairs on surfaces; leaf margins toothed; petals to 2.5 cm long, pink to violet; anthers 5-11 mm long, curved; hypanthium 7.5-10 mm long, with dense bristly glandular hairs; capsules to 7 mm long. Moist sites. Infreq. CF, NF, WF. W to Tex., N to N.C. Spr-sum.

CUBAN MEADOW BEAUTY
Rhexia cubensis Griseb.

Perennial, to 60 cm tall; stems with glandular hairs; leaves 2-4 cm long, with glandular hairs on surfaces; leaf margins toothed; petals to 2 cm long, violet or pink; anthers 7-12 mm long, curved; hypanthium 10-14 mm long, with glandular hairs; hypanthium neck longer than body; capsules 6-7 mm long. Low, wet sites. Freq. all Fla. W to Miss., N to N.C. Spr-sum.

YELLOW MEADOW BEAUTY
Rhexia lutea Walt.

Perennial, 10-60 cm tall; stems hairy, sometimes having additional glandular hairs; leaves 1-3 cm long; sepal lobes 4; petals 4, 9-15 mm long, yellow; stamens 8; hypanthium enclosing capsule, to 4 mm long. Low, wet sites. Common NF, WF. W to Tex., E to N.C. Spr-sum.

PALE MEADOW BEAUTY
Rhexia mariana L.

Perennial, 20-80 cm tall; stem with glandular hairs; stem with flat or concave surface narrower than convex; leaves 2-4 cm long, hairy, linear, lanceolate, elliptic or ovate; leaf margins toothed; petals 12-15 mm long, white to lavender; anthers 6-8 mm long; capsules 6-7 mm long. Low sites. Common all Fla. W to Tex., N to Mo., E to N.Y. Spr-fall.

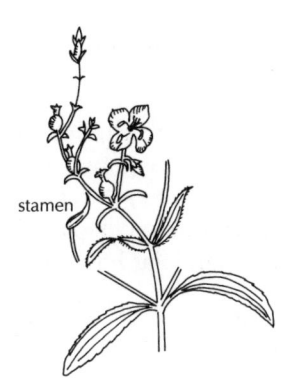

stamen

CLUSTERED MEADOW BEAUTY
Rhexia nashii Small

Perennial, to 1.5 m tall; stems and leaves covered with dense, coarse, yellow hairs; leaves to 7 cm long, with toothed margins; petals to 2.5 cm long, violet; anthers 5-11 mm long, curved; hypanthium to 1 cm long, sparsely hairy; capsules to 6 mm long. Wet sites. Freq. all Fla. Sum.

NUTTALL'S MEADOW BEAUTY
Rhexia nuttallii James

Perennial, 5-30 cm tall; stems with few hairs; leaves to 1 cm long, with toothed margins; petals to 1.2 cm long, pink to violet; anthers to 2 mm long, straight; hypanthium 5-7 mm long, with few glandular hairs; capsules to 4 mm long. Wet sites. Freq. all Fla. W to Miss., N to Ga. Spr-sum.

COASTAL PLAIN MEADOW BEAUTY
Rhexia petiolata Walt.

Perennial, 20-60 cm tall; stems smooth; leaves 1-2 cm long, with teeth on margins; petals 1-2 cm long, pink to violet; anthers to 2 mm long, straight; hypanthium to 12 mm long, smooth; capsules 4-5 mm long. Moist to wet sites. Freq. CF, NF, WF. W to Tex., N to N.C. Spr-sum.

COMMON MEADOW BEAUTY
Rhexia virginica L.

Perennial, to 1 m tall; stems 4-winged, hairy; leaves to 7 cm long, with toothed margins; petals to 2 cm long, pink, violet or purple; anthers 5-6 mm long, curved; hypanthium 8-10 mm long, with few glandular hairs; hypanthium neck shorter than body; capsules to 6 mm long. Low, wet sites. Freq. NF, WF. W to Tex., N to Canada. Spr-fall.

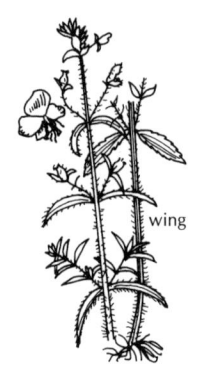

wing

BOGBEAN FAMILY   *MENYANTHACEAE*

Aquatic or wet terrestrial herbs; leaves alternate; flowers bisexual or unisexual; sepals 5; petals 5; ovary superior; fruit a capsule.

FLOATING HEARTS
Nymphoides aquatica (J. F. Gmel.) Kuntze

Perennial, aquatic and rooting in mud; leaves 5-20 cm wide, green above, rough and purple beneath, floating; flowers 1-2 cm wide. Quiet water. Common all Fla. W to Tex., N to N.J. Spr-sum.

NORTHERN FLOATING HEARTS
Nymphoides cordata (Ell.) Fern.

Perennial aquatic; leaves 2-6 cm long, floating, heart- to oval-shaped, mottled purple above, may be red beneath; petals to 1 cm in diam., white; capsules 4-5 mm long. Quiet water. Infreq. CF, NF, WF. N to Canada. Sum-fall.

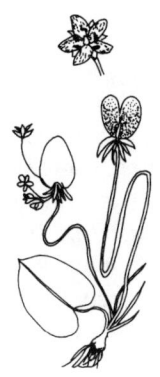

BAYBERRY FAMILY   *MYRICACEAE*

Trees or shrubs; leaves alternate; flowers unisexual (monoecious); floral parts 0; ovary superior; fruit a drupe.

WAX MYRTLE
Myrica cerifera L.

Perennial shrub to small tree, 0.3-12 m tall; leaves
3-10 cm long, evergreen, aromatic, with punctate
glands on both sides; leaves shorter at branch tips;
fruits to 2.5 mm in diam., covered with white
wax, in clustered spikes. Various dry to wet sites.
Common all Fla. W to Tex., N to N.J. Spr-fall.

NORTHERN BAYBERRY
Myrica heterophylla Raf.

Perennial shrub to small tree, to 3 m tall; leaves
to 12 cm long, evergreen, aromatic, with punctate
glands on lower surface only; leaves same length
even at branch tips; fruits 2-4.5 mm in diam.,
covered with white wax, in clustered spikes.
Wet sites. Infreq. CF, NF, WF. W to Tex.,
N to Va. Spr-fall.

(not shown)

MYRSINE FAMILY   *MYRSINACEAE*

Trees or shrubs; leaves alternate; flowers bisexual or unisexual (dioecious); sepals 4-6; petals 4-6; ovary superior or inferior; fruit a drupe.

MARLBERRY
Ardisia escallonioides Schlecht. & Cham.

Perennial shrub, to 4 m tall; bark white; leaves 4-18 cm long; sepals to 2 mm long; petals 5, white, recurved; flowers in panicles, to 15 cm long; drupes to 7.5 mm in diam., black. Pinelands, hammocks. Freq. SF, CF, NF. Spr.

MYRSINE
Rapanea punctata (Lam.) Lundell
[*R. guianensis* Aubl.]

Perennial shrub or small tree; leaves to 15 cm long, near end of branches; petals to 2 mm long, white; drupes to 8 mm long. Hammocks. Freq. CF, SF. Fall.

MYRTLE FAMILY *MYRTACEAE*

Evergreen trees or shrubs; leaves opposite or alternate, aromatic when crushed; flowers bisexual; sepals 4 or 5; petals 4 or 5; ovary inferior; fruit a berry, capsule, drupe or nut.

WHITE STOPPER
Eugenia axillaris (Sw.) Willd.

Perennial shrub, to 7 m tall; bark scaly; leaves 3-5 cm long; petals 4, 5-6 mm wide, white, in short axillary racemes; berries to 11 mm wide, black. Coastal hammocks. Infreq. SF, CF. All yr.

TWINBERRY
Myrcianthes fragrans (Sw.) McVaugh

Perennial tree, to 20 m tall; leaves 1-8 cm long; sepals minute, persistent; flowers to 1 cm wide, white; berries 6-10 mm long, red-brown. Hammocks. Infreq. SF, CF, NF. All yr.

NAIAD FAMILY   *NAJADACEAE*

Aquatic herbs; leaves opposite or nearly whorled; flowers unisexual (dioecious); floral parts consisting of spathes enclosing each flower; ovary superior; fruit an achene.

BUSHY POND WEED
Najas guadalupensis (Spreng.) Magnus

Annual, to 50 cm long; stems and leaf surfaces smooth; leaves to 3 cm long, to 2 mm wide; leaf margins with teeth to 0.5 mm long; *flowers monoecious; *achenes to 2 mm long. Shallow fresh water. Freq. all Fla. W to Tex. Sum-fall.

SPINY NAIAD
Najas marina L.

Annual; stems and lower leaf surfaces spiny; leaves to 4 cm long, to 3 mm wide; leaf margins with teeth to 1 mm long; flowers dioecious; achenes to 4 mm long. Fresh and salt water. Infreq. all Fla. W to Calif. Spr-fall.

(not shown)

> ## FOUR-O'CLOCK FAMILY NYCTAGINACEAE
>
> Herbs, shrubs or trees; leaves opposite or alternate; flowers bisexual or unisexual; sepals 5; petals 0; ovary superior; fruit an achene.

RED-SPIDERLING
Boerhavia diffusa L.
[*B. coccinea* Mill]

Perennial, to 1.2 m long; stems spreading; leaves opposite, with blades 2-8 cm long; *sepals to 1.5 mm long, white to lavender; *achenes to 4 mm long, rounded at apex, with glandular ribs. Disturbed sites, turf. Infreq. all Fla. Tex., Calif., N.C. Spr-sum, all yr S.

SPIDERLING
Boerhavia erecta L.

Annual, to 1.2 m long; stems sprawling; leaves 2-8 cm long; lower leaf surfaces white with dark dots; flowers to 1.2 mm long, white, pink or purple; achenes to 4 mm long, smooth, with one end flat. Disturbed sites. Infreq. CF, NF, WF. W to Tex., E to S.C. Spr-fall, all yr S.

COMMON FOUR-O'CLOCK
Mirabilis jalapa L.

Perennial, 0.4-1.4 m tall; stems smooth; leaf blades 3-15 cm long; sepals 3-5 cm long, red, purple, pink, white, yellow or a variation of several colors; flowers open late afternoon; achenes 7-10 mm long. Native to tropical America. Disturbed sites. Infreq. all Fla. Sum-fall.

WATER-LILY FAMILY   *NYMPHAEACEAE*

Rhizomatous aquatic herbs; leaves alternate, on long petioles; flowers bisexual; sepals 3-9; petals numerous; ovary superior or inferior; fruit a berry.

YELLOW LOTUS
Nelumbo lutea (Willd.) Pers.

Perennial, usually in water, emergent 0.1-1 m above water; peltate leaf 20-60 cm wide; flowers 10-30 cm wide, yellow; fruiting receptacle 7-10 cm wide. Quiet water. Infreq. all Fla. W to Tex., N to Canada. Sum-fall.

EGYPTIAN LOTUS
Nelumbo nucifera Gaertn.

Similar to *Nelumbo lutea* except: flowers to 40 cm in diam., pink or white. Native to Old World. Cultivated and escapes locally. Ponds. Rare CF, NF, WF. W to La., N to Mo., E to N.C. Sum.

(not shown)

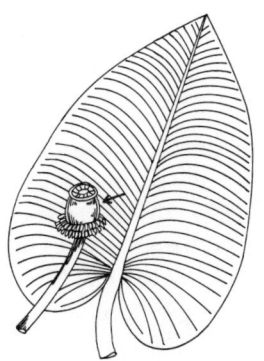

SPATTER-DOCK
Nuphar luteum (L.) Sibth. & Sm.
ssp. *macrophyllum* (Small) Beal

Perennial; rhizomes large; leaves floating or erect and emergent; blades 25-40 cm long, ovate; flowers 3-5 cm wide; sepals 6; petals yellow; peduncles 1-2 m long; *fruits 3-5 cm long. Quiet water. Common all Fla. W to Tex., N to Wis. and Maine. Spr-fall. Other varieties with various leaves occurring NF and WF to Va.

CAPE BLUE WATER-LILY
Nymphaea capensis Thunb.

Similar to *Nymphaea elegans* except: leaves 20-30 cm wide, sinuate-dentate, reddish beneath; sepals green; petals blue, rose, or purple. Native to Africa. Cultivated and escapes. Ditches. Infreq. CF. Spr-sum.

(not shown)

BLUE WATER-LILY
Nymphaea elegans Hook.

(not shown)

Similar to *Nymphaea odorata* except: leaves 8-20 cm in diam., purple beneath; sepals with dark purple streaks or dots; petals blue or violet. Ditches. Infreq. SF, CF. W to La. and Tex. Spr-sum.

YELLOW WATER-LILY
Nymphaea mexicana Zucc.

Perennial; leaves to 20 cm long, 6-20 cm wide, green above, crimson to purple below; petals 6-10 cm long, yellow; berries to 3 cm long. Ponds, lakes, slow streams. Infreq. all Fla. W to Tex., E to N.C. Sum-fall.

WHITE WATER-LILY
Nymphaea odorata Ait.

Perennial, with large rhizomes; leaves 15-30 cm in diam., green, occasionally purple beneath; sepals 4, green; petals 8-15 cm wide, white to pink, fragrant, floating; berries 2-3 cm in diam. Fresh water. Freq. all Fla. W to Tex., N to Canada. Spr-fall.

XIMENIA FAMILY   *OLACACEAE*

Trees, shrubs or vines; leaves alternate; flowers bisexual or unisexual (dioecious); sepals 4-6; petals 4-6; ovary appearing inferior; fruit a drupe or nut.

HOG-PLUM
Ximenia americana L.

Perennial shrub or small tree, to 7 m tall; thorns 1-2 cm long; leaves 3-10 cm long; *petals to 1 cm long, yellow to white; drupes to 1.7 cm long, edible. Dry to moist sites. Freq. SF, CF, NF. Spr-fall.

OLIVE FAMILY   *OLEACEAE*

Trees or shrubs; leaves opposite; flowers bisexual or unisexual (dioecious); sepals 4 or 0; petals 0, 4 or many; ovary superior; fruit a berry, capsule, samara, drupe or nut.

PIGMY FRINGE TREE
Chionanthus pygmaea Small

(not shown)

Similar to *Chionanthus virginicus* except: to 2 m tall; stems underground; branches 20-40 cm high; leaves 3-9 cm long; petals to 1 cm long; drupes to 2 cm long. Scrub. Infreq. CF, NF. Spr-sum.

OLD MAN'S BEARD
Chionanthus virginicus L.

Perennial shrub, to 10 m tall; leaves to 20 cm long, deciduous; sepals 4, minute; petals 4, 2-3 cm long, white; drupes 1-1.5 cm long. Moist woods. Infreq. CF, NF, WF. W to Tex., N to Ohio, E to N.J. Spr-sum.

PRIVET
Forestiera ligustrina (Michx.) Poir.

(not shown)

Perennial shrub, to 4 m tall; leaves to 5 cm long, deciduous, not visibly punctate; leaf margins toothed; sepals minute or 0, white; petals 0; drupes 7-8 mm long, round to oval. Dry woods, river banks. Infreq. CF, NF, WF. W to Tex., E to Ga. and Tenn. Sum-fall.

FLORIDA PRIVET
Forestiera segregata (Jacq.) Krug & Urban

Perennial shrub, 1-2 m tall; leaves 1-5 cm long, evergreen, simple, persistent, punctate; sepals minute or 0, white; petals 0; drupes 7-9 mm long, ellipsoid to globular. Hammocks, coastal marshes, coastal pinelands. Freq. all Fla. N to coastal Ga. Spr.

YELLOW JASMINE
Jasminum mesnyi Hance

Perennial evergreen shrub, to 3 m tall, with top recurved and viny; leaves to 7 cm long, divided into 2-3 leaflets; petals 7, to 2.5 cm long, yellow with dark center. Native to China. Escapes from cultivation. Disturbed sites. Rare CF, NF, WF. All yr.

WILD OLIVE
Osmanthus americanus (L.) Gray

Perennial shrub or small tree, to 15 m tall; bark pale; leaves 5-15 cm long, leather-like; petals 4, to 4 mm long, white; drupes to 1.5 cm in diam., blue or purple; seeds oval to elliptic. Hammocks, swamps. Freq. CF, NF, WF. W to La. Spr.

LARGE FRUIT WILD OLIVE
Osmanthus megacarpus (Small) Small ex Little

Similar to *Osmanthus americana* except: bark pale green; drupes to 2.5 cm in diam.; seeds round. Scrub. Infreq. CF, NF. Spr.

EVENING PRIMROSE FAMILY   *ONAGRACEAE*

Herbs or shrubs; leaves opposite or alternate; flowers bisexual or unisexual (dioecious); sepals 2-7; petals 2-7 or 0; ovary inferior; fruit nut-like or a capsule.

SOUTHERN GAURA
Gaura angustifolia Michx.

Perennial, 0.7-2 m tall; stems hairy; leaves 1-12 cm long, oblanceolate to lanceolate; petals 3-6 mm long, unequal, white to pink; stamens 8 or 6; capsules 5-10 mm long, indehiscent, elliptic. Drier pinelands. Freq. all Fla. W to Miss., E to N.C. Spr-fall.

WINGED WATER-PRIMROSE
Ludwigia alata Ell.

Perennial, 30-90 cm tall; stems erect, smooth,
*winged and angled; leaves 1-10 cm long, alter-
nate; *sepals 4, greenish white; petals 0; stamens
4; capsules to 4 mm long, 4-angled. Moist to wet
sites. Freq. all Fla. W to Miss., E to Va. Spr-fall.

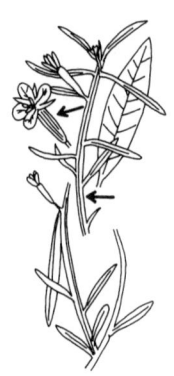

SEEDBOX
Ludwigia alternifolia L.

Perennial, 0.6-1.2 m tall; leaves 5-10 cm long,
alternate; sepals 4; petals 4, to 6 mm long, yellow;
stamens 4; *capsules 4-6 mm long, angled. Moist
to wet sites. Infreq. CF, NF, WF. W to Tex., N to
Canada. Spr-fall.

LONG-STALKED SEEDBOX
Ludwigia arcuata Walt.

Perennial; stems covered with short hairs;
leaves 8-20 mm long, opposite; flowers solitary,
with stalks longer than leaves; sepals 4, 4-8 mm
long; petals 4, 8-10 mm long, yellow; stamens 4;
capsules 6-10 mm long. Moist to wet sites.
Common all Fla. W to Ala., E to Va. Spr-fall.

LITTLE SEEDBOX
Ludwigia curtissii Chapm.
[*L. simpsonii* Chapm.]

Perennial, to 80 cm tall; stems smooth; leaves 8-
25 mm long, alternate; flowers axillary; sepals 4,
2-2.5 mm long; petals none; stamens 4; capsules
2-4 mm long. Moist to wet sites. Infreq. SF, CF,
NF. Endemic. Spr-fall.

WINGED WATER-PRIMROSE
Ludwigia decurrens Walt.
[*Jussiaea decurrens* (Walt.) DC.]

Annual, 0.2-2.5 m tall; stems smooth; *internodes 2-winged; leaves 2-18 cm long, alternate; flowers solitary; sepals 4; petals 4, 6-12 mm long, yellow; stamens 8; capsules 1-2.5 cm long. Moist to wet sites. Infreq. all Fla. W to Tex., E to Va. Spr-fall.

CREEPING WATER-PRIMROSE
Ludwigia grandiflora (Michx.) Zardini & Peng
[*Jussiaea grandiflorus* Michx.]

Perennial, creeping, floating; leaves 3-12 cm long; petals 1-2.5 cm long, yellow; capsules 1-4 cm long. Wet sites. Tex., Ariz, West Indies, Mex., S into South America. Spr-fall.

SPINDLE-ROOT
Ludwigia hirtella Raf.

Perennial, 30-60 cm tall; stems, leaves and fruits covered with long hairs; leaves 2-3 cm long, alternate; *flowers axillary; *sepals 4, 5-10 mm long; petals 4, 1-1.5 cm long, yellow; stamens 4; capsules 5-8 mm long. Moist sites. Infreq. all Fla. W to Tex., E to N.J. Spr-fall.

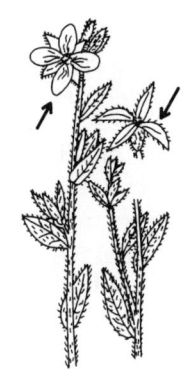

HAIRY PRIMROSE-WILLOW
Ludwigia leptocarpa (Nutt.) Hara

Perennial herb or shrub, 0.1-2 m tall; stems covered with long hairs; leaves 3-18 cm long; sepals and petals 5-7; petals 5-10 mm long, yellow; stamens 10-14; capsules 3-6 cm long. Low, wet sites. Freq. CF, NF, WF. W to Tex., N to N.C. Spr-fall, all yr S.

SWAMP SEEDBOX
Ludwigia linearis Walt.

Perennial, 0.3-1 m tall; leaves 2.5-7 cm long;
*sepals 4, persistent; flowers inconspicuous;
petals 4, 3-4 mm long, yellow; *capsules 3-4 mm
wide, to 7 mm long, usually sessile. Moist to wet
sites. Infreq. CF, NF, WF. W to Tex., E to N.J.
Sum-fall.

LITTLE SWAMP SEEDBOX
Ludwigia linifolia Poir

(not shown)

Perennial, 10-40 cm tall; leaves 1-3 cm long;
petals 4, 4-5 mm long, yellow; capsules to 1.2 mm
wide, 8-9 mm long. Wet sites. Freq. all Fla. W to
Miss., N to N.C. Spr-fall.

COASTAL PLAIN SEEDBOX
Ludwigia maritima Harper

Perennial, to 1 m tall; stems and leaves smooth
to hairy; leaves 2-8 cm long, alternate; flowers
axillary; sepals 4, 6-9 mm long; petals 4, 7-15
mm long, yellow; stamens 4; capsules 6-10 mm
long. Wet sites. Freq. all Fla. W to La., E to N.C.
Spr-fall.

TINY SEEDBOX
Ludwigia microcarpa Michx.

Perennial, 10-50 cm tall; stems erect or sprawling;
leaves 0.5-2 cm long, alternate; sepals to 1.5 mm
long; petals none; capsules to 1.5 mm long.
Moist to wet sites. Common all Fla. W to La.,
N to N.C. Spr-fall.

LONG-FRUITED PRIMROSE-WILLOW
Ludwigia octovalvis (Jacq.) Raven
[*Jussiaea angustifolia* Lam.]

Annual or perennial, to 3 m tall; leaves 2-15 cm
long, alternate; sepals 4; petals 4, 1-2.5 cm long,
yellow; stamens 8; capsules 1-6 cm long. Low,
wet sites. Common all Fla. W to Tex., E to N.C.
Spr-fall, all yr S.

WATER PURSLANE
Ludwigia palustris (L.) Ell.

Perennial, to 60 cm long; stems prostrate or
creeping; branches floating if in water; leaves
to 2.5 cm long, opposite; sepals to 1 mm long,
green or red; petals none; capsules to 4 mm
long. Moist to wet sites. Freq. CF, NF, WF.
W to Calif., N to Canada. Spr-fall.

COMMON PRIMROSE-WILLOW
Ludwigia peruviana (L.) Hara
[*Jussiaea peruviana* L.]

Perennial, 1-4 m tall; stems woody below,
herbaceous and hairy above; leaves 4-15 cm long;
sepals 4 or 5, persistent; petals 4 or 5, 1-3 cm
long; *capsules 1-3 cm long. Moist to wet sites.
Freq. all Fla. N to Ga. Spr-fall, all yr S.

FLOATING WATER-PRIMROSE
Ludwigia repens Forst.

Perennial, to 1 m long; stems prostrate; leaves
to 5 cm long, opposite; sepals to 3.5 mm long;
petals 4, yellow; capsules to 8 mm long. Wet
sites. Common all Fla. W to Tex., N to N.C.
Spr-fall.

POND WATER-PRIMROSE
Ludwigia spathulata Torr. & Gray

Perennial, to 20 cm long; stems prostrate, hairy; leaves 1-2 cm long, opposite; sepals to 1 mm long; petals none; capsules to 3 mm long. Dry ponds, lake bottoms. Infreq. CF, NF, WF. N to Ala., E to S.C. Sum.

(not shown)

HEADED SEEDBOX
Ludwigia suffruticosa Walt.

Perennial, 30-80 cm tall; stems smooth; leaves 2-10 cm long, alternate; *flowers in dense head-like terminal spikes; sepals 4, 8-14 mm long, white; petals minute or 0; stamens 4; *capsules 4-5 mm long. Low, wet sites. Common all Fla. W to Miss., E to N.C. Spr-fall.

LONG-PETALED SEEDBOX
Ludwigia virgata Michx.

Perennial; leaves 2-3 or 7.5 cm long, sessile, oblong; sepals 4; petals 4, 10-16 mm long, yellow, deciduous. Moist to wet sites. Freq. CF, NF, WF. W to Ala., E to Va. Spr-fall.

(not shown)

TALL EVENING PRIMROSE
Oenothera biennis L.

Biennial, 1.3-2 m tall; stems erect, covered with short stiff hairs; leaves 2.5-15 cm long; sepals 4; petals 4, 1-3 cm long, yellow; capsules 2-2.5 cm long. Dry sites. Freq. all Fla. W to Tex., N to Canada. Sum-fall.

SEASIDE EVENING PRIMROSE
Oenothera humifusa Nutt.

Perennial, 20-50 cm long; stems prostrate,
covered with appressed silky gray hairs; leaves
1-3.5 cm long, covered with densely appressed
hairs; sepals 4; petals 4, to 1 cm long, yellow;
capsules 1-3.5 cm long, with appressed hairs.
Coastal sites. Freq. all Fla. W to La., E to N.J.
Spr-fall, all yr S.

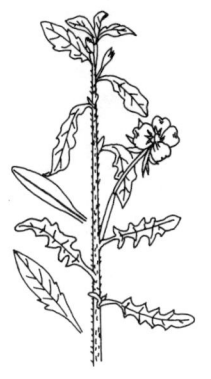

CUT-LEAVED EVENING PRIMROSE
Oenothera laciniata Hill

Biennial, up to 80 cm tall; stems lying flat or
ascending, smooth to slightly hairy; leaves 2.5-8 cm
long, with margins irregularly lobed or entire; sepals
4; petals 4, 0.5-1.8 cm long, 12-45 mm wide, yellow; capsules 2-4 cm long, hairy. Open disturbed
sites. Common all Fla. W to Tex., N to Vt. Spr-
sum, all yr S.

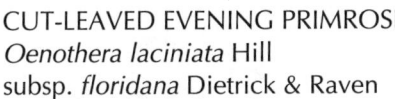

CUT-LEAVED EVENING PRIMROSE
Oenothera laciniata Hill
subsp. *floridana* Dietrick & Raven

Similar to *Oenothera laciniata* except: stems with
spreading hairs; leaf margins entire or shallow-
toothed. Disturbed sites. Infreq. CF, NF. All yr.

SHOWY PRIMROSE
Oenothera speciosa Nutt.

Perennial, 20-70 cm tall; stems erect, branching,
hairy; basal leaves to 7.5 cm long; petals 2.5-5 cm
long, white or pink; capsules 1-2 cm long, pointed.
Native to northwestern states. Dry disturbed sites.
Infreq. all Fla. W to Tex., N to N.C. Spr-fall.

ORCHID FAMILY ORCHIDACEAE

Epiphytic, terrestrial or saprophytic herbs; lower portion of stems in most species inflated into pseudobulb from which leaves arise; leaves mostly alternate or sheathing at base; flowers bisexual; sepals 3, petal-like; petals 3, with 2 alike and 1 a lip differing from others in shape, color or size and usually forming spur; stamens and style forming column; ovary inferior; fruit a capsule.

CORAL ROCK ORCHID
Basiphyllaea corallicola Ames

Perennial, terrestrial, 20-40 cm tall; leaves 1 or 2, 2-10 cm long, basal; *flowers greenish; sepals and petals to 8 mm long, yellow-green; lip to 7 mm long, yellowish; *capsules to 10 mm long. Rocky pinelands. Infreq. SF. Fall.

RAT-TAIL ORCHID
Bulbophyllum pachyrachis (A. Rich.) Griseb.

Perennial, epiphytic; pseudobulb to 2 cm long; leaves 2 at tip of pseudobulb, to 18 cm long; *flowers green with purple dots; sepals 6-7 mm long; petals to 2.5 mm long; lip to 3 mm long, red; *flower stems to 40 cm long; *capsules to 8 mm long. Hammocks. Rare SF. Fall-wint.

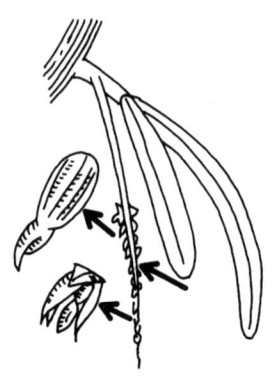

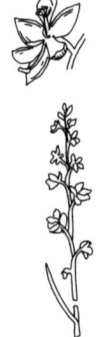

MANY-FLOWERED GRASS-PINK
Calopogon multiflorus Lindl.

Perennial, 10-40 cm tall; leaves 1, 2 or 0, 4-19 cm long; petals to 1.3 cm long, maroon to purple, spurless. Woods, often after fire. Freq. all Fla. W to Miss., E to Ga. Spr-sum.

PALE GRASS-PINK
Calopogon pallidus Chapm.

Perennial, terrestrial, 20-50 cm tall; leaves 1 or 2, to 20 cm long, to 5 mm wide, basal; flowers 1-12, pink or white; lip to 9 mm long, uppermost, 3-parted, pink or white; capsules to 10 mm long. Open marshy meadows. Infreq. all Fla. W to La., N to N.C. Spr-sum.

GRASS-PINK
Calopogon tuberosus (L.) BSP.

Perennial; leaves to 50 cm long, 5-50 mm wide; flower stalks 20-90 cm tall; flowers pink, rose, purple, red or white; sepals 15-27 mm long; *lip 15-19 mm long, bearded; column 18 mm long, winged. Wet sites. Common all Fla. W to Miss., N to Canada. Spr-sum.

ROSE ORCHID
Cleistes divaricata (L.) Ames

Perennial, terrestrial, 20-80 cm tall; leaf 1, 5-15 cm long, to 2 cm wide; flowers occurring singularly, brown, pink or white; lip to 5.5 cm long, 3-parted, yellow to green, with purple venation; capsules to 3 cm long. Swamps, moist sites. Rare CF; infreq. NF, WF. W to Miss., N to N.J. Spr-sum.

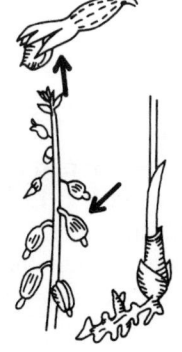

AUTUMN CORAL-ROOT
Corallorhiza odontorhiza (Willd.) Nutt.

Perennial, terrestrial, saprophytic; stem to 30 cm tall, yellow to purple; leaves absent; *flowers green with purple markings; sepals and petals 3-4 mm long; lip to 4 mm long, white with purple stripes; *capsules to 7 mm long. Dry woods. Rare WF. W to Tex., N to Maine. Sum-fall.

SPRING CORAL-ROOT
Corallorhiza wisteriana Conrad

Perennial, terrestrial, feeding off dead organic
matter (saprophytic); stem 20-35 cm tall, tan to
reddish purple; leaves absent; flowers green to
yellow, with purple markings; sepals and petals
6-8 mm long; *lip 5.5-10 mm long, white with
purple spots; *capsules to 10 mm long.
Rich deciduous woods. Infreq. CF, NF, WF.
W to Tex., N to Ill. Wint-spr.

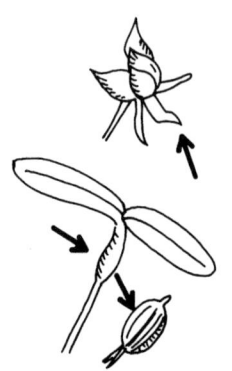

DWARF EPIDENDRUM
Encyclia pygmaea (Hook.) Dressler

Perennial, epiphytic, to 15 cm long; stems creeping; leaves 2 or 3, 2-8 cm long; *pseudobulb 2-
8 cm long; *flowers green; sepals to 5 mm long;
petals to 4 mm long; lip to 2 mm long, white
with red tip; flower stems short, between leaves;
*capsules 12-15 mm long, winged. Swamps.
Rare SF. Sum.

BUTTERFLY ORCHID
Encyclia tampensis (Lindl.) Small

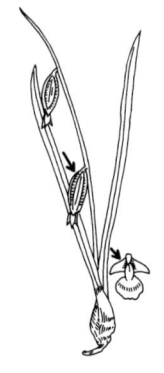

Perennial, epiphytic, 15-60 cm tall, from a
pseudobulb; leaves 1-3, 5-30 cm long; *flowers
yellow, green or brown, with purple markings;
sepals 12-22 mm long; petals 12-22 mm long; lip
3-lobed; *capsules to 30 mm long. Hammocks.
Freq. SF, CF, NF. Spr-sum.

GREEN-FLY ORCHID
Epidendrum conopseum R. Br.

Perennial, epiphytic, 5-30 cm tall; leaves 1-3,
3-9 cm long; *flowers green with purple markings; sepals 8-13 mm long; petals 8-12 mm long;
lip 5 mm long, 3-lobed; capsules 15-20 mm
long. Swamps. On live oaks, magnolias. Freq.
CF, NF, WF. W to La., E to N.C. Sum.

UMBELLED EPIDENDRUM
Epidendrum difforme Jacq.
[*Amphiglottis difformis* (Jacq.) Britt.]

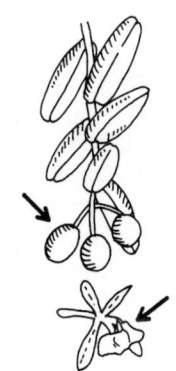

Perennial, epiphytic, 10-35 cm tall; stems
ascending; *leaves 5-10, 4-10 cm long; *flowers
green to yellow; sepals 12-15 mm long; petals to
12 mm long; lip to 7.5 mm long, green; *capsules
1.5-2 cm long. Swamps. Rare SF. Sum-fall.

RIGID EPIDENDRUM
Epidendrum rigidum Jacq.

Perennial, epiphytic, 10-50 cm long, creeping;
leaves 2-6, 3-8 cm long, elliptic; *flowers green,
inconspicuous; sepals to 6 mm long; petals to 5 mm
long; lip 2-3 mm long, uppermost; *capsules 15-
40 mm long. Swamps, hammocks. Infreq. SF, CF,
NF. Fall-spr.

WILD COCO
Eulophia alta (L.) Fawc. & Rendle

Perennial, 0.7-1.5 m tall; leaves 3-4, 0.2-1.2 m
long; flowers madder-purple; ovary 1.5-2.6 cm
long; *capsules 3-5 cm long. Low sites. Infreq.
CF, SF. Sum-fall.

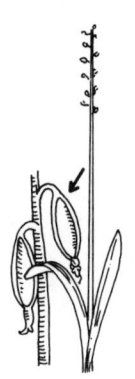

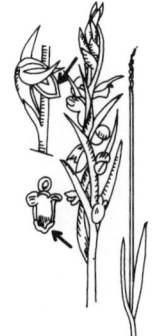

NORTHERN WILD COCO
Eulophia ecristata (Fern.) Ames

Perennial, terrestrial, 0.3-1.7 m tall; stems erect;
leaves to 70 cm long; *flowers yellow to green,
in racemes; *lip to 15 mm long, 3-lobed, purple;
capsules to 2 cm long. Dry sites, bogs. Infreq.
all Fla. N to N.C. Sum-fall.

SMALL GREEN WOOD ORCHID
Habenaria clavellata (Michx.) Spreng.

Perennial, 8-45 cm tall; leaves 1 or 2, 5-15 cm
long; spikes to 6 cm long; sepals and petals 3-5 mm
long; *flower yellowish green; *spur 8-12 mm long,
curved upward; capsules to 7 mm long. Swampy
sites. Infreq. WF. W to Tex., N to Canada. Sum.

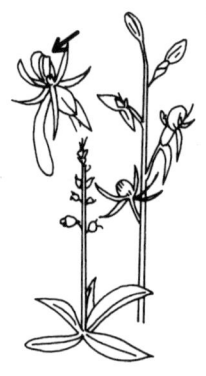

LONG-STALKED HABENARIA
Habenaria distans Griseb.

Perennial, terrestrial, 20-30 cm tall; leaves 2-6,
3-15 cm long, basal; *flowers yellow to green;
sepals 5-8 mm long; petals to 9 mm long, 2-part-
ed; lip 6-9 mm long, 3-parted, green; spur to
15 mm long; capsules 10-15 mm long. Moist
hammocks. Rare SF, CF. Sum-fall.

GREEN-CROSS ORCHID
Habenaria odontopetala Reichb. f.
[*Habenella strictissima*
var. *odontopetala* (Reichb. f.) L. O. Williams]

Perennial, terrestrial, 30-80 cm tall; leaves 5-20 cm
long; flowers yellow-green; sepals 3-8 mm long;
petals 3-5 mm long, forming *spur 12-18 mm long;
lip 6-12 mm long, slender; *capsules 9-13 mm
long. Hammocks, wet woods. Infreq. SF, CF, NF.
Fall-wint.

WATER-SPIDER ORCHID
Habenaria repens Nutt.

Perennial, terrestrial or aquatic, 10-90 cm tall,
leafy, often stoloniferous; leaves 5-30 cm long,
3-ribbed; flowers green to yellow; sepals 3-7 mm
long; petals 3-7 mm long, 2-parted; spur 9-14 mm
long; lip 9 mm long, 3-parted; capsules to 9 mm
long. Pools, watery ditches. Locally common all
Fla. W to Tex., N to N.C. Sum-fall.

TALL LIPARIS
Liparis elata Lindl.
[*L. nervosa* (Thunb.) Lindl.]

Perennial, 10-60 cm tall; leaves 3-5, 6-30 cm
long; *flowers 7-12 mm long; petals madder-purple; lip mauve-purple; column 3-5.5 mm long,
curved; capsules 1.5-2 cm long. Swamps, low
sites. Infreq. SF, CF. Sum, all yr S.

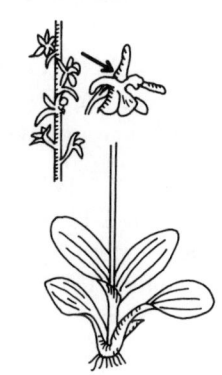

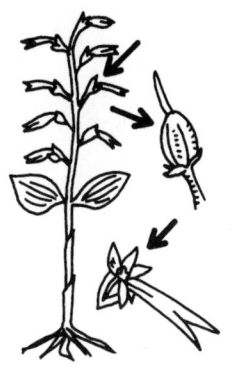

SOUTHERN TWAYBLADE
Listera australis Lindl.

Perennial, terrestrial, 10-30 cm tall; leaves 2, 1.5-
4 cm long; flowers green to red to purple; *sepals
to 2 mm long; petals to 1.5 mm long; lip 6-12 mm
long, 2-parted, linear; *capsules to 8 mm long.
Rich, moist sites. Infreq. CF, NF, WF. W to Tex.,
N to Canada. Wint-spr.

FLORIDA ADDER'S MOUTH
Malaxis spicata Sw.

Perennial, terrestrial to partially epiphytic on
stumps and logs, to 45 cm tall; leaves 2 or 3, to
10 cm long; *flowers few, small, green; sepals and
petals to 3.5 mm long; lip yellow, orange or red,
positioned at top of flower; capsules to 7.5 mm
long. Low, wet, calcareous sites. Infreq. all Fla.
N to Va. Sum-fall, sum-wint S.

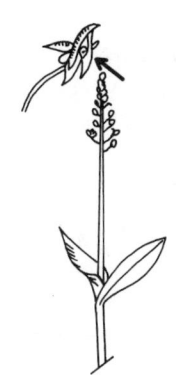

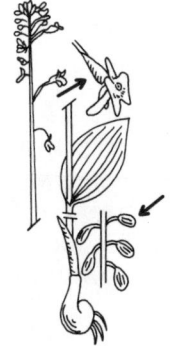

GREEN ADDER'S MOUTH
Malaxis unifolia Michx.

Perennial, terrestrial, to 25 cm tall; leaf 1, 1-9 cm
long; *flowers numerous, small, green; sepals and
petals 2-4 mm long; lip positioned at bottom of
flower; *capsules to 1 cm long. Moist, acidic
woods, meadows. Rare CF, NF, WF. W to Tex.,
N to Canada. Spr-sum.

LARGE WHITE FRINGED ORCHID
Platanthera blephariglottis (Willd.) Lindley
[*Habenaria blephariglottis* (Willd.) Hook.]

Perennial, 0.3-1.1 m tall; leaves 5-40 cm long;
sepals and petals 3-11 mm long, white; *lip 4-
13 mm long, fringed; *spur 10-25 mm long.
Moist sites. Infreq. CF, NF, WF. W to Tex.,
N to Canada. Sum-fall.

SMALL GREEN WOOD ORCHID
Platanthera clavellata (Michx.) Luer
[*Habenaria clavellata* (Michx.) Spreng.]

Perennial, 10-50 cm tall; leaves 1 or 2, 5-18 cm
long; flowers white or green; sepals and petals
2-7 mm long; spur 8-12 mm long, club-shaped;
capsules 7-8 mm long. Swamps, wet sites. Rare
NF, WF. W to Tex., N to Canada. Sum.

GYPSY-SPIKES
Platanthera flava (L.) Lindl.
[*Habenaria flava* (L.) R. Br.]

Perennial, terrestrial, 10-60 cm tall; leaves 1-4,
5-25 cm long; *flowers yellow to green; sepals
2-5 mm long; petals to 4 mm long; *lip to 5 mm
long, yellow to green, keeled; spur 4-11 mm long;
capsules to 8 mm long. Moist to wet sites. Infreq.
CF, NF, WF. W to Tex., N to Md. Spr-fall.

YELLOW FRINGELESS ORCHID
Platanthera integra (Nutt.) Gray
[*Habenaria integra* (Nutt.) Spreng.]

Perennial, terrestrial, 30-60 cm tall; stems erect;
leaves 5-20 cm long; *flowers yellow-orange, in
racemes; lip margins 3-5 mm long, entire, yellow-
orange; spur to 5 mm long; capsules to 1 cm long.
Low, wet sites. Rare CF, NF, WF. W to Tex.,
N to N.J. Sum-fall.

SNOWY ORCHID
Platanthera nivea (Nutt.) Spreng.
[*Habenaria nivea* (Nutt.) Spreng.]

Perennial, 20-90 cm tall; lower leaves 2-3,
7-26 cm long; spikes to 12 cm long; sepals
and petals 3-6 mm long; flowers white, with lip
uppermost; spur 1-1.5 cm long, curved upward;
capsules to 1 cm long. Moist, wet sites. Freq.
all Fla. W to Tex., E to N.J. Sum.

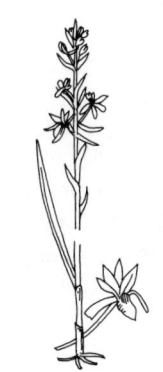

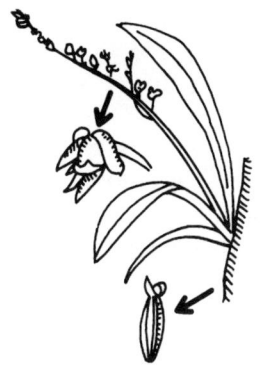

PALE FLOWERED POLYSTACHYA
Polystachya flavescens (Lindl.) J. J. Smith

Perennial, epiphytic, 10-60 cm tall; leaves 2-5,
10-20 cm long; *flowers waxy, yellow to green;
sepals 3-6 mm long; 2 sepals triangular; petals to
4 mm long; lip to 4 mm long, white; *capsules to
15 mm long. Hammocks, high in trees. Infreq.
CF, SF, Fla. Keys. Fall.

SHADOW WITCH
Ponthieva racemosa (Walt.) Mohr

Perennial, terrestrial; leaves 3-8, 2-15 cm long,
basal, quickly falling off; flower stem 20-30 cm
tall; flowers white to green; sepals 4-8 mm long;
petals 4-8 mm long; *lip to 7 mm long, white,
with green stripes; *capsules 8-13 mm long.
Moist to wet woods. Infreq. all Fla. W to Tex.,
N to Va. Wint.

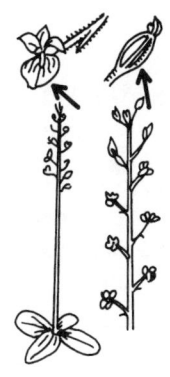

TEXAS LADIES'-TRESSES
Spiranthes brevilabris Lindl.

Perennial, terrestrial, to 40 cm tall; stems erect,
hairy; leaves to 7 cm long; *flowers densely
hairy, white-green or yellow-green, in spiraled
spikes; *lip tip to 5 mm long, appearing torn,
yellow; capsules to 5 mm long. Low, wet sites.
Rare CF. W to Tex. Spr.

NODDING LADIES'-TRESSES
Spiranthes cernua (L.) L. C. Rich.

Perennial, 10-90 cm tall; leaves 5-40 cm long,
mostly basal, deciduous; flower spike dense;
tepals 6-14 mm long, white; lip 6-12 mm long,
ovate-oblong, curved, yellow-green, spurless;
apex round. Moist to wet sites. Infreq. all Fla.
W to N.Mex., N to Canada. Fall-wint.

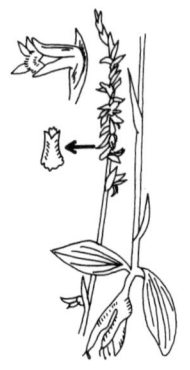

SLENDER LADIES'-TRESSES
Spiranthes gracilis (Bigel.) Beck

Perennial, 10-80 cm tall; basal leaves 1-6 cm
long, deciduous; spikes 3-15 cm long; sepals and
petals 3-5 mm long; flowers white with green
stripe on *lip; capsules to 5 mm long. Dry to wet
sites. Freq. CF, NF, WF. W to Tex., N to
Canada. Spr-fall.

LEAFLESS BEAKED ORCHID
Spiranthes lanceolata (Aubl.) Leon

Perennial, terrestrial, to 60 cm tall; leaves to 35 cm
long; petals to 2.2 cm long, orange-red; capsules to
1.5 cm long. Dry roadsides and woods, pastures,
vacant lots. Rare SF, CF, NF. Spr-sum.

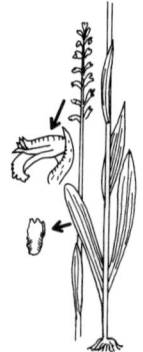

OVAL LADIES'-TRESSES
Spiranthes ovalis Lindl.

Perennial, 14-45 cm tall; basal leaves 2-4,
4-16 cm long; spikes 2-10 cm long; *sepals and
petals 4-5 mm long; flowers white, with *lip
tightly curved-recurved; capsules to 6 mm long.
Moist, woody sites. Infreq. CF, NF, WF. W to
Tex., N to Mo., E to Va. Sum-fall.

GREEN LADIES'-TRESSES
Spiranthes polyantha Reichenb. f.

Perennial, terrestrial, 15-70 cm tall; stems smooth; leaves 1-3, 3-6 cm long, basal, quickly falling off; flowers green to brown; sepals 4-5 mm long; petals 4-5 mm long; lip to 5 mm long, gray to green; *capsules 4-6 mm long. Hammocks, Keys, coastal sites. Infreq. SF, CF. Wint-spr.

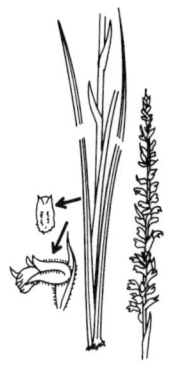

GIANT LADIES'-TRESSES
Spiranthes praecox (Walt.) S. Wats.

Perennial, terrestrial, 20-60 cm tall; leaves 5 or 6, 5-25 cm long, to 1 cm wide; *flowers to 40, spiraled, white; *lip to 9 mm long, white, sometimes with green veins; capsules to 8 mm long. Wet grassy sites. Freq. all Fla. W to Tex., N to N.J. Spr-sum.

SOUTHERN LADIES'-TRESSES
Spiranthes torta (Thunb.) Garay & Sweet
[*S. tortilis* (Swartz) L. C. Rich.]

Perennial, terrestrial, 20-50 cm tall; leaves 2 or 3, to 20 cm long, to 5 mm wide; flowers to 60, spiraled, white with green base; lip to 4 mm long, white with green center; capsules to 4 mm long. Marshes, moist pinelands. Rare SF. Spr-sum.

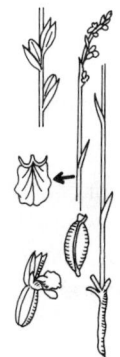

LITTLE LADIES'-TRESSES
Spiranthes tuberosa Raf.
[*S. grayi* Ames]

Perennial, 8-50 cm tall; leaves to 6.5 cm long; spikes slender, to 8 cm long; sepals and petals 2-3.5 mm long; flowers white, with *lip 2.5-4 mm long; capsules to 5 mm long. Dry, often woody sites. Infreq. CF, NF, WF. W to Tex., E to Mass. Sum.

SPRING LADIES'-TRESSES
Spiranthes vernalis Engelm. & Gray

Perennial, terrestrial, 15-65 cm tall; stems with
pointed hairs; leaves 4 or 5, to 25 cm long, to 1 cm
wide; flowers to 50, spiraled or secund, white with
yellow base; *lip to 7 mm long, white with yellow
center; capsules to 8 mm long. Marshes, low sites.
Freq. all Fla. W to N.Mex., N to Mass. Spr-fall.

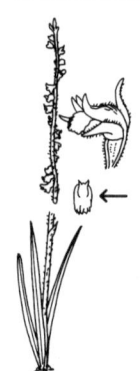

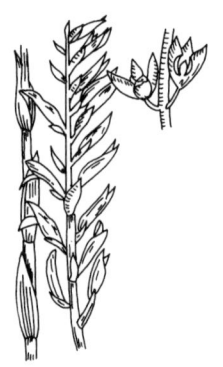

LACE-LIP LADIES'-TRESSES
Stenorrhynchos lanceolatus
(Aubl.) L. C. Rich. ex Spreng.
[*Spiranthes lanceolata* (Aubl.) Leon]

Perennial, to 1.2 m tall; lower leaves not persistent; flowers 2-3 cm long, red or yellow. Dry to
wet pinewoods. Infreq. SF, CF, NF. Sum.

SHORT STALKED NODDING CAP
Triphora gentianoides (Sw.) Ames & Schltr.

Perennial, 10-20 cm tall; leaves 10-18 mm long,
clasping; flowers 3-10, 8-11 mm long, yellow-
green with red-brown coloration, corymbose;
capsules 15-20 mm long. Pinelands. Infreq. SF,
CF. Sum-fall.

(not shown)

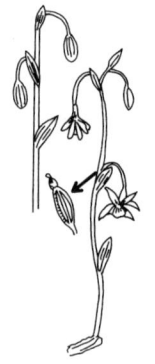

NODDING POGONIA
Triphora trianthophora (Sw.) Rydb.

Perennial, 8-30 cm tall; leaves 8-20 mm long,
slightly clasping; flowers 3 but can range from
1-6, 13-16 mm long, pink to white; *capsules
10-14 mm long. Rich woods. Infreq. CF, NF.
W to Tex., N to Wis., E to Maine. Sum-fall.

LAWN ORCHID
Zeuxine strateumatica (L.) Schltr.

Perennial, terrestrial, 6-25 cm tall; leaves 5-12,
2-9 cm long, to 8 mm wide; flowers to 50, white;
lip to 4 mm long, yellow; capsules to 7 mm long.
Native to tropical Asia. Naturalized in Fla.
Lawns, open moist sites. Infreq.-freq. all Fla.
Fall-wint.

BROOM RAPE FAMILY   *OROBANCHACEAE*

Parasitic herbs inhabiting tree roots and lacking chlorophyll;
leaves alternate; flowers bisexual; sepals 2-5; petals 4 or 5;
ovary superior; fruit a capsule.

SQUAW ROOT
Conopholis americana (L.) Wallr.

Perennial, 10-20 cm high, resembling fir cone,
brown-yellowish; leaves to 12 mm long, overlapping, scale-like; *petals to 15 mm long, yellow;
capsules to 1.5 cm long. Dry oak woods. Infreq.
CF, NF, WF. N to Maine. Spr-sum.

WOOD SORREL FAMILY *OXALIDACEAE*

Herbs; leaves alternate; flowers bisexual; sepals 5; petals 5; ovary superior; fruit a capsule.

CREEPING WOOD SORREL
Oxalis corniculata L.

Perennial, 5-40 cm long; stems creeping, hairy; stolons prostrate; roots fibrous; leaflets to 1 cm long, 5-12 mm wide, with fringed margins; flowers 5-10 mm long, yellow; capsules 8-30 mm long, with spreading or reflexed hairs; seeds brown. Disturbed sites. Infreq. all Fla. W to Calif., N to Canada. Spr-fall, all yr S.

PINK WOOD SORREL
Oxalis corymbosa DC.
[*Ionoxalis martiana* (Zucc.) Small]

Perennial, 8-30 cm tall; *bulblets sessile; *leaflets to 3.5 cm long, heart-shaped, with attachment to stem at pointed end, hairy; flower stems hairy; petals to 1.5 cm long, pink to purple, or white; capsules to 10 mm long. Native to tropical America. Cultivated and escapes. Disturbed sites. Infreq. all Fla. W to Tex. Spr-fall.

FLORIDA YELLOW WOOD SORREL
Oxalis florida Salisb.
[*O. filipes* (Small) Ahles]

Perennial, 10-40 cm tall; stems erect or lying flat, hairy; leaflets 10-16 mm wide; flowers 5-10 mm long, yellow; *capsules 7-14 mm long, smooth or with a few appressed or ascending hairs. Dry to moist, often disturbed sites. Common all Fla. W to La. and Mo., N to N.J. Spr-fall.

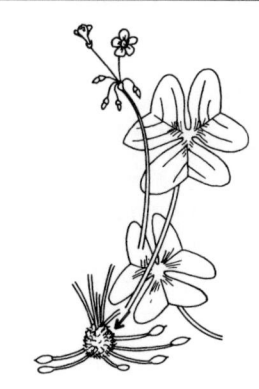

CUBAN PURPLE WOOD SORREL
Oxalis intermedia A. Rich.

Perennial, to 30 cm tall; *bulblets stalked;
leaflets to 4.5 cm long, triangular in shape, with
attachment to stem at pointed end; flower stems
smooth; petals to 1 cm long, violet; capsules
6-7 mm long. Native to West Indies. Disturbed
sites. Infreq. CF, NF. Spr-fall.

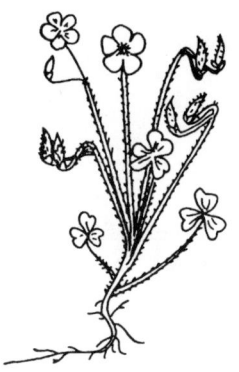

YELLOW WOOD SORREL
Oxalis stricta L.
[*O. dillenii* Jacq.]

Perennial, 10-25 cm tall; stems erect, lying flat or
inclining, hairy; leaflets 7-27 mm wide; flowers
9-12 mm long, yellow; capsules 8-10 mm long;
seeds brown with whitish ridges. Woods, disturbed sites. Common CF, NF. W to Tex.,
N to Canada. Spr-fall.

POPPY FAMILY   *PAPAVERACEAE*

Herbs or shrubs exuding milky, watery or yellow latex; leaves
alternate; flowers bisexual; sepals 2 or 3; petals 4-12; ovary
superior; fruit a capsule.

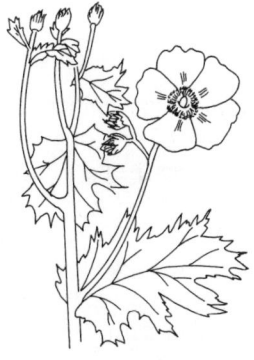

CAROLINA POPPY
Argemone albiflora Hornem.
[*A. alba* Lestib.]

Annual or biennial, 30-60 cm tall; stems spiny;
leaves 3-20 cm long; flowers 5-10 cm wide,
white; capsules 2.5-4 cm long. Disturbed sites.
Freq. CF, NF, WF. W to Tex., N to N.C. Spr-sum.

MEXICAN PRICKLY POPPY
Argemone mexicana L.

Annual or biennial; stems 30-90 cm tall, spiny; leaves to 25 cm long, with spiny margins; *petals 4-6, 2-3 cm long, 3-7 cm wide, yellow; *capsules to 3 cm long, spiny. Disturbed sites. Common all Fla. W to Tex., N to Mass. Spr-fall.

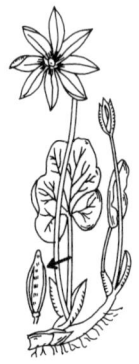

BLOOD ROOT
Sanguinaria canadensis L.

Perennial, 8-30 cm tall; stems smooth; rhizomes exuding red liquid; leaves 5-18 cm broad; petals to 3 cm long, white; *capsules 3-5 cm long. Rich woods. Infreq. WF. W to Ark., N to Canada. Spr. Roots used for dye.

PASSION-FLOWER FAMILY   *PASSIFLORACEAE*

Herbs, vines, shrubs or trees; leaves alternate; stipules present; flowers bisexual or unisexual (monoecious or dioecious); sepals 5; petals 5 or 0; sepals and petals forming crown-like extension (corona); ovary superior; fruit a berry or capsule.

PASSION-FLOWER
Passiflora incarnata L.

Perennial vine, often 2 m long; stems smooth to slightly hairy; leaf blades 5-15 cm long, 3-lobed, with toothed margins; flowers 1 or 2 per stalk; sepals to 3 cm long, white; petals 5, 3-4 cm long, blue, white or lavender; corona white, lavender or purple; berries 4-10 cm long, yellow or green, edible. Dry, disturbed sites. Infreq. to freq. all Fla. W to Tex., N to Va. Spr-sum.

YELLOW PASSION-FLOWER
Passiflora lutea L.

Perennial vine climbing to 5 m long; stems sparsely covered with long hairs; leaf blades to 7 cm
long, 3-lobed, with entire margins; flowers 1-3 per
stalk; sepals to 1 cm long, green; petals 9-15 mm
long, yellowish; corona green with dark purple
base; berries 1.2 cm long, black. Woody sites.
Infreq. NF, WF. W to Tex., N to Va. Spr-fall.

(not shown)

CORKY-STEMMED PASSION-FLOWER
Passiflora suberosa L.

Perennial, trailing or climbing; older stems corky-
winged; leaves 4-10 cm long, oval to 3-lobed;
*petioles glandular; sepals green; *berries 0.6-1 cm
in diam., purple to black. Pinelands, hammocks.
Freq. all Fla. All yr.

POKEWEED FAMILY   *PHYTOLACCACEAE*

Herbs, vines, shrubs or trees; leaves alternate; flowers bisexual
or unisexual (monoecious or dioecious); sepals 4 or 5; petals
0; ovary mostly superior; fruit a berry, nut or capsule.

INDIA CARPETWEED
Gisekia pharnaceoides L.

Annual, 20-45 cm long; stems diffuse; leaves to
2.5 cm long, fleshy; sepals to 2 mm long, pink;
stamens 5; styles 5; fruits to 3 mm long. Native
to India. Disturbed sites and orange groves.
Rare CF. Sum.

GUINEA-HEN WEED
Petiveria alliacea L.

Perennial, 0.3-1.2 m tall; crushed foliage having skunk-like odor; leaves 4-12 cm long; *sepals 4, 3-4 mm long, white, pink or greenish; achenes 6-10 mm long. Disturbed, woody sites. Infreq. CF, SF, NF. W to Tex. All yr.

POKEWEED
Phytolacca americana L.
[*P. rigida* Small]

Perennial, 1-3 m tall; stems smooth; leaf blades 7-30 cm long; sepals 5, 2-3 mm long, white or pink; berries 7-12 mm in diam., purple-black. Disturbed moist to wet, rich sites. Freq. all Fla. W to Tex., N to Canada. Spr-fall.

ROUGE PLANT
Rivina humilis L.

Perennial, 30-70 cm tall or clambering; leaves 3-15 cm long, ovate; sepals 4, 2-3 mm long, white or pink; berries 2-4 mm long, red. Disturbed, woody sites. Freq. SF, CF, NF. W to Tex. All yr S.

PLANTAIN FAMILY   *PLANTAGINACEAE*

Herbs; leaves alternate, opposite or basal; flowers bisexual; sepals 4; petals 4; ovary superior; fruit a capsule.

LARGE BRACTED PLANTAIN
Plantago aristata Michx.

(not shown)

Annual; leaves to 25 cm long, to 8 mm wide, linear; flower stalks to 50 cm tall; male and female parts in same flowers (perfect); floral bracts twice as long as flowers; capsules to 3.5 mm long. Dry sites. Rare CF, NF. W to N.Mex., N to Canada. Spr-fall.

ENGLISH PLANTAIN
Plantago lanceolata L.

Perennial, 10-60 cm tall; leaves 10-30 cm long; flower stalks 10-60 cm tall; flower spikes 1-8 cm long; petals to 2.5 mm long, white; capsules to 3 mm long. Disturbed sites. Infreq. all Fla. N to Canada. Spr-sum.

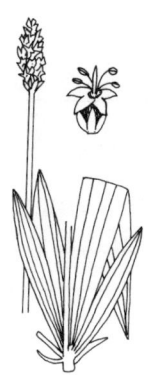

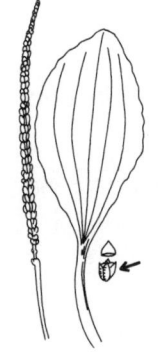

COMMON PLANTAIN
Plantago major L.

Perennial; stemless; leaves 5-35 cm long, 7-13 cm wide; leaves and flower stalks smooth to short-hairy; spikes 12-60 cm tall; *capsules 2.5-3 mm long. Native to Europe. Escapes from cultivation. Disturbed sites. Infreq. all Fla. All U.S. Spr-fall.

NARROWLEAF PLANTAIN
Plantago purshii Roem. & Schult.

Annual; stemless; leaves 3-15 cm long, 1-7 mm
wide, basal, linear, covered with white silky
hairs; capsules 2-2.5 mm long. Dry sites. Infreq.
NF. W to Tex., N to Canada. Spr-sum.

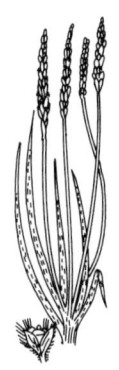

SOUTHERN PLANTAIN
Plantago virginica L.

Annual; stemless; leaves to 15 cm long, to 4 cm
wide, lance-shaped; flower stalks to 35 cm tall,
hairy; male and female flowers on separate plants
(dioecious); floral bracts shorter than or as long as
flowers; capsules to 2 mm long. Dry, disturbed
sites. Common all Fla. W to Tex., N to R.I.
Spr-sum.

LEADWORT FAMILY *PLUMBAGINACEAE*

Herbs, vines or shrubs; leaves alternate or basal; flowers
bisexual; sepals 5; petals 5; ovary superior; fruit a utricle.

SEA LAVENDER
Limonium carolinianum (Walt.) Britt.

Perennial; leaves 0.5-40 cm long, basal; flower
stems 20-80 cm tall; flowers to 6 mm long, lavender or violet; capsules to 5.5 mm long. Coastal
marshes. Freq. all Fla. W to Tex., N to Canada.
Spr-fall, all yr S.

WILD PLUMBAGO
Plumbago scandens L.

Perennial; stems erect, lying flat or climbing; leaves 2-10 cm long; bracts acuminate; *sepals with glandular hairs; petals 3-4 cm long, white; capsules to 7 mm long. Coastal hammocks, woody borders. Infreq. CF, NF, WF. All yr.

PHLOX FAMILY   *POLEMONIACEAE*

Herbs, shrubs, vines or trees; leaves alternate or opposite; flowers bisexual; sepals 5; petals 5; ovary superior; fruit a capsule.

STANDING CYPRESS
Ipomopsis rubra (L.) Wherry
[*Gilia rubra* (L.) Heller]

Biennial, to 1.7 m tall; leaves to 4 cm long, finely divided; petals to 4 cm long, red or yellow; capsules to 1 cm long. Dry open sites. Infreq. CF, NF, WF. W to Tex., N to N.C. Sum-fall.

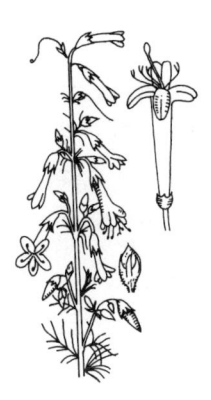

HAIRY PHLOX
Phlox amoena Sims

Perennial, 10-30 cm tall; stems lying flat, matting, with non glandular hairs; leaves to 5 cm long, to 0.8 cm wide; floral tube to 1.6 cm long, smooth, violet, purple or white; capsules 4-6 mm long. Open woods. Infreq. CF, NF, WF. W to Miss., N to N.C. Spr.

BLUE PHLOX
Phlox divaricata L.

Perennial, to 50 cm tall, rhizomatous; shoots prostrate or erect; leaves 4-5 cm long, to 2 cm wide;
floral tube to 1.6 cm long, smooth, lavender, blue
or white; *capsules 4-6 mm long. Rich woods.
Infreq. WF. W to Tex., N to Canada. Spr.

ANNUAL GARDEN PHLOX
Phlox drummondii Hook.

Annual, 10-70 cm tall; stems with glandular
hairs; leaves to 9 cm long, bristle-tipped; lower
leaves opposite; upper leaves alternate; floral
tube 12-17 mm long, white, pink, purple or red;
capsules 4-6 mm long. Open disturbed sites.
Freq., locally common CF, NF, WF. W to Tex.,
E to Va. Spr-sum.

FLORIDA PHLOX
Phlox floridana Benth.

Perennial, to 80 cm tall; lower stems smooth;
leaves to 10 cm long, smooth; upper leaves sub-
opposite; petals to 20 mm long, purple; capsules
4-6 mm long. Woody sites. Infreq. CF, NF, WF.
W to Ala., E to Ga. Sum-fall.

(not shown)

SMOOTH PHLOX
Phlox glaberrima L.
[*P. carolina* L.]

Perennial, to 1 m tall; stems erect, smooth; leaves
to 15 cm long; *sepals 7-10 mm long; floral tube
15-26 mm long, purple or white; stamens exserted;
capsules 4-6 mm long. Woody, open sites. Infreq
NF, WF. W to Tex.,   N to Mo., E to Va. Spr-fall.

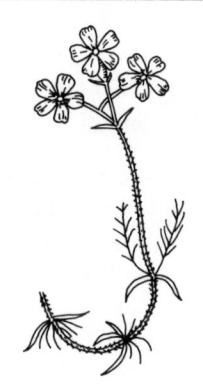

TRAILING PHLOX
Phlox nivalis Lodd. ex Sweet

Perennial evergreen shrub, 10-20 cm tall; stems lying flat; leaves 1-25 mm long, opposite, fascicled; flowers pink to rose or lavender to white; capsules 4-6 mm long. Dry sites. Infreq. NF, WF. W to Ala., E to Va. Spr.

DOWNY PHLOX
Phlox pilosa L.

Perennial, 20-60 cm tall; stems tufted, erect, with glandular hairs; leaves 3-9 cm long, with ciliate hairs; leaves nearly opposite on upper stem; *sepals 6-12 mm long; floral tube 10-18 mm long, violet, pink or white; capsules 4-6 mm long. Dry, open woods. Infreq. CF, NF, WF. W to Tex., N to N.D., E to Conn. Spr-sum.

MILKWORT FAMILY   *POLYGALACEAE*

Herbs, shrubs, vines or trees; leaves alternate, whorled or opposite; flowers bisexual; sepals 5 or 4-7, with inner 2 petal-like (wings); petals 3; ovary superior; fruit a capsule.

WHITE BACHELOR'S BUTTON
Polygala balduinii Nutt.

Annual or biennial, 10-70 cm tall; stems erect; leaves 5-25 mm long, variable, spatulate to obovate, alternate; rosette leaves not persistent; *racemes with white to green flowers; wings 3-4 mm long; capsules less than 1 mm wide. Moist to wet sites. Freq. all Fla. W to Tex., E to Ga. Sum.

SLENDER LEAFY MILKWORT
Polygala boykinii Nutt.
[*P. flagellaris* Small]

Perennial, 30-60 cm tall; stems lying flat; leaves
elliptic to suborbicular or linear; lower leaves
whorled; racemes 4-25 cm long; flowers white or
green; wings and keel 2-3 mm long; capsules to
3 mm long. Flatwoods. Infreq. all Fla. W to La.,
N to Ga. Spr-sum, all yr S.

Polygala brevifolia Nutt.

(not shown)

Similar to *Polygala cruciata* except: racemes on
stalks 2-8 cm long; bracts to 1 mm long; wings 3-
4 mm long, with sharply pointed tips. Swampy
sites. Infreq. WF. W to Miss., N to R.I. Spr-fall.

DRUMHEAD
Polygala cruciata L.

Annual, 5-40 cm tall; stems smooth; leaves to
4 cm long, whorled; racemes on stalks to 3 cm
long or sessile; bracts 2-3 mm long, persisting
after fruits fall; flowers red, purple or purple-
green; wings 1.5-3 mm long, with tapering tips;
petals shorter than wings; capsules to 2 mm long.
Moist to wet sites. Freq. all Fla. W to Tex.,
N to Mass. Sum-fall.

TALL MILKWORT
Polygala cymosa Walt.

Biennial, 0.4-1.2 m tall; basal leaves numerous,
3-14 cm long, linear, forming rosettes; racemes
in cyme-like panicles, yellow; capsules to 1 mm
wide. Moist sites. Freq. all Fla. W to La.,
N to Del. Spr-sum.

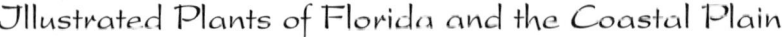

LARGE-FLOWERED POLYGALA
Polygala grandiflora Walt.

Perennial, 20-60 cm tall; stems hairy; leaves 1.5-5 cm long, alternate; *flowers purple, green or pink; wings 3-7 mm long; petals shorter than wings; capsules to 5 mm long. Dry to moist sites. Common all Fla. W to La., N to N.C. Spr-fall.

PROCESSION FLOWER
Polygala incarnata L.

Annual, 20-70 cm tall; stems covered with a white film; leaves 0.5-1.7 cm long, fleshy, alternate; flowers purple, pink or white; wings to 3.5 mm long; petals to 7 mm long; capsules 4 mm long. Dry sites. Freq. all Fla. W to Tex., N to Canada. Spr-sum.

SCRUB MILKWORT
Polygala lewtonii Small

Perennial, 10-20 cm tall; stems smooth; leaves to 2 cm long, alternate; *flowers pink or lavender; wings 4-5 mm long; petals to 3 mm long; capsules to 5 mm long. Dry scrub. Infreq. CF, NF. Spr-sum.

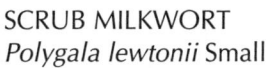

CANDY WEED
Polygala lutea L.

Annual, 5-40 cm tall; stems erect; leaves 1.5-6 cm long; racemes 1-4 cm long, orange or yellow; wings 5-7 mm long; capsules to 1.5 mm long. Wet, acid flatwood sites. Common all Fla. W to La., N to N.Y. Spr-fall.

MARYLAND MILKWORT
Polygala mariana Mill.

Annual, 10-50 cm tall; stems erect, smooth;
leaves to 3 cm long, alternate; racemes on stalks
to 4 cm long; flowers pink, white or purple;
wings and petals to 3 mm long; capsules to
2 mm long. Moist to wet sites. Infreq. NF, WF.
W to Tex., N to N.J. Spr-fall.

WILD BACHELOR'S BUTTON
Polygala nana (Michx.) DC.

Biennial or perennial, 2-15 cm tall; leaves 1-5.5 cm
long; flowers in head-like racemes to 4 cm long;
wings 6-8 mm long, yellow to green; *capsules to
2 mm long. Moist sites. Freq. all Fla. W to Tex.,
N to S.C. Spr-sum.

RACEMED MILKWORT
Polygala polygama Walt.

Biennial, 10-60 cm tall; stems lying flat, smooth;
leaves to 3 cm long, variable, obovate to linear,
alternate; racemes with *flowers white, pink,
rose, or purple; petals fringed; wings 3-6 mm
long; *cleistogamous flowers underground;
capsules less than 2x as long as wide. Dry
and coastal sites. Freq. all Fla. W to Tex.,
N to Canada. Spr-sum.

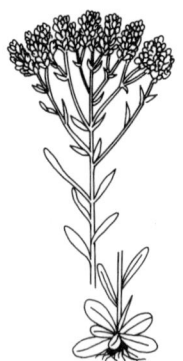

SHORT MILKWORT
Polygala ramosa Ell.

Annual, 10-30 cm tall; stem leaves to 2.5 cm
long; basal leaves 0.7-2.5 cm long, spoon-shaped,
quickly falling off; racemes 1-3 cm long, yellow;
wings 3-4 mm long; capsules to 1 mm wide.
Moist to wet sites. Freq. all Fla. W to Tex.,
N to N.C. Spr-fall.

YELLOW BACHELOR'S BUTTON
Polygala rugelii Shuttlw. ex Chapm.

Annual, biennial or perennial, 25-75 cm tall; stem
leaves 3-4 cm long; basal leaves 3-6 cm long;
racemes 1.5-3 cm long, yellow; wings 5-8 mm
long; capsules to 2 mm wide. Moist to wet sites.
Common all Fla. Spr-fall.

Polygala sanguinea L.

(not shown)

Similar to *Polygala mariana* except: wings 5-6 mm
long; petals 2-3 mm long; capsules to 3 mm long.
Dry to wet sites. Ala., W to Tex., N to Canada.
Sum.

SLENDER MILKWORT
Polygala setacea Michx.

Perennial, 10-35 cm tall; leaves scale-like;
racemes 5-15 mm long; flowers white, pink or
cream; wings and keel to 2 mm long; capsules to
1.5 mm long. Pinelands. Freq. all Fla. W to La.,
E to N.C. Spr-sum, all yr S.

WHORLED MILKWORT
Polygala verticillata L.

Annual, 5-30 cm tall; stems erect, branching,
smooth; leaves 6-30 mm long, variable, spatulate
to linear, whorled, some alternate; racemes with
white to pinkish-purple flowers; wings 1-1.5 mm
long; capsules to 1 mm long. Disturbed sites.
Infreq. CF, NF. W to Tex., N to Canada. Spr-sum.

BUCKWHEAT FAMILY   *POLYGONACEAE*

Herbs, vines, shrubs or trees; stem nodes swollen; leaves alternate; stipules forming membranous tube (ocreae); flowers bisexual or unisexual; sepals 3-6; petals 0; ovary superior; fruit a drupe, capsule, berry or nutlet.

JUMPSEED
Antenoron virginiana (L.) Raf.

(not shown)

Perennial, 0.3-1.2 m tall; leaves 5-15 cm long, 2-10 cm wide, ovate; racemes 15-60 cm long, green, white or pink; styles 2; achenes 3.5-4 mm long. Rich woods, thickets. Infreq. CF, NF, WF. W to Tex., N to Canada. Sum-fall.

CORAL VINE
Antigonon leptopus Hook. & Arn.

Perennial climbing vine; leaves to 15 cm long; *sepals 5-6, to 16 mm long, rose or purple, wing-like; *achenes 8-9 mm long, 3-winged. Native to Mexico. Escapes from cultivation. Disturbed sites. Infreq. SF, CF, NF. N to Ga. All yr.

PIGEON PLUM
Coccoloba diversifolia Jacq.

(not shown)

Perennial shrub or small tree; leaves 5-22 cm long, ovate, longer than wide, thin; sepals to 3.5 mm long, green to yellow; achenes to 1 cm in diam., enclosed by juicy hypanthium; fruits in clusters. Coastal hammocks. Infreq. SF, CF, NF. Spr-fall.

SEA GRAPE
Coccoloba uvifera (L.) L.

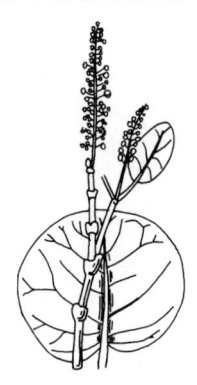

Perennial shrub or small tree, to 15 m tall; leaves 4-27 cm long, heart-shaped to rounded, broader than long, thick; sepals to 3 mm long, green to yellow; achenes to 2 cm in diam., enclosed by juicy hypanthium; fruits in grape-like clusters. Coastal hammocks and strands. Freq. SF, CF, NF. Spr-fall.

SCRUB BUCKWHEAT
Eriogonum longifolium Nutt.
[*E. floridanum* Small]

Perennial, 0.6-1 m tall; stems covered with soft silver hairs; basal leaves 8-20 cm long, with silver hairs on under surface; sepals 6, 5-6 mm long, white, covered with silver hairs; achenes to 6 mm long, beaked, with long white silky hairs. Dry sites. Infreq. CF. Spr-fall.

WILD BUCKWHEAT
Eriogonum tomentosum Michx.

Perennial, 0.4-1.2 m tall; leaves whorled, 3-5 per whorl, white- or tan-hairy beneath; rosette leaves 7-12 cm long; stem leaves smaller than rosette leaves; sepals 6, 3-4 mm long, white or pink; petals 0; stamens 9; achenes to 5 mm long. Dry sites. Freq. CF, NF, WF. W to Ala., E to S.C. Sum-fall.

SANDHILL WIREWEED
Polygonella fimbriata (Ell.) Horton

Perennial, 15-60 cm tall; leaves 1-3 cm long, persistent; sepals 5, to 2.5 mm long, white to pink; inner sepal fringed; *ocreae fringed with few long bristles; achenes to 2 mm long, to 1 mm wide. Pinelands, scrub. Infreq. NF. W to Ala., N to Ga. Sum-fall.

WIREWEED
Polygonella gracilis (Nutt.) Meisn.

Annual, 0.3-1.7 m tall; leaves 2-3 cm long, early deciduous; sepals 5, to 2 mm long, appressed, white to pink; ocreae entire; achenes to 2.2 mm long, 1-1.3 mm wide. Dry sites. Freq. CF, NF, WF. W to La., E to S.C. Sum-fall.

WOODY WIREWEED
Polygonella myriophylla (Small) Hort.

Perennial shrub, 0.2-2.5 m long; stems prostrate, branching; leaves 3-12 mm long; *sepals 2-3 mm long, white, pink or yellow; inner sepals longer than outer; *achenes to 3 mm long. Dry scrub. Infreq. CF. Endemic. All yr.

JOINTWEED
Polygonella polygama (Vent.) Engelm. & Gray

Perennial, 30-60 cm tall; leaves to 3 cm long, persistent; sepals 5, to 2 mm long, reflexed, white to pink; ocreae entire to pointed; capsules 1-2 mm long, 1-3 mm wide. Pinelands, scrub. Freq. all Fla. W to Tex., E to Va. Sum-fall.

BIG-FLOWERED SANDHILL WIREWEED
Polygonella robusta (Small) Nesom & Bates
[*Polygonella fimbriata* (Ell.) Horton
var. *robusta* (Small) Horton]

Perennial, 60-90 cm tall; leaves 2-6 cm long, persistent; sepals 5, 3.5-4 mm long, white to pink; inner sepal fringed; *ocreae fringed with many bristles; achenes over 2 mm long, to 1 mm wide. Sandhills, dry sites. Freq. CF, NF, WF. Sum-fall.

GIANT SMARTWEED
Polygonum densiflorum Meisn.

Perennial, to 1.5 m tall; stems lying flat to
ascending; leaves to 25 cm long, to 5 cm wide;
ocreae entire; sepals 5, to 3 mm long, green,
white or pink; achenes to 3 mm long. Wet sites.
Infreq. all Fla. W to Tex., N to N.Y. Sum-fall.

(not shown)

HAIRY SMARTWEED
Polygonum hirsutum Walt.

Perennial, to 1 m tall; stems lying flat to erect,
hairy; leaves 3-10 cm long, to 2 cm wide, hairy;
ocreae with long hairs; sepals 5, 2.5-3 mm long,
pink or white; achenes 2.5-3 mm long. Wet
sites. Infreq. CF, NF, WF. N to N.C. Spr-fall.

MARSH PEPPER SMARTWEED
Polygonum hydropiper L.

Annual, to 1 m tall; leaves to 10 cm iong, to 2 cm
wide; ocreae fringed; sepals 4, to 4 mm long,
white or green, glandular punctate; *achenes to
3 mm long. Native to Eurasia. Low wet sites.
Canada, S to N.C., W to Calif. Spr-fall.

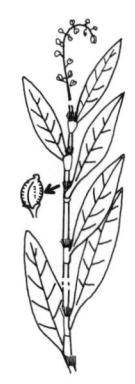

MILD WATER-PEPPER
Polygonum hydropiperoides Michx.

Perennial, 60-80 cm tall; stem or leaf sheaths
with appressed hairs; leaves 7-10 cm long, less
than 1 cm wide, lanceolate with appressed hairs;
ocreae fringed; sepals to 3 mm long, pink, white
or green, not glandular; achenes to 2.5 mm long,
lustrous. Wet areas. Freq. all Fla. W to Tex.,
N to Canada. Spr-fall.

PALE SMARTWEED
Polygonum lapathifolium L.

Annual, to 1.5 m tall; leaves 5-25 cm long, to 5 cm wide; *ocreae entire; *sepals 5, to 3 mm long, white or pink; *achenes to 2.5 mm long. Wet sites. Rare all Fla. Spr-fall.

OPELOUSAS SMARTWEED
Polygonum opelousanum Ridd. ex Small

Perennial, to 50 cm tall; leaves to 10 cm long, to 1 cm wide; ocreae fringed; sepals 5, to 3.5 mm long, mostly rose- or purple-tinted; *achenes to 2.5 mm long, protruding. Low wet sites. Freq. all Fla. W to Tex., N to Mass. Spr-fall.

PRINCESS FEATHER
Polygonum orientale L.

Annual, to 3 m tall; stems hairy; leaves to 25 cm long, to 7 cm wide; upper portion of ocreae with broad membranous sheath; sepals 5, 4-4.5 mm long, dark rose; achenes to 3.5 mm long. Native to Asia. Disturbed areas. Rare CF. N to N.C. Sum-fall.

(not shown)

DOTTED SMARTWEED
Polygonum punctatum Ell.

Perennial, 30-90 cm tall; stems erect to ascending, smooth; leaves to 15 cm long; ocreae bristles long-fringed; flowers bisexual; *sepals 5, glandular-dotted, persistent, pink; achenes to 3 mm long. Moist to wet sites. Common all Fla. Throughout U.S. Spr-fall, all yr S.

STUBBLE SMARTWEED
Polygonum setaceum Baldw.

Perennial, to 1.5 m tall; stem or leaf sheaths with appressed hairs; leaves to 20 cm long, to 1.5 cm wide, lanceolate, with appressed hairs; ocrea bristles to 1 cm long; sepals to 3 mm long, white, green or pinkish, rarely glandular; achenes 2-2.5 mm long, lustrous. Swamps. Freq. all Fla. W to Tex. and Mo., E to N.J. Spr-fall.

(not shown)

SHEEP SORREL
Rumex acetosella L.

(not shown)

Perennial, 20-40 cm tall, with slender rhizomes; lower leaves with petioles as long as leaves; leaf blades 2-5 cm long, hastate; flowers dioecious; valves of fruits ovate, longer than wide, not winged; achenes to 1 mm long. Disturbed sites. All U.S. except Fla. Spr-fall.

CURLY DOCK
Rumex crispus L.

Perennial, 0.3-1.5 m tall; leaves 1.5-40 cm long; margins crisped, wavy; *sepals 4-5 mm long, winged; *achenes 1.4-3 mm long. Native to Europe. Disturbed moist to wet sites. Infreq. all Fla. All U.S. Sum.

HEART-WING SORREL
Rumex hastatulus Baldw. ex Ell.

Annual, 0.15-1.3 m tall, with slender taproot; lower leaves with petioles as long as leaves; leaf blades to 8 cm long, hastate; *flowers unisexual with male and female flowers on separate plants (dioecious); valves of fruits kidney-shaped, wider than long, broadly winged; *achenes to 1.5 mm long. Old fields, disturbed sites. Common CF, NF, WF. W to Tex., N to Mass. Spr-sum.

TROPICAL DOCK
Rumex obovatus Danser

Annual, 40-70 cm tall, with taproot; lower leaves
with petioles as long as leaves; leaf blades 4-15 cm
long, oblong, elliptic, obovate, grayish green; flowers bisexual; valves of fruits ovate-triangular, with
tubercles; achenes to 2 mm long, winged. Native
to South America. Moist to wet sites. Rare CF, NF.
W to La. Spr-fall.

BITTER DOCK
Rumex obtusifolius L.

(not shown)

Perennial, to 1.2 m tall, with large taproot; lower
leaves with petioles shorter than leaves; leaf
blades 10-35 cm long, to 15 cm wide, heart-
shaped, elliptic or lanceolate; flower stalks much
longer than sepals; flowers bisexual; valves of
fruits ovate-triangular, with tubercles; achenes to
2 mm long. Moist to wet sites. Infreq. NF, WF.
Throughout U.S. Sum-fall.

FIDDLE DOCK
Rumex pulcher L.

Perennial, to 80 cm tall, with large taproot; lower
leaves with petioles shorter than leaves; leaf blades
to 20 cm long, fiddle-shaped; flower stalks as long
as sepals; flowers bisexual; valves of fruits ovate,
with tubercles; achenes to 2.5 mm long. Wet sites.
Infreq. all Fla. W to Tex., N to N.Y. Spr.

(not shown)

SWAMP-DOCK
Rumex verticillatus L.

Perennial, to 1 m tall, with fibrous roots; lower
leaves with petioles shorter than leaves; leaf
blades 5-20 cm long, linear, elliptic, or lanceolate; flowers green, bisexual; *valves of fruits
ovate, with tubercles; *achenes to 3 mm long.
Wet sites. Freq. all Fla. N to Canada, W to Tex.
Sum-fall.

PICKEREL WEED FAMILY   *PONTEDERIACEAE*

Aquatic herbs; leaves opposite or whorled; flowers bisexual; floral parts 6, subtended by sheaths; ovary superior; fruit a capsule or achene.

WATER HYACINTH
Eichhornia crassipes (Mart.) Solms

Perennial, to 1.2 m tall; stems free-floating; leaves 2-15 cm long, with inflated petioles; flowers in spikes, 4-7.5 cm wide, blue; capsules to 10 mm long. Wet sites. Common all Fla. W to Tex., N to Va. Spr-fall.

PICKEREL WEED
Pontederia cordata L.
[*P. lanceolata* Nutt.]

Perennial, 0.4-2 m tall; leaves including petioles to 2 m long; leaf bases varying from rounded to cordate; flowers to 1.5 cm long, blue or rarely white, with yellow marks; flowers with glandular hairs; capsules to 8 mm long. Wet sites. Common all Fla. W to Tex., E to Va. Spr-sum.

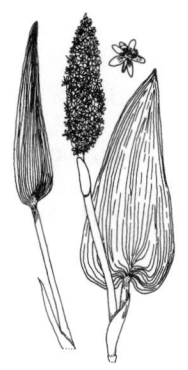

PURSLANE FAMILY *PORTULACACEAE*

Succulent or slightly woody herbs; leaves alternate or opposite; flowers bisexual; sepals 2; petals 4, 5 or 6; ovary superior; fruit a capsule.

BROADLEAF PINK PURSLANE
Portulaca amilis Speg.

Annual, to 15 cm tall; stems sprawling, succulent; leaves to 1 cm long, to 7 mm wide, flat; leaf axil hairs tufted; *petals 5, 1.5-2 cm wide, red; capsules to 6 mm in diam. Native to South America. Disturbed sites. Freq. CF, NF, WF. N to N.C. Spr-fall.

ROSE MOSS
Portulaca grandiflora Hook.

Annual, to 25 cm long; stems prostrate to ascending, branching; leaves to 2.5 cm long, linear or rounded; leaf axils hairy; petals 4-5, 15-25 mm long, yellow, red, pink or white; capsules 4-5 mm long. Disturbed sites. Cultivated. Rare CF, NF. Spr-fall.

(not shown)

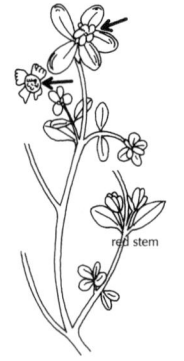

PURSLANE
Portulaca oleracea L.

Annual, to 45 cm long, edible; stems prostrate to ascending, many-branched, smooth, red to purple; leaves 3-5 cm long, spoon-shaped or almost round; leaf axils smooth or with few hairs; *petals 4-5, 3-8 mm long, yellow or orange; capsules 5-9 mm high. Disturbed sites. Freq. all Fla. All U.S. Spr-fall, all yr S.

PINK PURSLANE
Portulaca pilosa L.

Annual, 5-20 cm tall; stems erect to diffuse, succulent; leaves 1-2 cm long, to 3 mm wide, rounded; leaf axils with tufted hairs; petals 5, to 1 cm
wide, pink to purple; *capsules 3-4 mm in diam.
Disturbed sites. Freq. all Fla. W to La., N to
N.C. Spr-fall.

FAME FLOWER
Talinum paniculatum (Jacq.) Gaertn.

Perennial, 0.3-1.5 m tall, slightly woody; leaves
3-11 cm long, flat, spoon-shaped or obovate;
petals 5, pink; capsules to 3.5 mm in diam. Dry
open sites. Infreq. all Fla. W to Tex., N to N.C.
Sum-fall, all yr S.

Talinum teretifolium Pursh

Perennial, 10-30 cm tall; leaves 2-6 cm long,
nearly rounded; petals red or purple; capsules
4-5 mm in diam. Rocky sites. Ga., W to Ala.,
N to Va. Spr-fall.

(not shown)

PRIMROSE FAMILY   *PRIMULACEAE*

Rhizomatous or tuberous herbs; leaves basal, alternate or opposite; flowers bisexual; sepals 5 or 6; petals 4, 5 or 0; ovary superior or inferior; fruit a capsule.

SCARLET PIMPERNEL
Anagallis arvensis L.

Annual, 5-30 cm long; stems smooth, low, branching; leaves opposite, 5-20 mm long; *flowers axillary, stalked, solitary; sepals and petals 4 or 5; petals scarlet, white or blue; sepals 3-4 mm long; *capsules to 4 mm in diam. Native to Eurasia. Moist, disturbed sites. Rare CF, NF, WF. W to Tex., N to Canada. Spr-sum.

CHAFFWEED
Anagallis minima (L.) Krause
[*Centunculus minimus* L.]

Annual, 2-15 cm tall; stems low, branching, smooth; leaves 3-8 mm long, alternate; *flowers subsessile; sepals 2-2.5 mm long; petals pink to white; *capsules to 2 mm in diam. Moist sites. Occasional all Fla. W to Calif., N to Ohio and Canada. Spr-fall.

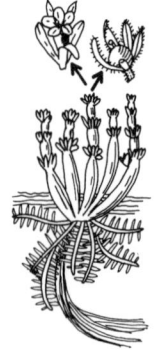

FEATHERFOIL
Hottonia inflata Ell.

Annual aquatic plant, 20-60 cm long; leaves 4-15 cm long, pectinate; *flowers whorled; petals 5, 3-4 mm long, with white lobes; *capsules to 3 mm in diam. Wet sites. Rare WF. W to Tex., N to Mo., E to Maine. Spr-sum.

STALKLESS LOOSESTRIFE
Lysimachia lanceolata Walt.
var. *lanceolata*
[*Steironema heterophyllum* (Michx.) Raf.]

Perennial, to 70 cm long; stems erect to reclining; leaves 3-18 cm long; flowers to 1 cm long, yellow, 5 per head; sepals and petals 5-lobed; capsules 3-4.5 mm in diam. Moist to wet woods, meadows. Rare to infreq. NF, WF. W to Ark., N to N.C. Spr-fall.

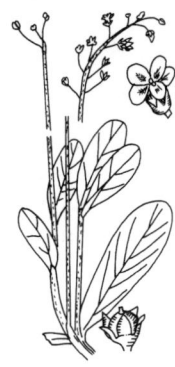

WATER PIMPERNEL
Samolus ebracteatus HBK.

Perennial, 10-20 cm tall; leaves 3-10 cm long; flowers pinkish; sepals 5; petals 5; stamens 5; capsules 3-4 mm wide. Wet sites. Freq. all Fla. W to Tex. All yr.

PINELAND PIMPERNEL
Samolus valerandi L.
subsp. *parviflorus* (Raf.) Hulten
[*S. parviflorus* Raf.]

Perennial, 10-60 cm tall; leaves 3-15 cm long; *flowers 2-3 mm wide, white; *capsules 2-3 mm in diam. Moist to wet sites. Freq. all Fla. W to Calif., N to Canada. Spr-sum.

CROWFOOT FAMILY	*RANUNCULACEAE*

Herbs or woody vines; leaves alternate or opposite; flowers bisexual or unisexual; sepals 3-5, petal-like or spurred; petals 0 or if present, spurred; ovary superior; fruit a capsule.

PINE HYACINTH
Clematis baldwinii Torr. & Gray

Perennial, 20-60 cm tall; stems erect, hairy; leaves 1.5-10 cm long, simple; leaf margins entire to lobed; flowers bisexual, solitary; sepals 2.5-5.5 cm long, purple; petals 0; styles to 10 cm long; achenes to 5 mm long. Moist to wet sites. Freq. SF, CF, NF. Spr-fall.

LEATHER FLOWER
Clematis crispa L.

Perennial; stems climbing to ascending; leaves divided into 3-5 leaflets each to 7 cm long; flowers bisexual, solitary, on bractless stems; sepals 3-5 cm long, pink, blue to violet, or white; petals 0; styles to 3 cm long; achenes 6-9 mm in diam. Wet sites. Freq. CF, NF, WF. W to Tex., N to Va. Spr-sum.

(not shown)

VASE VINE
Clematis reticulata Walt.

Perennial climbing vine, slightly hairy; leaves opposite; leaflets 2-6 cm long, variously lobed, with veins raised on both surfaces; *flowers 16-24 mm long, 4-lobed, solitary, on bracted stems; sepals 1-2 cm long, purple-red or pink-lavender; petals 0; *achenes 4-6 mm wide; style covered with long gold hairs. Dry woody sites. Freq. CF, NF, WF. W to Tex., N to S.C. Spr-sum.

AUTUMN CLEMATIS
Clematis terniflora DC.

Perennial; stems climbing; leaves divided into
5 leaflets each to 8 cm long; flowers bisexual, in
cymes; sepals 0.6-2 cm long, white; petals 0;
styles to 5 cm long; achenes 2-5 cm long. Native
to Japan. Dry to wet sites. Rare CF, NF, WF.
N to N.Y. Sum-fall.

(not shown)

VIRGIN'S BOWER
Clematis virginiana L.

(not shown)

Perennial; stems climbing; leaves divided into
3 leaflets each 2-10 cm long; flowers unisexual,
in cymes; sepals 6-12 mm long, white; petals 0;
styles to 4 cm long; achenes 1-3 cm long.
Wooded sites. Infreq. CF, NF, WF. W to Tex.,
N to Canada. Sum-fall

KIDNEY-LEAF BUTTERCUP
Ranunculus abortivus L.

Annual, 10-50 cm tall; stems smooth; basal
and upper leaves to 10 cm long, to 4 cm wide;
*flowers stalked; petals 5, 2-3 mm long, yellow;
achenes to 1.5 mm long, smooth. Floodplains.
Infreq. WF. W to Tex., N to Canada. Spr.

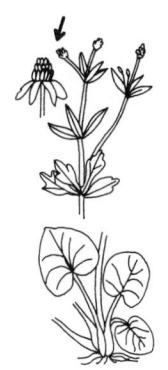

SWAMP BUTTERCUP
Ranunculus carolinianus DC.

Perennial, 10-65 cm long; stems weak, sprawling, rooting at nodes, forming dense mats; basal
leaves to 4 cm long, 3-lobed or ternately divided;
petioles to 15 cm long; *petals 0.5-1.5 cm long,
yellow; achenes to 3 mm long. Wet sites. Infreq.
NF, CF, WF. W to Tex., N to Md. Spr-fall.

SMALL-FLOWERED BUTTERCUP
Ranunculus parviflorus L.

Annual, 10-30 cm long; stems diffuse, hairy; basal leaves to 3 cm long, to 6 cm wide; upper leaves to 4 cm long, to 0.5 cm wide; flowers stalked; *petals 5, 1-2 mm long, yellow; achenes to 1.5 mm long, hairy. Native to Europe. Wet banks. Infreq. WF. Spr-sum.

TROPICAL BUTTERCUP
Ranunculus platensis Spreng.

(not shown)

Annual, 5-30 cm tall; stems hairy; basal and upper leaves to 8 cm long, to 1.5 cm wide; flowers sessile; petals 3, minute, yellow; achenes to 2 mm long, smooth to hairy. Native to South America. Moist sites. Infreq. WF. W to Tex. Spr.

SPEARWORT
Ranunculus pusillus Poir.

Annual, 10-40 cm tall; leaves 1-4 cm long; *flowers minute; petals to 3 mm long, yellow; achenes to l mm long, numerous. Low, wet sites. Infreq. NF, WF. W to Tex., N to N.Y. Spr-sum.

BUCKTHORN FAMILY   *RHAMNACEAE*

Vines, shrubs or trees; leaves alternate or opposite; stipules present; flowers bisexual or unisexual (dioecious); sepals 4 or 5; petals 4 or 5; ovary superior; fruit a drupe or capsule.

RATTAN VINE
Berchemia scandens (Hill) K. Koch

Perennial, scrambling and high climbing vine; leaves 3-8 cm long, simple, varying from oval to elliptic to lanceolate; flowers to 2 mm wide, greenish yellow; drupes to 6 mm long, black. Moist to wet woods. Freq. all Fla. W to Tex., E to Va., N to Mo. Spr.

NEW JERSEY TEA
Ceanothus americanus L.
[*C. intermedius* (Pursh) K. Koch]

Perennial shrub, 0.2-1 m tall; leaves 2-8 cm long; sepals 5, 1-1.5 mm long; petals 5, 1.5-2 mm long, white; stamens 5; drupes 2-4 mm long, 3-lobed. Dry woody sites. Infreq. CF, NF, WF. W to Tex., N to Canada. Spr-sum.

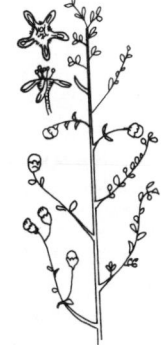

LITTLE LEAF RED ROOT
Ceanothus microphyllus Michx.

Perennial diffuse shrub, 30-60 cm tall; leaves 3-10 mm long; petals 5, to 1.5 mm long, white; drupes 4-5 mm wide, 3-lobed. Dry sites. Infreq. CF, NF, WF. W to Ala., N to Ga. Spr.

RED MANGROVE FAMILY   *RHIZOPHORACEAE*

Shrubs, vines or trees; leaves mostly opposite; flowers mostly bisexual; sepals 3-16; petals 3-16; ovary inferior or superior; fruit a berry or drupe.

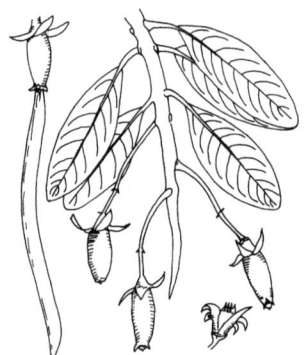

RED MANGROVE
Rhizophora mangle L.

Perennial shrub or tree, to 20 m tall; stilt roots many; leaves 4-15 cm long, evergreen, leathery; flowers to 20 cm wide; sepals 4 or 5; petals 4, yellow; stamens 4-12; fruits 2-3 cm long, growing to 30 cm long, developing root while still on parent tree and upon breaking off, floating root end down until reaching ground. Coastal sites in salty or brackish water. Freq. SF, CF; infreq. NF, WF. All yr.

ROSE FAMILY   *ROSACEAE*

Herbs, shrubs, vines or trees; leaves mostly alternate; stipules present; flowers bisexual; sepals mostly 5; petals mostly 5; ovary superior or inferior; fruit an achene, pome, drupe or follicle.

RED CHOKEBERRY
Aronia arbutifolia (L.) Ell.
[*Pyrus arbutifolia* (L.) L. f.]

Perennial shrub, to 3.5 m tall; leaves 2.5-7.5 cm long, elliptic to oval, hairy on lower surface; *petals to 8 mm long, white; *pomes to 6 mm in diam., red. Swampy sites. Freq. CF, NF, WF. W to La., N to Canada. Spr.

BLACK CHOKEBERRY
Aronia melanocarpa (Michx.) Ell.

Similar to *Aronia arbutifolia* except: leaves and cymes smooth; leaves oval; pomes to 8 mm in diam., black-purple. Swamps, woods. Ga., N to Mich. and N.Y. Spr.

SUMMER HAW
Crataegus flava Ait.

Perennial shrub, to 8 m tall; stems thorny; leaves 1-5 cm long, with marginal glands from base of petiole upwards; *petals to 15 mm wide, white; pomes 8-16 mm in diam., red, yellow or orange. Dry woody sites. Infreq. CF, NF, WF. W to Miss., N to N.C. Spr.

CHICKASAW PLUM
Prunus angustifolia Marsh.

Perennial shrub or small tree, to 4 m tall, growing in thick groupings; some twigs with thorns; leaves 2-8 cm long, deciduous; leaf margins with teeth; flowers 6-9 mm wide, fragrant, white; drupes 1-2.3 cm in diam., red to yellow. Fields, woody sites. Infreq. CF, NF, WF. W to Tex., N to N.J. Spr.

COMMON PEACH
Prunus persica (L.) Batsch

Perennial shrub or small tree; leaves 6-19 cm long, lanceolate; flowers present before leaves; petals to 1.5 cm long, pink; drupes to 6 cm wide, covered with fine hairs. Native to China. Rarely escapes. Disturbed sites. Rare CF, WF. E to Va. Spr.

FLATWOODS PLUM
Prunus umbellata Ell.

Perennial small tree, to 8 m tall, growing solitary; twigs mostly without thorns; leaves to 6 cm long; leaf margins with teeth; flowers 1-2.5 cm wide, odorless, white; drupes 1-2 cm in diam., purplish black or red. Well-drained woods. Freq. CF, NF, WF. W to Tex., N to S.C. Spr.

CHEROKEE ROSE
Rosa laevigata Michx.

Perennial, 2-5 m long, high climbing, thorny; leaflets 3 or 5, 1-8 cm long, smooth, evergreen; petals 3-4 cm long, white; stamens many; hips 3-4 cm long. Roadsides, woody sites. Infreq. CF, NF, WF. W to Tex., N to S.C. Spr.

SWAMP ROSE
Rosa palustris Marsh.
[*R. floridana* Rydb.]

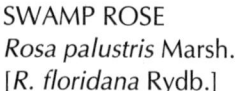

Perennial, 0.3-2 m tall; stems covered with curved prickles; leaflets 3-9, 2-6 cm long; sepals, petals and stamens united into tube (hypanthium) housing achenes (rose hips); hips to 11 mm in diam.; petals 2-3 cm long, pink; achenes to 3 mm long. Swampy sites. Freq. CF, NF, WF. W to Miss., N to Canada. Spr-sum.

HIGHBUSH BLACKBERRY
Rubus argutus Link

Perennial, to 3 m tall; stems erect or arching, covered with straight or curved prickles; first canes with 5 leaflets; each leaflet 6-10 cm long; *several flowers on each branchlet; petals to 2.5 cm long, rose, pink or white; *fruits to 1.2 cm long, black, edible. Various moist to wet sites. Infreq. CF, NF, WF. N to Md. and Mo. Spr-sum.

SAND BLACKBERRY
Rubus cuneifolius Pursh

Perennial, 0.3-1.5 cm tall; stems usually erect,
covered with straight or curved prickles; first
canes with 5 leaflets; each leaflet to 5 cm long;
leaves with white or grayish hairs beneath; petals
to 1.5 cm long, white; fruits 1-1.2 cm long,
edible. Various dry to wet sites. Freq. all Fla.
N to Conn. Spr-sum.

SWAMP DEWBERRY
Rubus hispidus L.

(not shown)

Perennial, to 2 m long; stems prostrate to trailing,
covered with straight or curved prickles; first
canes usually with 3 leaflets; each leaflet to 4 cm
long; several flowers on each branchlet; petals
to 1 cm long, white; fruits to 1 cm in diam.,
red to purple, edible. Wet sites. S.C., Ky.,
N to Canada. Spr-sum.

SOUTHERN DEWBERRY
Rubus trivialis Michx.

Perennial, to 2 m long; stems prostrate to trailing,
covered with curved prickles; first canes usually
with 5 leaflets; each leaflet to 6 cm long; flowers
solitary on branchlets; petals to 1.2 cm long,
white, pink or rose; fruits to 1.5 cm long, black,
edible. Dry to wet disturbed sites. Freq. all Fla.
N to Mo. and Va. Spr-sum.

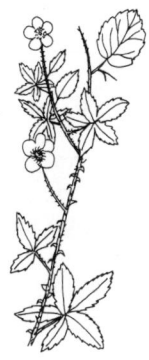

MADDER FAMILY   *RUBIACEAE*

Herbs, shrubs or trees; leaves opposite or whorled; stipules present; flowers bisexual; sepals 4 or 5; petals 4 or 5; ovary inferior; fruit a drupe, capsule, berry or nutlet.

BUTTON BUSH
Cephalanthus occidentalis L.

Perennial shrub, to 3 m tall; leaves 4-20 cm long; flowers in heads 2-4 cm wide; petals 6-10 mm long, white; styles exserted; nutlets 4-8 mm long. Low, swampy sites. Common all Fla. W to Tex., N to Canada. Sum-fall, all yr S.

SNOWBERRY
Chiococca alba (L.) A. Hitchc.

Perennial shrub, to 3 m tall; stems erect, reclining or spreading, smooth; leaves 1-6 cm long; sepals 5; petals 5, to 4 mm long, white to yellow; drupes 4-7 mm in diam., white. Coastal hammocks. Freq. SF, CF, NF. All yr.

POOR JOE
Diodia teres Walt.
[*D. rigida* (Willd.) Cham. & Schlecht.]

Annual, 10-80 cm tall; stems erect, hairy; leaves 1-4 cm long; *flowers 4-6 mm long, in sessile axillary few-flowered clusters, purple to pink to white; stamens 4; *fruits 3-4 mm long. Dry, disturbed sites. Common all Fla. W to Tex., N to Mich., E to Conn. Sum-fall, all yr S.

BUTTONWEED
Diodia virginiana L.
[*D. hirsuta* Pursh]

Perennial, to 1.5 m long; stems smooth, lying flat;
leaves to 10 cm long; sepals 2, to 6 mm long;
*petals 4, to 1 cm long, white; *nutlets 5-9 mm
long, with 6 vertical ridges. Low, wet sites.
Common all Fla. W to Tex., N to N.J. Spr-fall,
all yr S.

BEACH CREEPER
Ernodea littoralis Sw.

Perennial, to l m tall, prostrate, spreading, weakly
shrubby; stems 4-angled, smooth; leaves 2-4 cm
long, fleshy; sepals 4-5; petals 4-5, 7-11 mm
long, white, pink or red; drupes to 6 mm long.
Coastal dunes. Infreq. SF, CF. All yr.

CATCHWEED
Galium aparine L.

Annual, 0.1-1.5 m long; stems reclining,
scabrous, prickly; leaves 6 or 8 in a whorl, 1-8 cm
long, hairy; sepals 0; *petals 4, white; stamens 4
or 3; *fruits 3-5 mm in diam., with hooked hairs.
Disturbed, shady sites. Infreq. NF, WF. W to
Tex., N to Canada. Spr-sum.

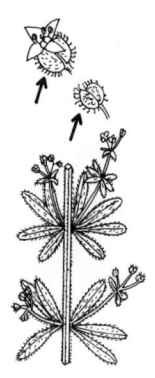

(not shown)

PURPLE GALIUM
Galium hispidulum Michx.

Perennial, 10-60 cm long; stems evergreen;
leaves in 4's, 3-15 mm long, elliptic, hairy below;
flowers 2 per cyme; petals to 2 mm long, greenish white; berries 3-5 mm wide, smooth, fleshy,
purple to blue. Dry, woody sites and dunes.
Freq. throughout Fla. W to La., E to N.J. Spr-fall.

ERECT BEDSTRAW
Galium obtusum Bigel.

Perennial, to 80 cm tall; stems diffuse to erect;
leaves in 4's, 5-20 mm long, oblanceolate to linear;
*flowers 2 or 3 in terminal clusters; petals to 1 mm
long, white; *fruits 2.5-4 mm wide, smooth, black.
Moist, shady sites. Ga., W to Tex., N to Canada.
Spr-sum.

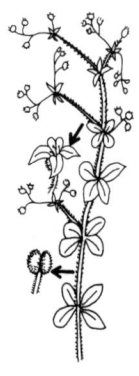

HAIRY FRUITED BEDSTRAW
Galium pilosum Ait.

Perennial, 30-80 cm tall; stems ascending or erect;
leaves in 4's, 10-25 mm long, oval to elliptic,
hairy; flowers paniculate; *petals to 1.5 mm long,
greenish white to purple; *fruits to 4 mm wide,
dry, bristly, brown or black. Woody sites. Infreq.
CF, NF, WF. W to Tex., N to Canada. Sum.

DYE BEDSTRAW
Galium tinctorium L.

Annual, to 50 cm long; stems ascending or reclining; leaves in 5's, 6's or 4's, to 25 m long, linear to
lanceolate; flowers 3 per cyme or single, petals 3
to 4, less than 2 mm broad, white; fruits 2-3 mm
wide, dry, smooth, black. Moist sites. Freq. all
Fla. W to Tex., N to Canada. Spr-sum.

(not shown)

(not shown)

BEDSTRAW
Galium uniflorum Michx.

Perennial, to 30 cm long; stems erect or lying flat,
smooth; leaves in 4's, 15-28 mm long, linear,
smooth below; flower 1 per cyme; petals to 2 mm
long, white; fruits 2-4 mm wide, black. Dry woods.
Rare CF, NF, WF. W to Tex., E to Va. Sum.

FIRE BUSH
Hamelia patens Jacq.

Perennial shrub or small tree, 1.5-4 m tall; leaves
8-15 cm long; flowers 1-2 cm long, red; berries
to 6 mm long, red or bluish black. Coastal sites,
hammocks. Freq. SF, CF. All yr.

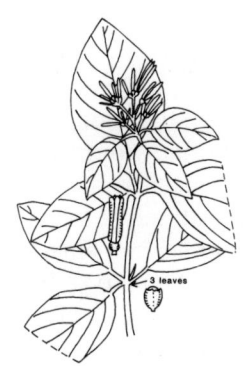

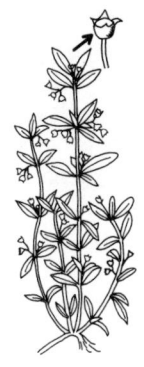

OLD WORLD DIAMOND-FLOWER
Hedyotis corymbosa (L.) Lam.

Annual, 20-50 cm long; stems diffuse, smooth;
leaves 1-3 cm long; sepals 4, to 1 mm long; petals
4, to 2 mm long, white; *capsules over 1 mm
long. Introduced pantropical weed. Moist to wet
disturbed sites. Freq. all Fla. W to Tex., N to S.C.
Spr-fall.

NARROW-LEAVED DIAMOND-FLOWER
Hedyotis nigricans (Lam.) Fosb.

Perennial, 8-70 cm tall; stems erect to diffuse;
leaves 12-40 mm long, linear; sepals 4; *petals 4,
white to purplish; stamens 4; *capsules 2.5-3 mm
long, exserted. Dry sites. Infreq. all Fla.
W to Ariz., N to Mich. Spr-sum.

INNOCENCE or FAIRY FOOTPRINTS
Hedyotis procumbens (Gmel.) Fosb.

Perennial, 5-40 cm long; stems prostrate, creeping to erect; leaves 0.5-1.5 cm long; sepals 4,
to 1.5 mm long; *petals 4, to 6 mm long, white;
capsules to 3 mm long. Moist disturbed sites.
Common all Fla. W to La., N to S.C. Spr-fall.

CLUSTERED DIAMOND-FLOWER
Hedyotis uniflora (L.) Lam.

Annual, 10-60 cm long; stems lying flat to
ascending, hairy to smooth; leaves 5-15 mm
long, lanceolate, ovate-elliptic; flowers axillary
and terminal, sessile or subsessile; sepals 4;
petals 4, to 1.5 mm long, white; stamens 4;
capsules to 2 mm long, hairy to smooth. Moist
to wet sites. Freq. all Fla. W to Tex., N to Mo.,
E to N.Y. Spr-fall.

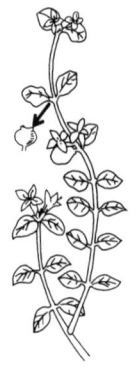

PARTRIDGE BERRY or TWINBERRY
Mitchella repens L.

Perennial, 15-30 cm long; stems creeping,
matting; leaves 0.6-2 cm long; sepals 4, to 1 mm
long; petals 4, to 1.2 cm long, white or pink,
occurring in pairs; *drupes 4-6 mm in diam., red
or white. Moist, woody sites. Freq. CF, NF, WF.
W to Tex., N to Canada. Spr-fall.

MITRACARPUS
Mitracarpus villosus (Sw.) Cham. & Schlecht.
[*M. hirtus* (L.) DC.]

Annual, to 50 cm tall; stems scabrous or hairy;
flowers minute, clustered in axils, white; petals 4;
capsules hairy, circumscissile. Dry sites. Rare all
Fla. W to La. Sum-fall.

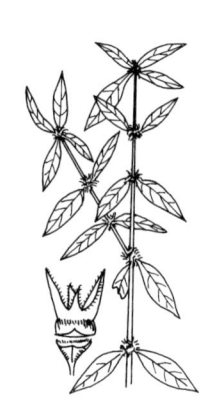

SKUNK VINE
Paederia foetida L.

Perennial twining vine; leaf blades 3-11 cm long;
petals 7-9 mm long, lilac; *berries to 1 cm long.
Native to East Indies. Escapes from cultivation.
Disturbed woody sites. Infreq. SF, CF, NF.
N to S.C. Spr-fall.

WILD COFFEE
Psychotria nervosa Sw.

Perennial shrub, 0.3-3.6 m tall; twigs and leaves
in the inflorescence smooth; leaves 4.5-18 cm
long; sepals 5, minute; petals 5, 2.5-4 mm long,
white; drupes 5-9 mm long, red or yellow.
Hammocks, woody sites. Freq. SF, CF, NF. All yr.

DULL-LEAF WILD COFFEE
Psychotria sulzneri Small

Perennial shrub, to 2 m tall; stems branching,
hairy; leaves 8-15 cm long, dull blue-green; sepals deltoid and conspicuous; petals to 2.5 mm
long, green; drupes to 5 mm long, red, orange or
yellow. Woods, hammocks. Freq. SF, CF, NF.
Spr-sum.

WHITE INDIGO BERRY
Randia aculeata L.

Perennial shrub, 0.3-3 m tall; stems spiny; leaves
1.5-5 cm long; sepals 4, to 1 mm long; petals 5,
5-7 mm long, white; berries 8-10 mm long,
white. Coastal hammocks. Infreq. SF, CF, NF.
All yr.

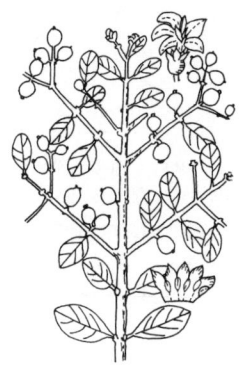

BRAZIL PUSLEY
Richardia brasiliensis (Moq.) Gomes

Perennial, 10-70 cm long; stems spreading, lying
flat, hairy; leaves 1.5-4 cm long; *petals 3-4 mm
long, white; fruits 3-4 mm long, with stiff hairs.
Native to South America. Disturbed sites. Freq.
all Fla. W to Tex., E to Va. Sum-fall, all yr S.

FLORIDA PUSLEY
Richardia scabra L.

Annual, to 80 cm tall; stems covered with very
fine hairs; leaves to 8 cm long, smooth; sepals 3,   (not shown)
to 2.5 mm long; petals 3, 5-6 mm long, white;
capsules to 4 mm long, covered with wart-like
projections. Sandy, disturbed sites. Common all
Fla. W to Ark., N to Va. All yr.

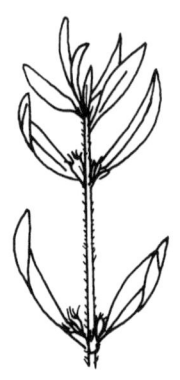

LARGE LEAF BUTTONWEED
Spermacoce assurgens Ruiz & Pavon
[*Borreria laevis* of authors]

Annual, 20-50 cm tall; stems ascending or erect;
leaves 2-4 cm long; flowers in dense axillary
clusters; sepals 4; petals 2-3 mm long, white;
stamens 4; fruits to 2 mm long. Moist, woody
sites. Freq. SF, CF. W to La. All yr.

SLENDER BUTTONWEED
Spermacoce prostrata Aubl.
[*Borreria ocimoides* (Burm.) DC.]

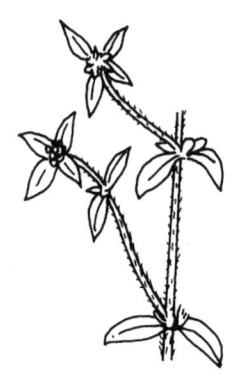

Annual, 10-60 cm tall; stems prostrate to ascending; leaves 10-40 mm long, elliptic or linear-
elliptic; flowers in compact cymes, to 1 mm long;
sepals and petals 4; petals as long as sepals,
white; stamens 4; capsules 1-2 mm long, smooth
to hairy. Pinelands. Freq. all Fla. All yr.

WHITE HEAD BROOM
Spermacoce verticillata L.
[*Borreria terminalis* Small]

Perennial, 5-30 cm tall; leaves 1-3 cm long;
*petals to 3 mm long, white; *capsules to 2 mm
long. Flatwoods, disturbed dry sites. Infreq. SF,
CF. All yr.

RUE FAMILY   *RUTACEAE*

Trees, shrubs or herbs; leaves alternate, aromatic when crushed; flowers mostly bisexual; sepals 3-7; petals 3-7; ovary superior; fruit a berry, samara or capsule.

WAFER ASH TREFOIL
Ptelea trifoliata L.

Perennial low shrub, to 8 m tall; leaflets 3 or 5, 2-15 cm long; petals 4-6 mm long, white or green; *samaras 1.5-2.5 cm long, cymose. Rich woods, low, rocky sites. Occasional CF, NF, WF. W to Ariz., N to Canada. Spr.

HERCULES'-CLUB
Zanthoxylum clava-herculis L.

Perennial shrub or tree, to 17 m tall; stems very prickly; leaflets 5-19, 3-7 cm long; sepals 5; petals 5, white; follicles to 5 mm long. Hammocks, woods. Freq. all Fla. W to Tex., E to Va. Spr-sum.

(not shown)

WILD LIME
Zanthoxylum fagara (L.) Sarg.

Perennial shrub or small tree, to 10 m tall; stems prickly; leaflets 9 or 11, 1-3 cm long; rachis wing-margined; sepals 4; petals 4, white; follicles to 4 mm long. Hammocks. Freq. SF, CF, NF. W to Tex. All yr.

SALVINIA FAMILY   *SALVINIACEAE*

Aquatic, floating herbs; leaves numerous, small; spores borne in sorus on leaves.

MOSQUITO FERN
Azolla caroliniana Willd.

Annual, 5-15 mm long, green or red, forming mats; stem branching; sporocarps on submerged leaves. Quiet fresh water. Freq. all Fla. W to Ariz., N to Canada. Sum-fall.

SOAPBERRY FAMILY   *SAPINDACEAE*

Trees, shrubs or vines; leaves mostly alternate; flowers unisexual (dioecious); sepals 4-5; petals 4-5 or 0; ovary superior; fruit a capsule, berry, nut, samara or drupe.

SMALL-FRUITED BALLOON-VINE
Cardiospermum microcarpum HBK.

Annual climbing vine; leaves to 12 cm long, divided into several leaflets; *petals 2-3 mm long, white; *capsules to 2 cm long. Flatwoods, hammocks. Infreq. SF, CF, NF. W to Tex., N to Del. All yr.

VARNISH LEAF
Dodonaea viscosa (L.) Jacq.

Perennial shrub or tree, 1-5 m tall; leaves 2-15 cm
long, evergreen; flowers pale yellow; sepals 3-5;
petals 0; capsules 2-3 cm long, 3-winged, papery;
seeds to 3.5 mm long. Pinewoods, hammocks.
Infreq. SF, CF, NF. Sum-fall.

SAPODILLA FAMILY   *SAPOTACEAE*

Shrubs or trees; stems producing milky sap; leaves mostly
alternate or spiraled; flowers bisexual; sepals 4, 5 or 12, in 1
to several whorls; petals equal in number to sepals, in 1
whorl; ovary superior; fruit a berry.

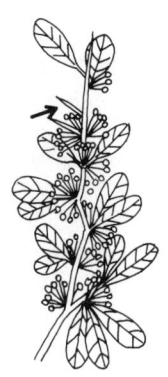

SHRUBBY BUCKTHORN
Bumelia reclinata (Michx.) Vent.

Perennial shrub, to 3 m tall; stems erect to diffuse,
*thorny; leaves 1-7 cm long, evergreen; flowers in
umbels; sepals to 2 mm long; petals to 4 mm wide,
white; berries 4-7 mm long. Moist to wet flatwoods. Freq. all Fla. W to La., E to Ga. Spr-sum.

PITCHER PLANT FAMILY   *SARRACENIACEAE*

Carnivorous, rhizomatous herbs; leaves mostly basal and hollow, modified into tubes (pitchers) expanding at top to form hoods; pitcher openings concealed or not by hoods; carnivory taking place within pitchers after insects become trapped inside, drown in liquid held in pitchers and later digested by enzymes; flowers bisexual; sepals 4, 5 or 6; petals 5; ovary superior; fruit a capsule.

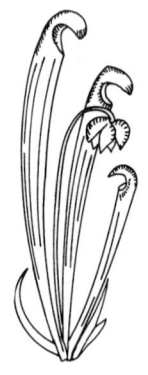

HOODED PITCHER PLANT
Sarracenia minor Walt.

Perennial; leaves 20-50 cm tall, erect; leaf hood whitish blotched, covering pitcher opening; petals 3-4 cm long, yellow; capsules to 1.5 cm wide. Wet, acid bogs. Common CF, NF, WF. N to N.C. Spr.

PARROT PITCHER PLANT
Sarracenia psittacina Michx.

Perennial; leaves 8-30 cm long, lying flat; leaf hood white blotched, strongly arched and covering pitcher opening; petals 3-4 cm long, maroon; capsules 9-14 mm in diam. Flatwoods, bogs. Common NF, WF. W to La., N to Ga. Spr. Hybridizes with *S. minor*.

(not shown)

SIDE-SADDLE
Sarracenia purpurea L.

(not shown)

Perennial; leaves 10-40 cm long, evergreen, lying flat, sprawling, urnlike; leaf hood green to purple, not covering pitcher opening; petals 4-5 cm long, purple; capsules 10-20 mm in diam. Bogs. Common WF. W to La., N to Canada. Spr.

SWEET PITCHER PLANT
Sarracenia rubra Walt.

Perennial; leaves 8-50 cm tall, erect; leaf hood
green to red, not arched over pitcher opening;
flowers fragrant; petals 3-4 cm long, red to purple; capsules to 1.5 cm wide. Bogs. Rare WF.
W to Miss., E to N.C. Spr.

(not shown)

LIZARD'S-TAIL FAMILY   *SAURURACEAE*

Rhizomatous herbs; leaves alternate; flowers bisexual; sepals
0; petals 0; flowers consisting of stamens only; ovary superior
or inferior; fruit a capsule.

LIZARD'S-TAIL
Saururus cernuus L.

Perennial, 0.3-1.2 m tall; leaves alternate; leaf
blades 5-15 cm long; racemes 10-20 cm long,
white, drooping at first; perianth 0; stamens 6-8;
capsules 2-3 mm long. Low, wet sites. Common
all Fla. W to Tex., N to Canada. Spr-sum.

SAXIFRAGE FAMILY   *SAXIFRAGACEAE*

Herbs, vines or shrubs; leaves alternate, opposite or basal; flowers mostly bisexual; sepals 3-10; petals 3-10 or 0; ovary superior or inferior; fruit a capsule or berry.

VIRGINIA WILLOW
Itea virginica L.

Perennial shrub, 1-3 m tall; leaves 5-10 cm long, alternate; sepals 5, minute; petals 5, 4-7 mm long, white; style persistent; capsules 6-10 mm long. Swampy sites. Freq. all Fla. W to Tex., N to Mo., E to N.J. Spr.

SAXIFRAGE FAMILY   *SCROPHULARIACEAE*

Herbs, shrubs, vines or trees; stems round to 4-angled; leaves alternate, whorled or opposite; flowers bisexual; sepals 4 or 5; petals 4 or 5; ovary superior; fruit a capsule.

LEAFLESS GERARDIA
Agalinis aphylla (Nutt.) Raf.

(not shown)

Annual, 0.5-1.2 m tall; stems smooth; branches several, spreading to erect; leaves 0.5-2 mm long, to 0.5 mm wide, slightly hairy; flower stems 1-2 mm long; petals to 1.5 cm long, pink; capsules to 4 mm long. Moist to wet sites. Infreq. NF, WF. W to La., N to N.C. Fall.

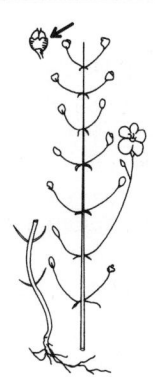

LITTLE GERARDIA
Agalinis divaricata (Chapm.) Penn.

Annual, 30-80 cm tall; stems branching; leaves
1.5-2.5 cm long; petals 1-1.5 cm long, pink;
*capsules to 3 mm long. Dry scrub, pinelands.
Infreq. CF, NF, WF. W to Miss., E to Ala. Fall.

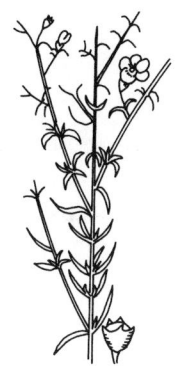

CLUSTER-LEAF GERARDIA
Agalinis fasciculata (Ell.) Raf.

Annual, to 1.2 m tall; stems with short, rough-textured hairs; branches many, spreading to upright;
leaves to 4 cm long, to 1·cm wide, with short
rough hairs above; flower stems to 4 mm long;
petals to 3.5 cm long, purple; capsules to 6 mm
long. Flatwoods. Common all Fla. W to Tex.,
N to N.C. Sum-fall.

FINE-LEAF GERARDIA
Agalinis filifolia (Nutt.) Raf.

Annual, to 80 cm tall; stems smooth; branches
numerous, spreading to erect; leaves 1-2 cm long,
to 0.5 mm wide, fleshy; flower stems to 3 cm
long; petals to 3.5 cm long, pink to purple; capsules to 4 mm long. Dry to moist sites. Freq. all
Fla. N to Ga. and Ala. Fall.

FLAX-LEAVED GERARDIA
Agalinis linifolia (Nutt.) Britt.

Perennial, 0.8-1.6 m tall; stems smooth, rounded;
leaves 3-5 cm long, opposite; petals 3-4 cm long,
pink to purple; *capsules to 7 mm long. Wet sites.
Common all Fla. W to La., N to Del. Sum-fall.

SEASIDE GERARDIA
Agalinis maritima (Raf.) Raf.

Annual, 5-40 cm tall; stems fleshy, smooth;
branches few, spreading; leaves 2-3 cm long, to
3 mm wide, smooth; flower stems to 8 mm long;
petals to 2.5 cm long, pink, lavender or purple;
capsules to 6 mm long. Coastal marshes. Infreq.
all Fla. W to Tex., N to Canada. Sum-fall.

Agalinis obtusifolia Raf.

(not shown)

Annual, 30-80 cm tall; stems parallel-grooved;
leaves 1-1.5 cm long, mostly opposite; petals
1-1.5 cm long, purple to pink; capsules to
3 mm long. Dry to wet sites. Infreq. all Fla.
W to Miss., N to Pa. Sum-fall.

PINELANDS GERARDIA
Agalinis pinetorum Pennell

Annual, to 70 cm tall; stems smooth; branches
few, upright; leaves 2-2.5 cm long, to 1 mm
wide, with short, rough-textured hairs; flower
stems 1-3 mm long; petals to 2 cm long, pink
or purple; capsules to 5 mm long. Wet sites.
Infreq. WF. W to La., E to Ga. Sum-fall.

(not shown)

SMOOTH GERARDIA
Agalinis purpurea (L.) Pennell

Annual, 0.4-1.2 m tall; stems smooth; branches
numerous, spreading; leaves 2-4 cm long, to 2 mm
wide, with short, rough-textured hairs; flower
stems to 6 mm long; petals 2-4 cm long, pink or
purple; capsules to 6 mm long. Moist sites. Freq.
all Fla. W to Tex., N to Mass. Sum-fall.

FALSE FOXGLOVE
Agalinis setacea (J. F. Gmel.) Raf.

Annual, to 80 cm tall; leaves to 3.5 cm long; petals
to 3 cm long, purple; capsules to 3.5 mm long.
Dry sites. Infreq. CF, NF, WF. N to N.Y. Fall.

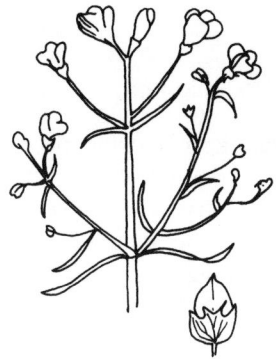

SLENDER AGALINIS
Agalinis tenuifolia (Vahl) Raf.

Annual, to 80 cm tall; leaves 2-5 cm long; petals
to 1.2 cm long, purple; capsules to 7 mm long.
Deciduous woods. Infreq. CF, NF, WF. W to La.,
N to Maine. Sum-fall.

YELLOW FOXGLOVE
Aureolaria flava (L.) Farwell

Perennial, 1.5-2.5 m tall; stems smooth, covered
with a white film; leaves to 14 cm long, entire to
lobed; petals 3.5-6 cm long, yellow; capsules to
2 cm long. Dry wooded sites. Infreq. CF, NF,
WF. W to Tex., N to Maine. Sum-fall.

HAIRY FOXGLOVE
Aureolaria pectinata (Nutt.) Penn.

Annual, 0.4-1.1 m tall; stems covered with glandular hairs; leaves to 4 cm long, narrowly lobed
like teeth in a comb (pectinate), with glandular
hairs; petals to 5 cm long, yellow; capsules to
1.2 cm long. Dry wooded sites. Infreq. all Fla.
W to Tex., N to Va. Spr-fall.

DOWNY FOXGLOVE
Aureolaria virginica (L.) Penn.

Perennial, 0.8-1.2 m tall; stems hairy; leaves to
15 cm long, downy, entire to lobed; petals 3-5 cm
long, yellow; capsules to 1.5 cm long. Dry wooded sites. Infreq. NF, WF. W to Ala., N to Mass.
Spr-fall.

BLUE HYSSOP
Bacopa caroliniana (Walt.) Robins.

Perennial, to 1 m long; stems creeping, floating,
covered with hairs; erect portion of stem 10-30 cm
tall; leaves 0.5-2 cm long, having lemon odor
when crushed; petals 9-11 mm long, blue to
violet; capsules 4-5 mm long. Moist to wet sites.
Freq. all Fla. W to Tex., E to Va. Spr-fall, all yr S.

ROUND-LEAF WATER-HYSSOP
Bacopa innominata (Gomez Maza) Alain

Perennial, 20-30 cm tall; stems hairy; leaves
5-15 mm long, rounded-ovate, palmately 3- to
5-veined; petals 5, 3-4 mm long, white; stamens
2; capsules to 2.3 mm long. Muddy ditches,
rivers and streams. Infreq. CF, NF, WF. Rare N
to N.C. Sum-fall.

(not shown)

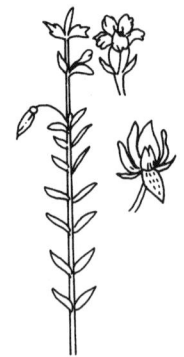

SMOOTH WATER-HYSSOP
Bacopa monnieri (L.) Penn.

Perennial, 10-30 cm tall; stems smooth, succulent,
lying flat to ascending; leaves 5-17 mm long,
oblanceolate or spatulate, 1-veined; leaf margins
entire; flowers solitary, axillary; sepals 5; petals
7-10 mm long, white or pink; stamens 4; capsules
4-5 mm long. Wet sites. Freq. all Fla. W to Tex.,
E to Va. Spr-fall, all yr S.

COMMON BLUE HEARTS
Buchnera americana L.
[*B. floridana* Gand.]

Perennial, 30-80 cm tall; larger leaves 2.5-8 cm
long; floral tube 6-12 mm long, purple or white;
lobes 2-9 mm long; capsules to 6 mm long.
Flatwoods, open disturbed areas. Common all
Fla. W to Tex., E to N.C., N to Canada. Spr-fall.

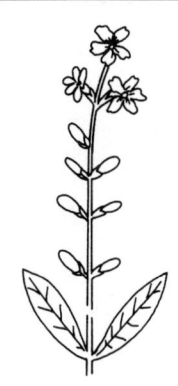

INDIAN PAINT-BRUSH
Castilleja coccinea Engelm.

Annual, 20-45 cm tall; stems hairy; leaves 3- to
5-parted, with segments to 4 cm long; bracts
scarlet; petals 1.8-2.5 cm long, yellowish green;
capsules to 1 cm long. Moist to wet sites. Rare
CF. W to Okla., N to N.H. Spr-sum.

GOLDEN HEDGE-HYSSOP
Gratiola aurea Muhl. ex Pursh

Perennial, 10-40 cm tall; leaves 1-2.5 cm long;
petals to 1.5 cm long, yellow; capsules to 3 mm
long. Wet barrens. Rare WF. N to Canada.
Spr-fall.

MATTED CREEPING HEDGE-HYSSOP
Gratiola brevifolia Raf.

Perennial, 20-40 cm tall; leaves 1-2 cm long;
*petals to 1.2 cm long, white; *capsules to 2 mm
long. Low wet sites. Infreq. WF. W to Tex.,
N to Tenn. Spr-sum.

SCRUB HEDGE-HYSSOP
Gratiola hispida (Benth.) Pollard
[*Sophroranthe hispida* Benth.]

Perennial, 5-20 cm tall; stems with stiff hairs; leaves 8-15 mm long, linear, smooth; sepals 3-6 mm long; flowers sessile; petals salverform; tubes 10-13 mm long, yellow; petal lobes white; stamens 2; capsules 4-5 mm long. Dry to moist sites. Freq. all Fla. W to Miss., N to Ga. Sum-fall.

HAIRY HEDGE-HYSSOP
Gratiola pilosa Michx.
[*Tragiola pilosa* (Michx.) Small & Penn.]

Perennial, 20-60 cm tall; stems soft; leaves 1-2 cm long, ovate to lanceolate, pebbly above, glandular-punctate below; sepals 5-7 mm long; flowers ses-sile or subsessile; petals tubular at base; floral tubes 5-9 mm long, yellow; petal lobes white; stamens 2; capsules 4-5 mm long. Dry to wet sites. Freq. all Fla. W to Tex., E to Tenn. and N.J. Sum-fall.

CREEPING HEDGE-HYSSOP
Gratiola ramosa Walt.

Perennial, 10-30 cm tall; stems and leaves with glandular hairs; leaf blades 0.7-2 cm long, with few teeth on top half, clasping, lance-subulate; sepals 3-6 mm long; flowers stalked; stalks 6-17 mm long; floral tubes yellow; petal lobes 1-1.4 cm long, white; stamens 2; capsules 1-2 mm long. Moist to wet sites. Freq. all Fla. W to Tex., E to N.C. Spr-sum.

CLAMMY HEDGE-HYSSOP
Gratiola virginiana L.

Annual, 10-40 cm tall; stems smooth; leaves 2-5 cm long, elliptic-lanceolate to ovate, glandular-punctate; sepals 4-6 mm long; flowers stalked; stalks 1-5 or 12 mm long; floral tubes yellow; petal lobes 9- 14 mm long, white; capsules 4-7 mm in diam. Wet, shady sites. Infreq. CF, NF, WF. W to Tex., N to Ohio, E to N.J. Spr-sum.

OLD FIELD TOADFLAX
Linaria canadensis (L.) Dum.

Biennial or annual, 15-70 cm tall; many stems
from base; leaves 0.3-2 cm long; flowers 6-10 mm
long, blue to violet; spur 2-6 mm long, curved;
flower stalks 1-5 mm long; capsules 2-3 mm long.
Dry disturbed areas. Common all Fla. W to Tex.,
N to Canada. Spr.

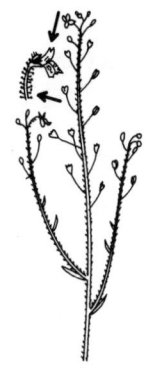

FLORIDA TOADFLAX
Linaria floridana Chapm.

Biennial or annual, 10-40 cm tall; leaves 1-2.5 cm
long; *flowers 5-6 mm long, blue-violet; spur 0.5
mm long, tapering; *flower stalks 5-10 mm long,
with glandular hairs; capsules 2-3.5 mm long.
Dry sites. Freq. all Fla. W to Miss., N to Ga. Spr.

TEXAS TOADFLAX
Linaria texana Scheele

Biennial or annual, to 70 cm tall; many stems
from base; leaves 1.5-3 cm long; flowers 10-
12 mm long, violet; spur 5-9 mm long; flower
stalks 5-10 mm long; capsules 2.5-3.5 mm long.
Old fields, pastures, roadsides. Infreq. CF, NF.
W to Calif. Wint-spr.

(not shown)

VARIABLE FALSE PIMPERNEL
Lindernia anagallidea (Michx.) Penn.

Annual, 5-20 cm tall; stems erect, smooth, 4-
angled; leaves 5-20 mm long, usually longer than
wide; flowers to 4 mm long, solitary, lavender
to white; flower stalks longer than subtending
leaves; *capsules to 3 mm long. Open wet sites.
Infreq. all Fla. W to Tex., N to N.D., E to N.H.
Sum-fall.

PURPLE FALSE PIMPERNEL
Lindernia crustacea (L.) F. Muell.

Annual; stems branching, lying flat, purplish; floral
stems 5-20 cm long, ascending; leaves 0.8-2 cm
long, ovate to elliptic; flowers 6-7 mm long, blue,
violet or white; stalks 1.5-2.5 cm long, longer than
leaves; *capsules 2-4 mm long. Moist sites. Freq.
all Fla. W to La., N to S.C. Fall, all yr S.

SHORT-STALKED FALSE PIMPERNEL
Lindernia dubia (L.) Penn.

(not shown)

Annual, 10-25 cm tall; stems ascending, branching; leaves 1-3 cm long, oblanceolate to elliptic;
flowers 7-10 mm long, lavender to white; flower
stalks 1-2.5 cm long, shorter than leaves; capsules 4-6 mm long. Wet sites. Infreq. NF, WF.
W to Tex., N to Canada. Sum-fall.

ROUND-LEAVED FALSE PIMPERNEL
Lindernia grandiflora Nutt.

Annual, to 40 cm tall; stems prostrate, creeping,
matting; branches ascending to erect; leaves 0.5-
1.5 cm long, round; flowers 8-10 mm long, blue
to purple, with mottled white lobes; flower stalks
15-40 mm long; capsules 4-6 mm long; seeds
winged. Moist to wet sites. Common all Fla.
N to Ga. Spr-fall, all yr S.

ERECT LINDERNIA
Lindernia monticola Muhl. ex Nutt.

Perennial, 10-30 cm tall; basal leaves 2-3 cm
long, to 1 cm wide, glandular; flower stems to
4 cm long; petals to 1 cm long, blue to purple;
capsules to 4 mm long. Moist sites. Rare CF, NF.
N to N.C. Spr-fall.

WHITE-FLOWERED MECARDONIA
Mecardonia acuminata (Walt.) Small

Perennial, 10-60 cm tall; stems erect to prostrate;
leaves 1-5 cm long; flowers 6-11 mm long,
white with purple veins; pedicels 7-30 mm long;
stamens 4; capsules 4-6 mm long. Moist sites,
partial shade. Freq. all Fla. W to Tex., N to Mo.,
E to Del. Spr-fall.

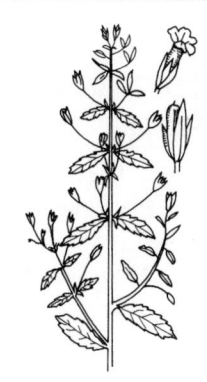

HEMIANTHUS
Micranthemum glomeratum (Chapm.) Shinners

Very similar to *Micranthemum umbrosum*
except: stems 2-5 (20 when submersed) cm long;
leaves 2-4 or 10 mm long, oblanceolate; petals
to 2 mm long, 1-lipped, 3-lobed, white to pink.
Wet sites. Infreq., endemic to peninsula Fla.
Spr-fall, all yr S.

GLOBIFERA
Micranthemum umbrosum (J. F. Gmel.) Blake
[*M. orbiculatum* Michx.]

Annual, 5-50 cm long; stems smooth, creeping,
branching; leaves 3-11 mm long, orbicular; leaf
margins entire; flowers 1-2 mm long, axillary,
solitary; sepals 4, to 1.5 mm long; petals 4, white
to purple; stamens 2; capsules to 1 mm long;
seeds numerous. Low, wet sites. Freq. all Fla.
W to Tex., E to Va. Spr-fall.

MONKEY FLOWER
Mimulus alatus Ait.

Perennial, 0.4-1.2 m tall; stems winged, smooth;
leaves 5-15 cm long, with margined petioles;
flowers 2.5-3.5 cm long, blue-lavender, pink
or white; flower stalks to 1 cm long; *capsules
to 1 cm long. Wet sites. Rare WF. W to Tex.,
N to Canada. Sum-fall.

CLASPING LEAF MONKEY FLOWER
Mimulus ringens L.

Perennial, 0.5-1.2 m tall; stems not winged; leaves 5-12 cm long, clasping (sessile); flowers 2.5-3 cm long, violet; flower stems 3-6 cm long; capsules 1-1.2 cm long. Wet areas, along rivers and streams. Ga., N Canada, W to Colo. Sum-fall.

(not shown)

SLENDER BEARD TONGUE
Penstemon australis Small

Perennial, 20-80 cm tall; stems covered with glandular hairs; leaves 5-14 cm long; petals 2-2.5 cm long, red to purple; capsules to 9 mm long. Dry sites. Freq. CF, NF, WF. N to Ala., E to Va. Spr.

BEARD TONGUE
Penstemon multiflorus Chapm.

Perennial, 0.8-1.5 m tall; leaves to 20 cm long; petals to 2.2 cm long, white; capsules to 8 mm long. Dry sites. Freq. all Fla. N to Ga. Spr-sum.

FIRECRACKER PLANT
Russelia equisetiformis Schlecht. & Cham.

Perennial, erect, arching; branches whorled; leaves 1-2 cm long, whorled; sepals 4; petals 2-2.5 cm long, red; *capsules to 6 mm long. Disturbed sites. Infreq. SF, CF. Spr-sum.

GOATWEED
Scoparia dulcis L.

Perennial, 30-80 cm tall; stems smooth; leaves
1-4 cm long, glandular-punctate; leaf margins
toothed; sepals 4; petals 4, 7-10 mm wide, white;
stamens 4; *capsules 1-2.5 mm long. Disturbed
dry to wet sites. Freq. all Fla. W to La., E to S.C.
Spr-fall, all yr S.

SENNA SEYMERIA
Seymeria cassioides (Walt.) Blake

Annual parasite with chlorophyll, 0.5-1 m tall;
stems slightly hairy; leaves 5-15 mm long, opposite, once or twice finely pinnately divided; flowers to 1 cm long, yellow; sepals and petals 5-
lobed, smooth; capsules to 4 mm long, smooth;
seeds wingless. Dry to wet sites on pine roots.
Infreq. CF, NF, WF. W to Tex., N to N.C. Fall.

STICKY SEYMERIA
Seymeria pectinata Pursh

Annual parasite with chlorophyll, 20-60 cm tall;
stems slightly hairy; leaves 15-30 mm long,
opposite, pinnately divided; flowers to 1 cm long,
yellow; sepals and petals 5-lobed, hairy; capsules
5-7 mm long, with glandular hairs; seeds winged.
Dry sites on pine roots. Freq. all Fla. W to Miss.,
N to N.C. Sum-fall.

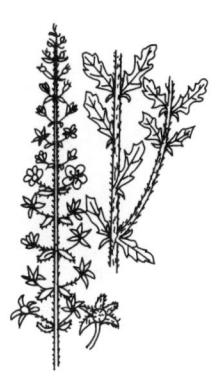

MOTH MULLEIN
Verbascum blattaria L.

Similar to *Verbascum thapsus* except: 0.4-1.2 m
tall; stems smooth or with glandular hairs; leaves
2-24 cm long, smooth; flowers 25-35 mm wide,
yellow or white; capsules 6-8 mm long.
Roadsides, disturbed areas. Infreq. CF, NF, WF.
W to Tex., N to Canada. Spr-fall.

WOOLLY MULLEIN
Verbascum thapsus L.

Perennial or biennial, 0.3-1.8 m tall; stems woolly; leaves 5-40 cm long, covered with star-shaped
hairs; flowers 15-22 mm wide, yellow; capsules
to 8 mm long. Roadsides, disturbed areas.
Infreq. SF, CF, NF. W to Calif., N to Canada.
Sum-fall.

PURPLE STAMEN MULLEIN
Verbascum virgatum Stokes

Perennial or biennial, 0.6-1.2 m tall; stems with
glandular hairs; leaves 7-25 cm long; flowers to
2.5 cm wide, yellow; pedicels 3-5 mm long;
sepals and petals 5-lobed; capsules 7-8 mm long.
Roadsides, old fields, disturbed areas. Infreq. CF,
NF, WF. W to Tex., N to N.C. Spr-sum.

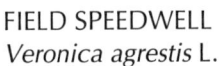

FIELD SPEEDWELL
Veronica agrestis L.

Annual, 5-20 cm long; stems spreading or erect,
branching, hairy; leaves 4-12 mm long; *flowers
to 3 mm wide, violet or blue; *capsules to 4 mm
long. Lawns, fields. Rare WF. E to Va. Spr.

CORN SPEEDWELL
Veronica arvensis L.

Annual, 10-30 cm tall; stems erect or lying flat,
hairy; leaves to 1 cm long; petals to 2.5 mm
wide, blue; capsules to 4 mm long. Native to
Europe. Dry or cultivated sites. Infreq. CF, NF,
WF. W to Okla., N to Canada. Wint.-spr.

PURSLANE SPEEDWELL
Veronica peregrina L.

Annual, 10-30 or 40 cm tall; stems smooth;
leaves 10-26 mm long, oblanceolate to linear,
smooth or with a few hairs; flowers axillary,
stalked; stalks to 2 mm long; sepals 4; petals
2-2.5 mm wide, white; stamens 2; capsules
2-4 mm long. Open disturbed areas. Infreq.
all Fla. W to Tex., N to Canada. Spr.

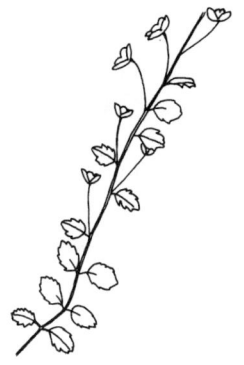

PERSIAN SPEEDWELL
Veronica persica Poir.

Annual, 10-30 cm tall; stems lying flat, hairy;
leaves to 3 cm long; petals to 5 mm long, blue;
capsules to 5 mm long. Native to Europe.
Disturbed sites. Rare CF, NF, WF. W to Calif.,
N to Canada. Spr-sum.

NIGHTSHADE FAMILY *SOLANACEAE*

Herbs, shrubs or trees; leaves alternate; flowers bisexual; sepals 5 or 3-10; petals 5; ovary superior; fruit a berry or capsule.

BIRD PEPPER
Capsicum annuum L.

Annual, 0.2-1 m tall; larger leaves 7-12 cm long;
petals to 8.5 mm long, white to pink; berries to
1 cm long, red. Coastal hammocks. Rare SF, CF,
NF. N to N.C. All yr S.

TABASCO PEPPER
Capsicum frutescens L.

Perennial, to 2 m tall; leaves 1-10 cm long; petals
to 4 mm long, white; berries to 2.5 cm long, red.
Hammocks. Infreq. SF, CF, NF. Fall.

(not shown)

ANGEL TRUMPET
Datura innoxia Mill.

(not shown)

Annual, 2-3 m tall; leaves to 20 cm long; flowers
10-20 cm long, white, fragrant; capsules to 3 cm
long, nodding, subglobose. Disturbed sites. Rare
CF, NF, WF. N to R.I. Sum-fall.

JIMSON WEED
Datura stramonium L.

Annual, 0.5-1.5 m tall, with rank odor; leaves
7-20 cm long; flowers 6-10 cm long, white,
lavender or violet; capsules 2.5-6 cm long, erect.
Escapes from cultivation. Disturbed sites. Freq.
all Fla. Widespread in U.S. Spr-fall.

CHRISTMAS BERRY
Lycium carolinianum Walt.

Perennial shrub, 0.3-3 m tall; branches recurved,
thorny; leaves 0.5-2.5 cm long, succulent; flowers blue, lavender or white, axillary; berries
8-15 mm long, red. Coastal sites. Freq. all Fla.
W to Tex., N to Ga. Spr-fall, all yr S.

CUT LEAF GROUND CHERRY
Physalis angulata L.

Annual, 0.3-1 m tall; stems smooth, diffuse; leaves 4-13 cm long, with several sharp teeth on margins; flowers 5-10 mm wide, yellow; anthers blue; sepals 1.5-3 cm long, papery, with purple veins; berries 8-11 mm in diam. Well-drained disturbed areas. Freq. all Fla. W to Tex., E to Pa. and Va. Sum, all yr S.

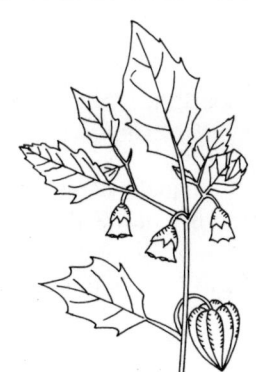

PUBESCENT GROUND CHERRY
Physalis arenicola Kearney

(not shown)

Perennial; stems diffuse, with simple and sparse hairs; leaves 1.5-8 cm long, with several rounded teeth on margins; flowers 15-20 mm wide, yellow; anthers yellow; sepals to 3 cm long, papery, reticulate; berries to 9 mm in diam. Pinelands, sandhills. Common all Fla. Sum.

DOWNY GROUND CHERRY
Physalis pubescens L.

Annual, 15-60 cm tall; stems smooth or covered with white shaggy hairs; leaves 3-9 cm long, with toothed or entire margins; flowers 5-10 mm wide, yellow, with 5-spotted center; anthers blue; sepals 2-4 cm long, papery, 5-angled; berries 5-8 mm in diam. Moist to dry woods. Infreq. CF, NF, WF. W to Calif., E to Pa. and Va. Sum.

(not shown)

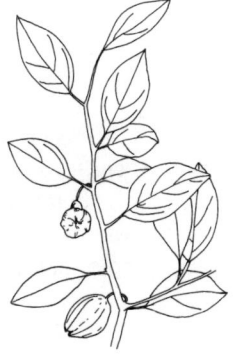

STICKY GROUND CHERRY
Physalis viscosa L.

Perennial, 20-60 cm tall; stems creeping, with star-shaped hairs; leaves 3-10 cm long; leaf margins entire; flowers 15-20 mm wide, green to yellow, with dark center; anthers yellow; sepals 2-4 cm long, papery, enclosing berry; berries 7-10 mm in diam. Sandhills, dry pinelands, coastal sites. Common all Fla. W to N.Mex., E to Va. Spr-fall.

AMERICAN BLACK NIGHTSHADE
Solanum americanum Mill.

Annual, to 1 m tall; stems smooth; leaves 2-10 cm
long; petals 6-9 mm wide, white; berries 0.5-1 cm
in diam., black. Various dry disturbed sites. Freq.
all Fla. W to Tex., N to Maine. Sum-fall.

HORSE-NETTLE
Solanum carolinense L.

Perennial, 20-80 cm tall; stems spiny; leaves
5-12 cm long, with star-shaped hairs on underside; flowers 5-lobed, 2-3 cm wide, white, yellow
or lavender; berries 10-15 mm in diam., orange-
yellow, poisonous. Disturbed sites. Freq. CF,
NF, WF. W to Tex., N to Canada. Spr-fall.

WHITE HORSE-NETTLE
Solanum elaeagnifolium Cav.

Perennial, 0.3-1.1 m tall; stems with silver hairs
and prickles; leaves 5-15 cm long, with silver
hairs on lower side; petals to 2.5 cm wide, violet
to white; berries to 1.2 cm in diam., yellow to
black. Native to Mexico. Disturbed sites. Infreq.
all Fla. W to Ariz., N to Kans. Spr-fall.

POTATO TREE
Solanum erianthum D. Don

Perennial shrub, 0.4-3.6 m tall; stems covered
with velvety hairs; leaves 8-30 cm long, with
*velvety hairs on upper surfaces; *petals 1.5-2 cm
wide, white; berries 1-2 cm in diam., yellow.
Hammocks, disturbed sites. Infreq. SF, CF, NF.
All yr.

LARGE-FLOWERED BLACK NIGHTSHADE
Solanum pseudogracile Heiser

Annual, to 70 cm long; stems trailing, climbing; leaves to 11 cm long; sepals to 1.5 m long, persistent; petals 5, white to purple; berries to 1 cm in diam., black. Disturbed sites. Infreq. CF, NF, WF. N to N.C. Spr-fall.

EASTERN BLACK NIGHTSHADE
Solanum ptycanthium Dun.

(not shown)

Annual, to 1 m tall; leaves to 12 cm long; petals to 7 mm wide, white; berries to 1.5 cm in diam., purple to black. Disturbed dry sites. Freq. all Fla. Throughout eastern North America into Quebec. All yr S, warm months N.

BUFFALO BUR
Solanum rostratum Dunal

Annual, 20-70 cm tall; stems, leaf veins, sepals and *fruits covered with long, thick golden spines to 1 cm long; leaves 5-22 cm long; sepals 5, persistent; petals 5, yellow; stamens 5; *berries to 1.5 cm long. Cultivated, escapes. Disturbed areas. Rare NF. W to N.Mex., N to Canada. Spr-fall.

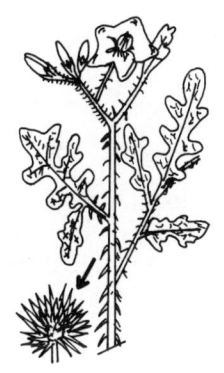

BRAZILIAN NIGHTSHADE
Solanum seaforthianum Andrews

Perennial vine, to 3 m long; some leaves to 13 cm long, pinnately divided; flowers to 2 cm wide, blue or lavender; berries to 1 cm wide, red. Native to West Indies. Escapes from cultivation. Disturbed sites. Infreq. SF, CF. All yr.

STICKY NIGHTSHADE
Solanum sisymbriifolium Lam.

Annual, 30-70 cm tall; stems covered with shaggy hairs and yellow spines; leaves 6-22 cm long;
leaf midribs covered with yellow spines; petals
to 3 cm wide, white to blue; *berries to 2 cm in
diam., red, covered with yellow spines.
Disturbed sites. Infreq. CF, NF, WF. W to La.,
E to Ga. Spr-fall.

CHOCOLATE FAMILY   *STERCULIACEAE*

Herbs, shrubs or trees; leaves alternate; stipules present; flowers
bisexual or unisexual; sepals 3-5; petals 5 or 0; ovary superior;
fruit a capsule.

RED WEED
Melochia corchorifolia L.

Annual, to 1.5 m tall; stems slightly hairy; leaves
to 7.5 cm long; flowers stalked at tips of branches; petals to 7 mm long, white, pink or purple;
*capsules to 5 mm in diam. Dry to wet sites.
Infreq. all Fla. W to Tex., N to N.C. Spr-fall.

HAIRY MELOCHIA
Melochia villosa (Mill.) Fawc. & Rendle

Annual, to 1.5 m long; stems lying flat to ascending, covered with long hairs; leaves to 5 cm long,
hairy on both surfaces; flowers sessile, in leaf
axils or at tips of branches; petals to 1 cm long,
violet; *capsules to 3 mm in diam. Disturbed
sites. Infreq. SF, CF. All yr.

LINDEN FAMILY   *TILIACEAE*

Trees, shrubs or herbs; leaves mostly alternate; stipules present; flowers mostly bisexual; sepals 5; petals 5 or 0; ovary superior; fruit nut-like or a capsule.

JUTE
Corchorus orinocensis HBK.

Perennial shrub, to 60 cm tall; leaves 2-8 cm long; *petals to 6 mm long, yellow; *capsules to 7 cm long. Disturbed sites. Rare SF. Ala., La., Tex., Ariz. All yr.

TURNERACEAE FAMILY   *TURNERACEAE*

Herbs, shrubs or trees; leaves alternate; flowers bisexual; sepals 5; petals 5; ovary superior; fruit a capsule.

PIRIQUETA
Piriqueta caroliniana (Walt.) Urban
[*P. tomentosa sensu* Small]

Perennial, 15-50 cm tall; stems and leaves with thick brown star-shaped hairs; leaves 1-7 cm long; flowers 15-22 mm wide, yellow; capsules 5-7 mm high, globose. Dry to moist sites. Freq. all Fla. N to S.C. Spr-fall, all yr S.

SMOOTH STEM PIRIQUETA
Piriqueta caroliniana (Walt.) Urban
var. *glabra* (DC.) Urban

Similar to *Piriqueta caroliniana* except: stems
and leaves smooth; leaves 1-8 cm long, 2-5 mm
wide, linear. Moist, low sites. Infreq. SF, CF, NF.
Spr-fall, all yr S.

CAT-TAIL FAMILY · *TYPHACEAE*

Aquatic herbs; leaves linear; flowers unisexual (monoecious);
flowers in terminal spikes with male flowers above female
flowers; floral parts present only as scales or threads; fruit an
achene.

NARROW LEAF CAT-TAIL
Typha angustifolia L.

Perennial, 1-1.5 m tall; white pith at base of
stem; leaves to 6 mm wide, convex on back at
tip; flower spikes to 20 cm long, to 2 cm in
diam.; male and female flowers separated on
spike. Wet sites. Throughout U.S. except for Fla.
Sum-fall.

(not shown)

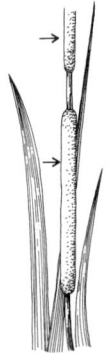

SOUTHERN CAT-TAIL
Typha domingensis Pers.

Perennial, 2-3 m tall; white pith at base of stem;
leaves 6-12 mm wide, convex on back at tip;
flower spikes to 45 cm long, to 2.2 cm in diam.;
*male and *female flowers separated on spike.
Marshes, shores, ditches. Common all Fla.
N to Del., W to Calif. Sum.

BLUE CAT-TAIL
Typha x *glauca* Godron
[*T. latifolia* L. x *T. angustifolia* L.]

A hybrid of *Typha latifolia* and *Typha angustifolia*.
Perennial, to 1 m tall; yellow pith at base of stem;
leaves to 12 mm wide, convex on back at tip;
flower spikes to 25 cm long, to 2.5 cm in diam.;
male and female flowers separated on spike.
Wet sites. Found in U.S. where both parents are
growing near each other. Sum-fall.

(not shown)

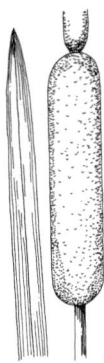

COMMON CAT-TAIL
Typha latifolia L.

Perennial, 1-2.5 m tall; white pith at base of stem;
leaves 8-15 mm wide, flat on back at tip; flower
spikes to 18 cm long, to 3 cm in diam.; male and
female flowers close together on spike. Shores,
wet sites. Freq. all Fla. All U.S. Sum-fall.

CARROT FAMILY   *UMBELLIFERAE*

Herbs or shrubs; leaves alternate, sometimes young leaves are
not true leaves, but flattened petioles resembling leaves; flowers bisexual or unisexual; flowers in flat-topped clusters;
sepals 5 or 0; petals 5; ovary inferior; fruit dry, splitting into
two parts.

BISHOP'S-WEED
Ammi majus L.

Annual, 0.3-1.5 m tall; stems smooth; upper
leaves to 20 cm long, finely divided; flowers 20-
30 per cluster, white; fruits to 1.8 mm long.
Native to Europe. Disturbed sites. Rare CF, NF,
WF. W to Tex., E to S.C. Spr-sum.

ANGELICA
Angelica dentata (Chapm.) Coult. & Rose

Perennial, to 1 m tall; stems smooth; leaves finely
divided, coarsely toothed; flowers 5-12 per cluster, white; fruits smooth. Dry scrub. Infreq. WF.
E to Ga. Sum-fall.

HAIRY ANGELICA
Angelica venenosa (Green.) Fern.

Perennial, 0.5-1 m tall; stems hairy near top;
leaves divided into leaflets; leaf margins finely
toothed; flowers 16-25 per cluster, white; *fruits
with stiff hairs. Dry, woody sites. Infreq. WF.
W to Okla., N to Mich., E to Mass. Sum.

MARSH PARSLEY
Apium leptophyllum (Pers.) F. Muell.

Annual, 70-80 cm tall; *lower leaves dissected,
with very narrow divisions; flower clusters stalkless or short-stalked; lower 1-3 clusters opposite
leaves; *fruits 1-2 mm long, smooth, slightly
bumpy, with very narrow ribs. Moist disturbed
sites. Occasional all Fla. W to Okla., N to Mo.,
E to N.Y. Spr-sum.

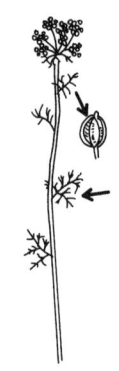

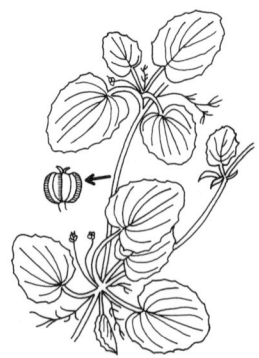

COINWORT
Centella asiatica (L.) Urban

Perennial; stems slightly hairy, prostrate, creeping; leaves 1.5-5 cm wide, clustered at nodes;
leaf blades round to heart-shaped or oblong;
flower clusters simple; sepals 0; petals white to
green, deciduous; *fruits 3-5 mm wide, ribbed.
Moist to wet sites. Common all Fla. W to Tex.,
E to Del. Spr-fall, all yr S.

PROCUMBENT WILD CHERVIL
Chaerophyllum procumbens (L.) Crantz

Annual, to 50 cm long; stems lying flat, spreading, smooth; leaves divided; flower clusters multiple, almost stalkless; mature bracts spreading; flowers minute; fruits to 8 mm long, beakless. Moist to wet woody sites. Infreq. WF. W to Okla., N to Mich., E to N.Y. Spr.

(not shown)

WILD CHERVIL
Chaerophyllum tainturieri Hook.

Annual, 20-70 or 90 cm tall; stems erect, hairy; leaves 2-12 cm long, divided; flower clusters multiple, almost stalkless; mature bracts bent down; flowers minute; *fruits 5-8 mm long, beaked. Disturbed sites. Infreq. CF, NF, WF. W to Tex., N to Kans., E to Va. Spr.

WATER HEMLOCK
Cicuta mexicana Coult. & Rose

Perennial, 1-2.5 m tall; stems purple-striped; leaves divided into 2 or 3 leaflets, each to 10 cm long, to 3.5 cm wide; flowers in clusters to 16 cm in diam.; petals white; fruits to 3 mm long. Low, wet sites. Freq. all Fla. N to Va. Sum-fall.

WILD CARROT
Daucus carota L.

Biennial, 0.4-1.2 m tall; stems hairy; leaves 5-20 cm long, finely divided; flowers few to several per cluster, white or pink; *fruits 2-4 mm long, with winged ribs; each wing having to 12 prickles. Disturbed sites. Infreq. NF, WF. N to Canada. Spr-fall.

SMALL QUEEN ANNE'S-LACE
Daucus pusillus Michx.

Annual, 5-80 cm tall; stems hairy; leaves 5-12 cm long, finely divided; flowers few to numerous per cluster, white; fruits 3-4 mm long; each wing of fruit having 1-8 *prickles. Disturbed sites. Infreq. CF, NF, WF. W to Calif., N to Canada. Spr-fall.

CORN SNAKEROOT
Eryngium aquaticum L.

Biennial, 0.3-2 m tall; lower leaves 15-40 cm long, falling prematurely; involucral bracts often bent down, blue; floral bracts rigid; petals to 1 mm long, blue; heads 1-1.5 cm long; fruits 1.5-2 mm long. Low sites. Infreq. CF, NF, WF. W to Ala., N to N.J. Spr-fall.

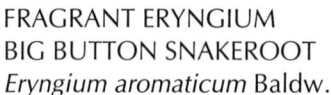

FRAGRANT ERYNGIUM
BIG BUTTON SNAKEROOT
Eryngium aromaticum Baldw.

Biennial or perennial, 10-60 cm long; stems spreading or lying flat; lower leaves 3-4 cm long, deeply lobed, spine-tipped; heads 8-10 mm long, bristly; fruits to 2 mm long, granular. Dry sites. Freq. all Fla. N to Ga. Spr-fall.

MATTED BUTTON SNAKEROOT
Eryngium baldwinii Spreng.

Biennial or perennial, prostrate, with creeping stems; leaves 2-9 cm long; some basal leaves lobed or divided; *heads to 5 mm long, usually blue or occasionally white; bractlets exserted; involucre little longer than radius of base; *fruits to 2 mm wide. Wet sites. Common all Fla. W to La., E to Ga. Spr-fall.

SAND PINE ERYNGIUM
Eryngium cuneifolium Small

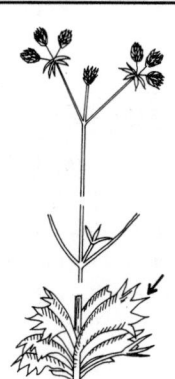

Perennial, 20-50 cm tall; leaves *spine-tipped;
basal leaves long-stalked; petals 1.5-2 mm long,
blue, white, greenish; heads 5-8 mm in diam.;
fruits to 1.5 mm long. Scrub. Rare Highlands
County area, CF. Sum-fall.

LANCE-LEAF BUTTON SNAKEROOT
Eryngium integrifolium Walt.

Perennial, 40-90 cm tall; stems erect; leaves 2-
7 cm long, entire to toothed; heads to 1 cm long;
bracts to 1.5 cm long; petals blue; fruits to 2 mm
long. Wet sites. Infreq. NF, WF. W to Okla.,
N to N.C. Sum-fall.

CREEPING ERYNGIUM
Eryngium prostratum Nutt.

Perennial, to 70 cm long; stems prostrate; leaves
to 9 cm long, entire to palmately lobed; *heads
5-7 mm long; bracts to 5 mm long, well extending past heads; petals blue; fruits to 2 mm long.
Wet sites. Common NF, WF. W to Tex., E to Va.
Spr-sum.

RATTLESNAKE MASTER
Eryngium yuccifolium Michx.

Perennial, 0.3-1.6 m tall; lower leaves to 1 m
long, 1-4 cm wide, with *bristly margins and
parallel veins; heads 8-25 mm long, somewhat
bristly, white, greenish; fruits to 2.5 mm wide.
Open woods, dry pinelands. Freq. all Fla.
W to Tex., N to Minn., E to N.J. Spr-fall.

FENNEL
Foeniculum vulgare Mill.

Biennial or perennial, 0.5-1.5 m tall; clasping leaf
sheaths 3-10 cm long; leaf segments 1-3 cm long,
to 1 mm wide; *flowers yellow; *fruits 3-4 mm
long. Native to Europe. Escapes from cultivation.
Disturbed sites. Rare SF. W to Tex., N to Mich.,
E to Conn. Sum-fall.

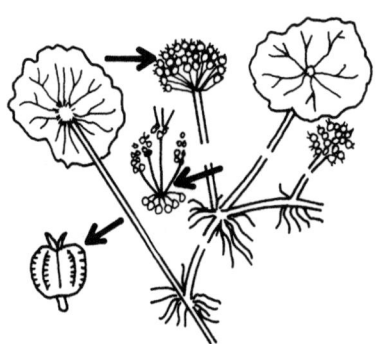

COASTAL PLAIN PENNYWORT
Hydrocotyle bonariensis Lam.

Perennial; stems smooth, creeping or floating;
leaf 1 per node, to 10 cm wide, rounded, petiole
attached near center of lower side of blade
(peltate); *flower clusters multiple, bearing off-
shoots; flower stalks longer than leaves; flowers
white; *fruits 3-4 mm wide, with corky ribs. Wet
sites. Freq. all Fla. W to Tex., E to N.C. Spr-fall.

FLOATING PENNYWORT
Hydrocotyle ranunculoides L. f.

Perennial, floating, creeping; leaves 2-5 cm wide,
with a deep split; petioles 0.05-1.3 m long; flowers 5-10 per stalk; *fruits to 3 mm in diam.
Shallow water. Freq. CF, NF, WF. W to Tex.
and Ark., E to Pa. and Del. Spr-sum.

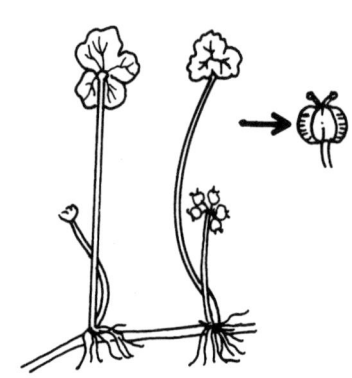

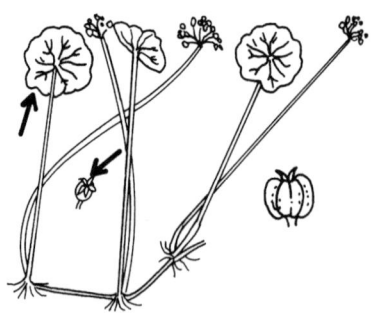

WATER PENNYWORT
Hydrocotyle umbellata L.

Perennial; stems smooth, creeping or floating;
leaf 1 per node, 2-7 cm wide, round; *petiole
attached near center of lower side of blade
(peltate); flower clusters simple; flower stalks
equal to or longer than leaves; *flowers white;
fruits 2-3 mm wide, with corky ribs. Moist to wet
sites. Common all Fla. W to Tex., N to Canada.
Sum, all yr S.

WHORLED PENNYWORT
Hydrocotyle verticillata Thunb.

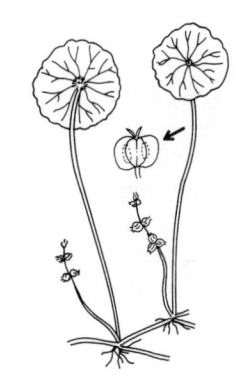

Perennial; stems smooth, creeping; leaf 1 per node, 3-6 cm wide, rounded to elliptic, petiole attached near center of lower side of blade (peltate); flowers of interrupted whorled spikes, white; flower stalks shorter than leaves; *fruits 3-4 mm wide, ribbed. Moist to wet sites. Freq. all Fla. W to Calif., N to Nev., E to Mass. Spr-sum.

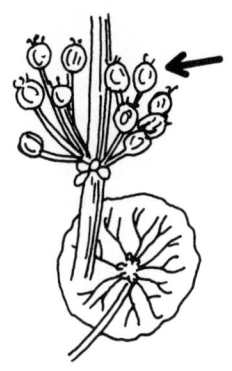

SWAMP PENNYWORT
Hydrocotyle verticillata Thunb. var. *triradiata* (A. Rich.) Fern.

Perennial; stems smooth, creeping; leaves rounded, petiole attached near center of lower side of blade (peltate); blades 1.5-5 cm wide; stalks to 12 cm long; flowers white, in whorls and spikes; stalks 1-10 mm long; *fruits 3-4 mm wide. Moist to wet sites. Infreq. all Fla. N to Mass., W to Nev. and Calif. Spr-fall.

CAROLINA FALSE-LILY
Lilaeopsis carolinensis Coult. & Rose

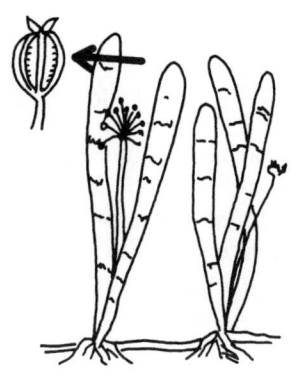

Perennial; stems smooth, creeping, matting, submerged; leaves 10-30 cm long, divided into 7-15 compartments; flower cluster stalks shorter than leaves; flowers 5-15 per cluster, white; *fruits to 3 mm in diam. Freshwater sites. Infreq. WF. W to La., E to Va. Sum.

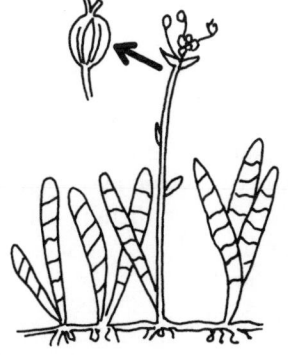

SALT MARSH FALSE-LILY
Lilaeopsis chinensis (L.) Kuntze

Perennial; stems smooth, forming thick mats; leaves 2-5 cm long, divided into 4-6 compartments; flower cluster stalks longer than leaves; flowers 4-10 per cluster, white; *fruits to 2 mm long. Saltwater sites. Infreq. CF, NF, WF. W to La., N to Canada. Spr-sum.

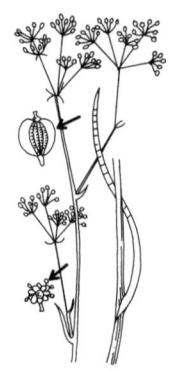

WATER DROPWORT
Oxypolis filiformis (Walt.) Britt.

Perennial, 0.5-1.8 m tall; leaves 3-60 cm long, hollow, round, slightly jointed; flowers 6-17 per cluster; flowers to 1.5 mm wide, white; *fruits 4-7 mm long, broadly winged. Open, wet sites. Freq. all Fla. W to Tex., N to N.C. Sum-fall, all yr S.

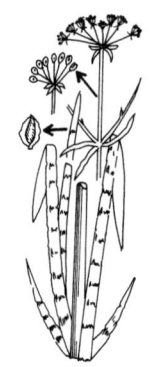

PURPLE DROPWORT
Oxypolis greenmanii Math. & Const.

Perennial, to 2 m tall; leaves chambered, hollow, almost round, mostly basal, much shorter upwards; flowers to 1.5 mm wide, maroon; *fruits to 9 mm long. Swamps, wet sites. Rare, endangered, locally abundant WF. Endemic. Sum.

PIG-POTATO
Oxypolis rigidior (L.) Raf.

Perennial, to 1.5 m tall; leaves to 30 cm long, compound with 5-11 leaflets; leaflets 6-12 cm long; flowers 12-25 per cluster; flowers to 3 mm wide, white; *fruits to 5 mm long, thinly winged. Moist to wet sites. Infreq. WF. W to Tex., N to Minn. and N.J. Sum-fall.

THREE LEAF DROPWORT
Oxypolis ternata (Nutt.) Heller

Perennial, 50-90 cm tall; leaves to 20 cm long, to 3 mm wide, simple or divided into 3 leaflets; flowers in clusters to 13 cm in diam.; petals white; fruits to 5 mm long. Low, wet sites. Rare WF. E to Va. Sum-fall.

MOCK BISHOP'S-WEED
Ptilimnium capillaceum (Michx.) Raf.

Annual, 10-80 cm tall; leaves dissected, very narrow, minute; petals less than 1 mm long, white; *fruits to 2 mm long, smooth, with prominent ribs. Wet, swampy sites. Freq. all Fla. W to Tex., N to Mo., E to Ky. and Mass. Spr-fall, all yr S.

SNAKEROOT
Sanicula canadensis L.

Biennial, 0.2-1.3 m tall; leaves to 6 cm long, divided into 3 sections, but giving appearance of 5 divisions; petals white or green; stamens and style shorter than flower; fruits 3-5 mm long, with hooked bristles. Woody sites. Freq. CF, NF, WF. W to Tex., N to Minn., E to N.H. Spr-sum.

LONG-STYLED SNAKEROOT
Sanicula gregaria Bickn.

Perennial, 30-90 cm tall; leaves 3-9 cm long, divided into 3 sections; flower clusters with 3 flowers per stalk; ovary and fruits with bristles; bristles not swollen at base; styles longer than fruit and flower bristles; flowers yellow; fruits 3-5 mm long. Wet sites. Infreq. WF. W to Kans., N to Wis., E to Vt. Spr.

SMALL'S SNAKEROOT
Sanicula smallii Bickn.

Perennial, 25-60 cm tall; leaves 3-9 cm long, divided into 3 sections, but giving appearance of 5 divisions; flower cluster of 3 sessile flowers; ovary and fruits with hooked hairs; bristles not swollen at base; styles equal to or shorter than bristles; flowers white or green; fruits 3-5 mm long. Rich, rocky woods. Infreq. WF. W to Miss., N to Mo., E to N.C. Spr.

FLORIDA WATER PARSNIP
Sium suave Walt.

Perennial, 0.6-1.9 m tall; leaves 10-25 cm long, finely divided; leaflets 7-17; flowers 8-21 per cluster, white; *fruits 2-3 mm long, with winged ribs. Low, wet sites. Rare CF, NF, WF. N to Canada. Sum-fall.

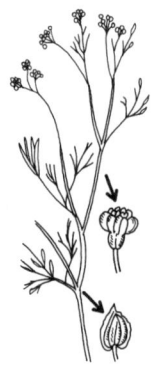

SPREADING SCALE SEED
Spermolepis divaricata (Walt.) Raf.

Annual, 10-70 cm tall, branching in upper half; leaves 1-8 cm long, dissected, with thin segments; sepals 0; *petals to 1 mm long, white; *fruits 1-1.5 mm long, with bumps. Moist to dry sites. Freq. CF, NF, WF. W to Tex., N to Kans., E to N.C. Spr.

BRISTLY SCALE SEED
Spermolepis echinata (Nutt). Heller

Annual, 10-50 cm tall, branching from near base; leaves to 6 cm long, dissected; sepals 0; petals to 1 mm long, white; *fruits to 2 mm long, bumpy, covered with hooked hairs. Dry sandy sites. Infreq. CF, NF, WF. W to Tex., N to Mo., E to S.C. Spr.

MEADOW-PARSNIP
Thaspium barbinode (Michx.) Nutt.

Perennial, 0.3-1.2 m tall; all leaves divided into 3 segments, sometimes subdivided into 3 additional segments each 3-6 cm long; flower clusters 3-6 cm wide; petals to 3 mm wide, yellow; fruits 5-6 mm long, winged. Woods, bluffs, stream banks. Infreq. WF. W to Miss., N to Canada. Spr-sum.

PURPLE MEADOW-PARSNIP
Thaspium trifoliatum (L.) A. Gray

Perennial, 20-70 cm tall; leaflets 3 or 1, to 7 cm
long; flower clusters 2-5 cm wide; petals small,
yellow to purple; *fruits to 4 mm long, winged.
Woodlands. Infreq. WF. W to Miss., N to Ill.,
E to R.I. Spr-sum.

GOLDEN ALEXANDER
Zizia aptera (Gray) Fern.

(not shown)

Perennial, 0.4-1 m tall; basal leaves 3-10 cm long,
simple; upper leaves with three leaflets, heart-
shaped to ovate; margins with 4-7 teeth per cm;
flower clusters multiple, yellow; fruits 3-4 mm
long. Moist to well-drained sites. Infreq. NF, WF.
N to Canada. Spr-sum.

THREE-LEAVED GOLDEN ALEXANDER
Zizia trifoliata (Michx.) Fern.

Perennial, 40-80 cm tall; basal and upper leaves
with three leaflets; leaflets 1-6 cm long; margins
with 2-4 teeth per cm; flower clusters multiple,
yellow; *fruits 2-3.5 mm long. Swamps. Rare
CF, NF, WF. N to Tenn. and Va. Spr-sum.

NETTLE FAMILY · URTICACEAE

Herbs, shrubs or trees; leaves alternate or opposite; stipules present; flowers unisexual (monoecious or dioecious) or bisexual; sepals 3-6; petals 0; ovary superior; fruit an achene.

BOG HEMP
Boehmeria cylindrica (L.) Sw.
[*B. drummondii* Wedd.]

Perennial, 0.2-1.2 m tall; stems hairy; leaves 2-18 cm long, opposite, green, hairy on lower surface; flowers green, monoecious; sepals 4; stamens 4; *achenes to 1 mm long, hairy. Low, woody sites. Common all Fla. W to Tex., N to Canada. Sum-fall.

RAMIE
Boehmeria nivea (L.) Gaud.

Perennial shrub, 1-3 m tall; leaves 8-30 cm long, alternate, with white hairs densely covering lower side; flowers pink to yellow, dioecious; achenes to 1 mm long, hairy. Native to Asia. Escapes from cultivation. Disturbed sites. Infreq. all Fla. W to Tex., N to S.C. Sum-fall.

(not shown)

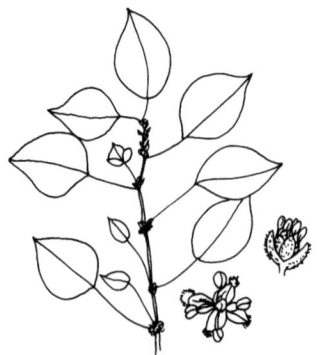

FLORIDA PELLITORY
Parietaria floridana Nutt.
[*P. nummularia* Small]

Annual, 10-30 cm long; stems prostrate to ascending; leaves to 2 cm long, longer than wide; achenes less than 1 mm long. Moist, shady, disturbed sites. Freq. all Fla. W to Tex., N to Del. Spr-sum.

EUROPEAN PELLITORY
Parietaria officinalis L.

Perennial, to 90 cm tall; stems hairy; leaf blades 1.5-3 cm long; achenes to 1 mm long. Disturbed sites. CF, NF. Sum-fall. Perhaps extirpated, not collected in recent yrs.

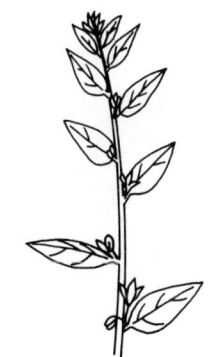

PENNSYLVANIA PELLITORY
Parietaria pensylvanica Muhl. ex Willd.

(not shown)

Annual, to 40 cm long; stems reclining, hairy; leaf blades 2-6 cm long; sepals 4, to 2 mm long, white; achenes to 1 mm long. Wet sites. Rare WF. W to Tex., N to Canada. Spr.

WHITE PELLITORY
Parietaria praetermissa Hinton

Annual, to 50 cm long; stems reclining, hairy; leaf blades 2-6 cm long; sepals 4, to 2 mm long; achenes to 1.3 mm long. Moist, shady, disturbed sites. Freq. all Fla. W to La., N to N.C. Spr-fall.

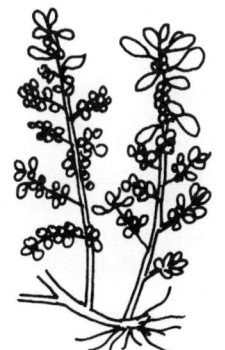

ARTILLERY PLANT
Pilea microphylla (L.) Liebm.

Annual or perennial, 2-15 cm long; stems diffuse, succulent; leaves to 8 mm long; sepals to 1 mm long, green or pink; achenes to 0.4 mm long. Moist sites. Infreq. all Fla. All yr.

BURNING NETTLE
Urtica urens L.

Annual, 10-40 cm tall; stems with stinging hairs;
leaf blades 1-4 cm long; sepals white or yellow;
achenes to 2 mm long. Disturbed sites, cultivated fields. Infreq. CF; rare NF, WF. W to Calif.,
N to Canada. Spr-fall.

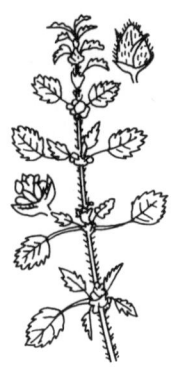

VALERIAN FAMILY   *VALERIANACEAE*

Mostly herbs; leaves opposite; flowers bisexual; sepals modified into numerous feather-like extensions, but mostly sepals
5; petals 5; ovary inferior; fruit a cypsela.

VALERIAN
Valeriana scandens L.

Perennial vine; leaves to 18 cm long including
long petioles, divided into 3 leaflets (compound);
flowers to 2 mm long, pink; achenes to 3 mm
long. Thickets, hammocks. Infreq. SF, CF, NF.
All yr.

CORN SALAD
Valerianella radiata (L.) Dufr.

Annual, 20-70 cm tall; leaves to 6 cm long,
simple, sessile; *flowers to 2 mm long, white;
*achenes to 2 mm long, tetragonal. Moist sites.
Infreq. WF. W to Tex., N to Minn. Spr-sum.

VERVAIN FAMILY   *VERBENACEAE*

Herbs, shrubs, vines or trees; leaves mostly opposite; flowers bisexual; sepals 2, 4, 5 or 7; petals 4, 5 or 7; ovary superior; fruit 2-4 nutlets or a drupe.

FRENCH MULBERRY
Callicarpa americana L.

Perennial shrub, 1-2 m tall; leaves to 15 cm long; upper leaf surfaces slightly hairy; lower leaf surfaces very hairy; flowers lavender, pink or white; drupes 4-5 mm in diam., violet, pink, purple or rarely white. Dry to moist, woody sites. Common all Fla. W to Tex., E to Va. Spr-fall.

STINKING GLORY-BOWER
Clerodendrum bungei Steud.

Perennial, 1-2 m tall; leaves 6-15 cm long, smooth; petals lilac to red-purple, in dense cymes. Disturbed sites. Rare all Fla. W to Tex. Sum-fall.

(not shown)

GLORY-BOWER
Clerodendrum fragrans R. Br.

(not shown)

Perennial shrub, 1-3 m tall; leaves 9-25 cm long, with short hairs on underside; floral tube 2-3 cm long, white, rose or blue; drupes to 10 mm long, blue-green. Native to Asia. Disturbed sites. Infreq. SF, CF, NF. Sum-fall.

TURK'S TURBAN or TUBE FLOWER
Clerodendrum indicum (L.) Kuntze

Perennial shrub, 1-4 m tall; leaves 5-16 cm long, smooth; floral tube 10-14 cm long, white; *drupes 3-12 mm long, red, purple or black. Native to East Indies. Cultivated. Disturbed sites. Infreq. all Fla. W to Tex., E to S.C. All yr.

GLORY-BOWER
Clerodendrum speciosissimum Van Geert

Perennial shrub, to 4 m tall; leaves to 30 cm long, densely hairy; sepals 5, 4-8 mm long; petals 5, 2-3 cm long, scarlet; drupes ovoid. Escapes. Disturbed sites. Rare SF, CF, NF. Sum-fall, all yr S.

MOSS VERBENA
Glandularia pulchella (Sweet) Troncoso
[*Verbena tenuisecta* Briq. - in many books]

Perennial, 10-30 cm tall; stems lying flat, ascending or branching, hairy; leaves 1-4 cm long, bipinnatifid; *flowers over 1 cm wide, white, purple or pink; nutlets to 3 mm long. Disturbed sites. Freq. all Fla. W to Tex., E to N.C. Spr-fall.

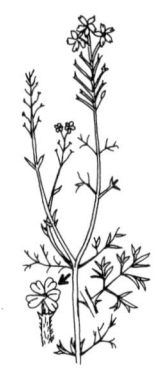

LANTANA
Lantana camara L.

Perennial, 1-1.5 m tall; stems with few weak prickles; leaves 2-12 cm long; petals to 7 mm long, cream, yellow, pink, orange or scarlet; drupes to 7 mm long. Native to Tropics. Disturbed areas. Common all Fla. W to Tex., E to Ga. Spr-fall, all yr S.

FLORIDA LANTANA
Lantana depressa Small
var. *floridana* (Mold.) Sanders

Similar to *Lantana camara* except: to 1.1 m long;
stems not prickly; leaves to 7 cm long; petals to
6 mm long, yellow; drupes to 3.5 mm long.
Hammocks. Infreq. SF, CF, NF. All yr.

(not shown)

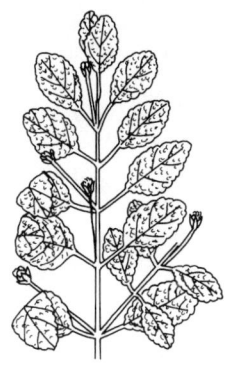

WHITE LANTANA
Lantana involucrata L.

Perennial shrub, to 2 m tall; leaves 1-4 cm long,
covered with very fine hairs; petals to 8 mm long,
white to purple; drupes 3-4 mm wide. Coastal
sites. Infreq. SF, CF, NF. All yr.

POLECAT GERANIUM
Lantana montevidensis (Spreng.) Briq.

Perennial shrub, 0.3-2 m high; stems lying flat to
trailing, covered with hairs or glands; leaves 1-3 cm
long; petals 5, 8-18 mm long, red or lilac; flowers
in heads; stamens 4; drupes to 5 mm long. Native
to South America. Naturalized in disturbed sites,
woods. Infreq. all Fla. Spr. Having characteristic
odor.

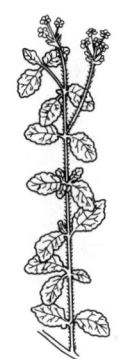

TALL MATCH-HEAD
Phyla lanceolata (Michx.) Greene

(not shown)

Perennial, to 60 cm long or 20-80 cm tall; stems
creeping and ascending; leaves 1.5-5 cm long; leaf
margins with 7-11 teeth; flower spikes 9-15 mm
long, round to cylindric; petals to 3 mm long,
purple, white, pink or blue; fruits to 1.5 mm long.
Wet sites. Infreq. WF. N to Canada. Sum-fall.

MATCH-HEAD
Phyla nodiflora (L.) Greene
[*Lippia nodiflora* (L.) Michx.]

Perennial, 0.2-1.3 m long or to 10 cm tall; stems
creeping and ascending, hairy; leaves 1-6 cm
long; leaf margins with 3-5 teeth; flower spikes 8-
25 mm long, round to cylindric; petals to 2.2 mm
long, white to purple; fruits to 1 mm long. Moist,
open sites. Common all Fla. W to Tex., N to Va.
Spr-fall.

ROADSIDE VERBENA
Verbena bonariensis L.

Perennial, 0.6-2 m tall; stems hairy on sides and
angles; leaves 5-15 cm long, largest at midstem,
nearly cordate and clasping; *spikes 5-6 mm
broad; flowers to 2 mm wide, purplish or occasionally white; floral tube to 4 mm long; nutlets
4, to 1 mm long. Disturbed, moist sites.
Occasional all Fla. W to Tex., E to Tenn. and
N.C. Spr-sum.

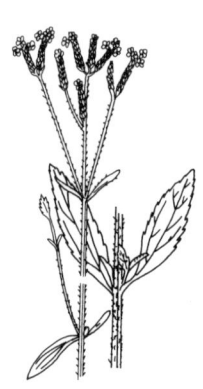

BRAZILIAN VERBENA
Verbena brasiliensis Vell.

Perennial, 1-2.5 m tall; leaves 4-10 cm long,
lanceolate; petals blue-purple; spikes to 4 cm long,
to 4.5 mm wide; fruits of mericarps to 2.2 mm
long. Disturbed sites. Infreq. all Fla. W to Tex.,
E to Va. Sum-fall.

ROSE VERBENA
Verbena canadensis (L.) Britt.

Perennial, 30-60 cm long; stems lying flat or
spreading; leaf blades 3-8 cm long; tip of sepal
lobes 2-3 mm long, bristle-like; nutlets to 3 mm
long. Disturbed areas. Infreq. CF, NF, WF.
W to Tex., E to Va. Spr-sum.

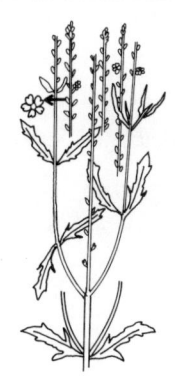

WIDE-MOUTH EUROPEAN VERVAIN
Verbena halei Small
[*V. officinalis* L. var. *halei* (Small) Barber]

Similar to *Verbena officinalis* except: leaves 3-
10 cm long, pinnately dissected; sepals to 3.5 mm
long; petals blue, lavender or white. Roadsides.
Infreq. CF, NF, WF. N to N.C., W to Miss. Spr.

BLUE VERVAIN
Verbena hastata L.

Perennial, 0.4-1.5 m tall; stems with rough hairs;
leaves 3-15 cm long; flowers to 3 mm wide, blue,
white or pink; nutlets 1.5-2 mm long. Disturbed,
moist to wet sites. Throughout most of U.S.
except Fla., to SE Canada. Sum.

SEASIDE VERBENA
Verbena maritima (Small) Small

Perennial, 0.2-2 m tall; stems diffuse or creeping;
leaf blades 1-6 cm long; floral tube to 2 cm long,
rose or purple; nutlets to 4 mm long. Dry to
moist sites. Infreq. SF, CF, NF. Endemic. All yr.

EUROPEAN VERVAIN
Verbena officinalis L.

Annual, 0.2-1 m tall; leaves 2-10 cm long, pinnately dissected; sepals to 2 mm long; petals 3-
5 mm long, 4-7 mm wide, white, blue or purple;
nutlets to 2 mm long. Low, disturbed sites.
Very infreq. CF. N to Va. Spr-fall.

STIFF VERBENA
Verbena rigida (L.) Spreng.

Perennial, 20-70 cm tall; stems lying flat, with
rough hairs; petals to 9 mm long, to 8 mm wide,
purple. Disturbed sites. Infreq. NF, WF. W to
Tex., E to N.C. Spr-fall.

(not shown)

WHITE VERVAIN or HARSH VERBENA
Verbena scabra Vahl
[*V. urticifolia* L.]

Annual or perennial, 40-50 cm tall; stems with
rough hairs; leaves 6-12 cm long; petals to 2 mm
long, 2-3 mm wide, white, pink or lavender;
*nutlets 1-5 mm long. Low, marshy sites. Infreq.
all Fla. W to Calif., E to Va. Spr-sum.

CHASTE TREE
Vitex agnus-castus L.

Perennial shrub, to 4 m tall; twigs with gray hairs;
leaflets 5-7, 4-8 cm long, grayish tomentose
beneath; *flowers blue; *drupes to 4 mm long.
Native to Eurasia. Escapes from cultivation.
Disturbed sites. Rare SF, CF, NF. W to Tex.,
E to N.C. Spr-fall.

THREE-LEAF VITEX
Vitex trifolia L. f.

Perennial shrub, to 3 m tall; stems prostrate; leaf
or leaflet blades 5-7 cm long, gray-white tomentose beneath; panicles to 10 cm long; flowers
blue. Native to Asia. Escapes from cultivation.
Disturbed sites. Infreq. all Fla. Sum.

VIOLET FAMILY   *VIOLACEAE*

Herbs or shrubs; leaves mostly alternate; stipules present; flowers bisexual; sepals 5; petals 5, with lowermost petal forming spur; ovary superior; fruit a capsule.

FLORIDA VIOLET
Viola affinis LeConte
[*V. floridana* Brainerd]

Perennial, having no stolons; stemless; leaf blades 3-8 cm long, oval to triangular, smooth; flowers 1.5-3.5 cm wide; petals blue or purple; spurred petal with hairs; capsules 0.6-1.6 cm long. Moist to wet woody sites. Common all Fla. W to Tex., N to Wis. Spr.

LONG-LEAF VIOLET
Viola lanceolata L.
subsp. *vittata* (Greene) Russell

Perennial, smooth, stoloniferous; leaves 5-20 cm long, linear to lanceolate; flower stems erect, longer or shorter than leaves; flowers 1-2 cm wide; petals white; *capsules 6-12 mm long. Open moist to wet sites. Freq. all Fla. W to Tex., E to Va., N to Canada. Spr.

BIRDFOOT VIOLET
Viola pedata L.

(not shown)

Perennial; stemless; leaves divided into 3; each leaf division again 3- to 5-cleft; leaf blades to 5 cm long, smooth; petals purple; capsules to 9 mm long. Dry sites. Rare NF. N to Ga., W to Tex. Spr.

PRIMROSE-LEAVED VIOLET
Viola primulifolia L.

Perennial, smooth to hairy, stoloniferous; leaves
1-10 cm long, ovate to elliptic; flower stems
erect, longer than or equal to leaves; flowers 1-
2 cm wide; petals white; capsules 7-10 mm long.
Open, wet sites. Freq. CF, NF, WF. W to Tex.,
N to Canada. Spr.

SEVEN-LOBED VIOLET
Viola septemloba LeConte

Perennial; stemless; leaf blades to 10 cm long,
heart-shaped to oval, 3-, 5- or 7-lobed, smooth;
petals blue or purple; spur petal bearded; capsules to 1.5 cm long. Low sites. Freq. all Fla.
W to La., N to Va. Spr.

THREE-LOBED VIOLET
Viola triloba Schwein.

Perennial; stemless; leaf blades to 10 cm long,
ovate, 3-lobed, with hairs on margins and veins;
petals blue or purple; capsules to 1.2 cm long.
Dry woods. Infreq. CF, NF, WF. W to Tex.,
N to Mass. Spr.

(not shown)

SOUTHERN DOWNY VIOLET
Viola villosa Walt.

(not shown)

Perennial; stemless; leaf blades to 6 cm long,
round to oval, densely hairy above; petals blue or
purple; capsules to 1 cm long. Dry sites. Infreq.
CF, NF, WF. W to Tex., N to Va. Spr.

WALTER'S VIOLET
Viola walteri House

Perennial; stems present, ascending; leaf blades
to 5 cm long, heart-shaped, red to purple on
lower surface, with dark venation on upper
surface; petals blue or purple; capsules to 6 mm
long. Moist woods. Infreq. CF, NF, WF.
W to Tex., N to Va. Spr.

(not shown)

GRAPE FAMILY   *VITACEAE*

Woody or herbaceous vines with tendrils, or shrubs or trees;
leaves mostly alternate; flowers bisexual or unisexual
(monoecious); sepals 4 or 5; petals 4 or 5; ovary superior;
fruit a berry.

PEPPER VINE
Ampelopsis arborea (L.) Koehne

Perennial; leaves 10-20 cm long, composed of
several leaflets (compound); *petals to 3 mm
long, green; berries to 1 cm in diam., black.
Moist to wet, woody sites. Freq. all Fla.
W to Tex., N to Md. Spr-fall.

FALSE GRAPE
Ampelopsis cordata Michx.

(not shown)

Perennial; leaves to 22 cm long, simple, grape-like;
petals to 3 mm long, green; berries 5-8 mm in
diam., blue. Wet woods. Infreq. WF. W to Tex.,
N to Va. Spr.

POSSUM GRAPE
Cissus sicyoides L.

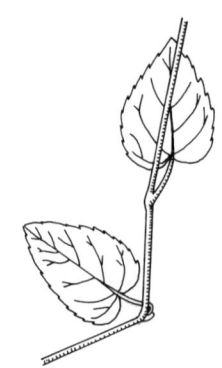

Perennial, fleshy to woody vine; stems with short
dense hairs; leaf blades 2-8 cm long; flowers in
peduncled cymes; sepals 4; petals 4, spreading;
berries to 1 cm in diam., black, inedible.
Hammocks. Infreq. SF, CF. Spr-fall, all yr S.

MARINE IVY
Cissus trifoliata L.
[*C. incisa* (Torr. & Gray) Desmoul.]

Perennial, partially woody vine; leaves fleshy,
divided into 3 leaflets each 3-8 cm long; *flowers
in cymes, to 5 cm long; *berries to 1.2 cm in
diam., inedible. Salt marshes, rocky sites. Infreq.
all Fla. W to Tex., N to Kans. Spr-fall.

VIRGINIA CREEPER
Parthenocissus quinquefolia (L.) Planch.

Perennial, climbing or trailing by *tendrils with
sticky discs; leaves divided like a hand into 3-5
(usually 5) leaflets; each leaflet to 15 cm long,
with toothed margins; flowers grouped in cymes;
petals 2-3 mm long, red with green margins;
berries 8-9 mm in diam., black to blue.
Poisonous. Woody sites. Common all Fla.
W to Tex., N to Canada. Spr, all yr S.

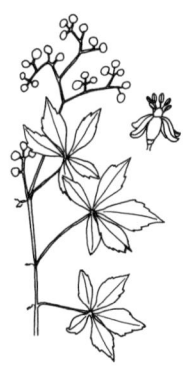

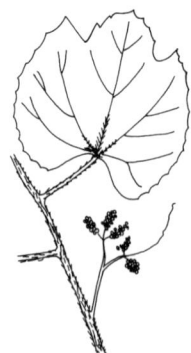

SUMMER GRAPE
Vitis aestivalis Michx.

Perennial, woody vine; stems rounded, covered
with red-brown or white woolly hairs; bark peeling
with age; leaves to 15 cm long, 3- to 5-lobed, with
red-brown or white woolly hairs on and in between
leaf axils or with white film on lower surface; flowers in panicles 10-15 cm long; petals 5; berries 5-
12 mm in diam., black or purple. Woody sites.
Freq. all Fla. W to Tex., N to Canada. Spr-sum.

≡≡≡≡ *Illustrated Plants of Florida and the Coastal Plain* ≡≡≡≡

PIGEON GRAPE
Vitis cinera Engelm. ex Millardet

Perennial, woody vine; stems angled; bark
peeling with age; branchlets, petioles and leaves
covered with red-brown or white woolly hairs;
leaves to 15 cm long, slightly lobed; flowers in
panicles 10-20 cm long; petals 5; berries 4-9 mm
in diam. Low, woody sites. Infreq. all Fla.
W to Tex., E to Va. Spr-sum.

(not shown)

RED GRAPE
Vitis palmata Vahl

(not shown)

Perennial, woody vine; stems angled to rounded,
smooth, purplish red; bark peeling with age;
leaves 4-8 cm long, 3-lobed, with hairs in leaf
axils on lower surface; flowers in panicles to
15 cm long; petals 5; berries to 8 mm in diam.,
black. Wet sites. Infreq. WF, NF. W to Tex.,
N to Ind. Spr.

SCUPPERNONG GRAPE
Vitis rotundifolia Michx.
[*V. munsoniana* Simpson ex Munson]

Perennial, climbing or trailing by simple tendrils;
leaves 5-12 cm long, 4-9 cm wide, simple, with
toothed margins; flowers grouped in panicles;
petals to 2.5 mm long, yellow; berries 1-2.5 cm
in diam., purple, black or bronze. Woody sites.
Common all Fla. W to Tex., N to Del. Spr-sum.

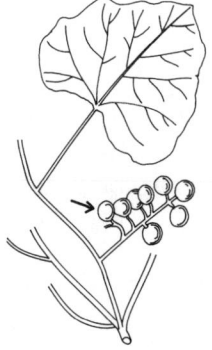

CALUSA GRAPE
Vitis shuttleworthii House

Perennial woody vine; leaf blades 4-12 cm long;
leaf petioles and underside silvery-felty; flowers in
panicles to 5 cm long; *berries to 1 cm in diam.
Well-drained sites. Infreq. SF, CF, NF. Spr.

FROST GRAPE
Vitis vulpina L.
[*V. cordifolia* Lam.]

Perennial, woody vine; stems rounded, smooth,
gray or brown; bark peeling with age; leaves 10-
18 cm long, slightly lobed, with hairs in leaf axils
on lower surface; flowers in panicles 10-15 cm
long; petals 5; berries 5-10 mm in diam., black.
Woody sites. Infreq. CF, NF, WF. W to Tex.,
N to N.Y. Spr-sum.

YELLOW-EYED-GRASS FAMILY **XYRIDACEAE**

Herbs; leaves basal; flowers bisexual; flowers subtended by
stiff scale; sepals 3, with outer sepal regular and inner 2
keeled; petals 3, separate or united at base with upper portion
lobed; ovary superior; fruit a capsule.

MORNING YELLOW-EYED-GRASS
Xyris ambigua Beyr. ex Kunth

Perennial; leaves to 60 cm long, 0.3-2 cm wide,
*with brown or pink bases; flower stems to 1 m
long; *sepals enclosed in bracts; petals to 8 mm
long, yellow, open in a.m.; seeds to 0.6 mm long.
Moist sites. Common all Fla. W to Tex., N to Va.
Spr-fall.

ST. MARY'S-GRASS
Xyris baldwiniana Roem. & Schult.

Perennial; leaves 10-30 cm long, to 1 mm wide,
filiform, often twisted, twice as long as or equal
in length to flower stem sheath; flower stems
20-40 cm tall; keel margins torn; petals 3-4 mm
long, opening in a.m.; seeds to 0.9 mm long.
Moist sites. Rare CF, NF, WF. W to Tex.,
E to Ga., N to N.C. Spr-fall.

SHORT-LEAVED YELLOW-EYED-GRASS
Xyris brevifolia Michx.

Perennial; leaves to 11 cm long, to 2 mm wide,
shorter than flower stem sheath; flower stems
8- 40 cm tall; keel margins entire; petals 2.5-3 mm
long, opening in a.m.; seeds to 0.4 mm long. Low
wet sites. Common all Fla. N to N.C. Spr-fall.

CAROLINA YELLOW-EYED-GRASS
Xyris caroliniana Walt.
[*X. pallescens* (C. Mohr) Small; *X. flexuosa* Muhl. ex Ell.]

Perennial; *leaves to 50 cm long, to 5 mm wide,
twisted, sinuous, with brown bases; flower stems
10-70 cm tall, with 2 smooth edges above; *sepals
exserted from bracts; petals 8-9 mm long, yellow or
white, open in p.m.; seeds to 1 mm long. Pinelands,
moist sites. Common all Fla. W to Tex., N to
Maine. Sum-fall.

PINK-LEAVED YELLOW-EYED-GRASS
Xyris difformis Chapm.

Perennial; leaves 10-50 cm long, to 1.5 cm wide;
leaf bases pink or purple; flower stems 15-70 cm
tall, smooth to ridged; spikes to 1 cm long; sepals
not exserted, light brown, having keel with
toothed margins; petals to 4 mm long, open in
a.m. Moist to wet sites. Freq. all Fla. W to Tex.,
N to Canada. Spr-fall.

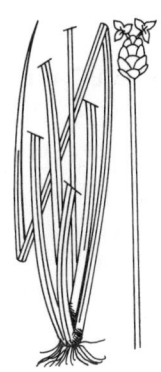

DRUMMOND'S YELLOW-EYED-GRASS
Xyris drummondii Malme

(not shown)

Perennial; leaves to 10 cm long, to 5 mm wide,
dark at base; leaves equal or nearly equal length
of flower stem sheath; flower stems 4-20 cm tall;
keel margins hairy; petals to 3 mm long, open in
a.m.; seeds to 0.3 mm long. Low wet sites.
Infreq. NF, WF. N to Ga., W to Miss. Sum.

THIN-LEAVED YELLOW-EYED-GRASS
Xyris elliottii Chapm.

Perennial; leaves 3-30 cm long, to 2.5 mm wide,
with brown bases; flower stems to 70 cm long,
with rough upper edges; sepals included or
slightly exserted; petals to 5 mm long, yellow,
open in a.m.; seeds to 0.6 mm long. Moist sites.
Common all Fla. W to Miss., N to S.C. Spr-fall.

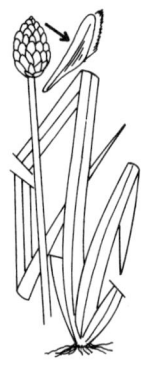

FRINGED YELLOW-EYED-GRASS
Xyris fimbriata Ell.

Perennial; leaves 4-70 cm long, to 2.5 cm wide;
leaf bases green, light brown or pink; flower
stems 0.6-1.2 m tall, rough, 2-edged; spikes 1.5-
2.5 cm long; *sepals exserted, yellow to brown,
with fringed keel; petals 5-6 mm long, open in
a.m. Low, wet sites. Freq. CF, NF, WF. W to
Miss., N to N.J. Spr-fall.

ONE-STEMMED YELLOW-EYED-GRASS
Xyris flabelliformis Chapm.

Perennial; leaves 1-4 cm long, to 4 mm wide,
shorter than flower stem sheath; flower stems to
30 cm tall; keel margins hairy; petals to 3 mm long,
open in a.m.; seeds to 0.3 mm long. Wet sites.
Common all Fla. W to La., N to N.C. Spr-sum.

(not shown)

COMMON YELLOW-EYED-GRASS
Xyris jupicai L. Rich.
[*X. communis* Kunth]

Perennial; leaves to 60 cm long, to 1 cm wide,
longer than flower stem sheath; flower stems
20-90 cm tall; keel margins torn; petals to 3 mm
long, open in a.m.; seeds to 0.5 mm long. Low
wet sites. Common all Fla. W to Tex., N to N.J.
Spr-fall.

BROAD SCALE YELLOW-EYED-GRASS
Xyris platylepis Chapm.

Perennial; leaves 20-50 cm long, 5-10 mm wide,
with pink bases; flower stems 0.5-1.1 m tall;
sepals enclosed in bracts; petals to 5 mm long,
yellow or white, open in p.m.; seeds to 0.6 mm
long. Moist to wet sites. Common CF, NF, WF.
W to Tex., N to Va. Sum-fall.

PURPLE-LEAF YELLOW-EYED-GRASS
Xyris smalliana Nash

Perennial; leaves 30-70 cm long, to 1.5 cm wide,
often purple near base; flower stems 0.5-1.5 m
tall; keel margins torn at tip, with remaining portion entire; petals 5-6 mm long, open in p.m.;
seeds to 0.7 mm long. Low wet sites. Freq. all
Fla. W to La. Sum.

GINGER FAMILY   *ZINGIBERACEAE*

Rhizomatous and sometimes tuberous-rooted herbs; leaves
arranged in 2 levels; flowers bisexual; sepals 3; petals 3; 2
staminodes fusing to form 2- or 3-lobed lip; ovary inferior;
fruit a capsule.

GARLAND FLOWER
Hedychium coronarium Koenig

Perennial, 1.2-1.6 m tall; leaves to 60 cm long,
to 8 cm wide; flowers in spikes, fragrant; petals
to 7.5 cm long, white. Native to tropical Asia.
Escapes from cultivation. Wet sites. Infreq. CF,
NF. Sum-fall.

CALTROP FAMILY *ZYGOPHYLLACEAE*

Herbs, shrubs or trees; leaves opposite or alternate; stipules present, mostly spiny; flowers bisexual; sepals 4-5; petals 4-5; ovary superior; fruit a capsule.

CALTROP
Kallstroemia maxima (L.) Hook. & Arn.

Annual, 20-60 cm long; stems prostrate; leaflets 6-8, to 2 cm long each; petals to 8 mm long, green to yellow; capsules to 1 cm long, smooth. Disturbed sites. Infreq. all Fla. W to Tex., E to Ga. All yr.

PUNCTURE VINE
Tribulus cistoides L.

Perennial, to 50 cm long; stems diffuse, covered with short dense hairs; leaves 2-6 cm long; leaflet blades to 2 cm long, to 5 mm wide; sepals 5; *petals 5, 1-2.5 cm long, to 4 cm wide; *capsules 8-9 mm long, spiny. Disturbed sites. Infreq. SF, CF. W to Tex., N to Ga. Spr-fall, all yr S.

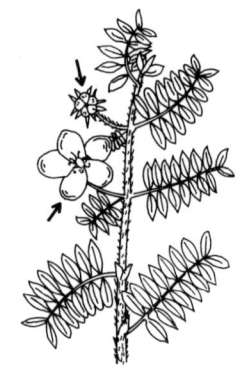

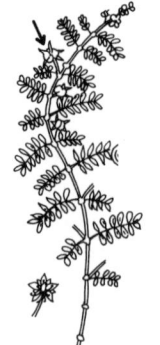

PUNCTURE WEED
Tribulus terrestris L.

Annual, 0.2-1.1 m long; stems prostrate, radiating, sparsely covered with coarse hairs; leaves to 4.5 cm long; leaflet blades to 1 cm long, to 5 mm wide; sepals 5; petals 5, 3-5 mm long, less than 1 cm wide; *capsules 6-7 mm long, spiny. Disturbed sites. Infreq. all Fla. W to Tex., N to N.Y. Spr-fall.

Species with flowers that may be BLUE ...

PHILIPPINE VIOLET .... *Barleria cristata*
SOUTHERN DOWNY VIOLET .... *Viola villosa*
TWIN FLOWER .... *Dyschoriste oblongifolia*
WILD PETUNIA .... *Ruellia caroliniensis*
WHITE THUNBERGIA .... *Thunbergia fragrans*
BLUE DOGBANE .... *Amsonia ciliata*
INDIAN HELIOTROPE .... *Heliotropium indicum*
CAPITATE BURMANNIA .... *Burmannia capitata*
TALL BELLFLOWER .... *Campanula americana*
PIEDMONT LOBELIA .... *Lobelia amoena*
BAY LOBELIA .... *Lobelia feayana*
COASTAL PLAIN LOBELIA .... *Lobelia glandulosa*
NUTTALL'S LOBELIA .... *Lobelia nuttallii*
DOWNY LOBELIA .... *Lobelia puberula*
POSSUM HAW .... *Viburnum nudum*
TROPICAL SPIDERWORT .... *Commelina benghalensis*
SPREADING DAY-FLOWER .... *Commelina diffusa*
WOODS DAY-FLOWER .... *Commelina virginica*
DOVE WEED .... *Murdannia nudiflora*
HAIRY SPIDERWORT .... *Tradescantia hirsuticaulis*
COMMON SPIDERWORT .... *Tradescantia ohiensis*
ZIG ZAG SPIDERWORT .... *Tradescantia subaspera.*
TIGHT-LEAVED ASTER .... *Aster adnatus*
CHAPMAN'S ASTER .... *Aster chapmanii*
BUSHY ASTER .... *Aster dumosus*
THISTLE-LEAF ASTER .... *Aster eryngiifolius*
SWAMP ASTER .... *Aster paludosus*
ARROW LEAVED ASTER .... *Aster sagittifolius*
ANNUAL MARSH ASTER .... *Aster subulatus*
DAISY FLEABANE .... *Erigeron strigosus*
MIST FLOWER .... *Eupatorium coelestinum*
WILD LETTUCE .... *Lactuca floridana*
BIG BLUE PROSTRATE MORNING-GLORY ...... *Bonamia grandiflora*
IVY-LEAF MORNING-GLORY .... *Ipomoea hederacea*
TALL MORNING-GLORY .... *Ipomoea purpurea*
SMALL-FLOWERED MORNING-GLORY ..... *Jacquemontia tamnifolia*
CATESBY GENTIAN .... *Gentiana catesbaei*
HYDROLEA .... *Hydrolea ovata*
ANGLEPOD BLUE-FLAG .... *Iris hexagona*
DWARF IRIS .... *Iris verna*
WIDE-WINGED BLUE-EYED-GRASS .... *Sisyrinchium angustifolium*
NARROW-WINGED BLUE-EYED-GRASS .... *Sisyrinchium atlanticum*
YELLOW BLUE-EYED-GRASS .... *Sisyrinchium exile*
SCRUB BLUE-EYED-GRASS .... *Sisyrinchium solstitiale*
MINTWEED .... *Hyptis pectinata*

BLUE SAGE .... *Salvia azurea*
LYRE-LEAVED SAGE .... *Salvia lyrata*
SOUTHERN SAGE .... *Salvia riparia*
SMALL WHITE SAGE .... *Salvia serotina*
COMMON LARGE SKULLCAP .... *Scutellaria integrifolia*
MARSH SKULLCAP .... *Scutellaria lateriflora*
HAIRY SKULLCAP .... *Scutellaria ovalifolia*
BLUE CURLS .... *Trichostema dichotomum*
BLUE CURLS .... *Trichostema suffrutescens*
BASTARD INDIGO .... *Amorpha fruticosa*
BUTTERFLY-PEA .... *Clitoria mariana*
CREEPING BUSH-CLOVER .... *Lespedeza repens*
SKY-BLUE LUPINE .... *Lupinus diffusus*
SUNDIAL LUPINE .... *Lupinus perennis*
BUCKROOT .... *Psoralea canescens*
SAND VETCH .... *Vicia acutifolia*
OCALA VETCH .... *Vicia ocalensis*
BLUE FLAX .... *Linum usitatissimum*
CAPE BLUE WATER-LILY .... *Nymphaea capensis*
BLUE WATER-LILY .... *Nymphaea elegans*
BLUE PHLOX .... *Phlox divaricata*
WATER HYACINTH .... *Eichhornia crassipes*
SCARLET PIMPERNEL .... *Anagallis arvensis*
LEATHER FLOWER .... *Clematis crispa*
DULL-LEAF WILD COFFEE .... *Psychotria sulzneri*
BLUE HYSSOP .... *Bacopa caroliniana*
OLD FIELD TOADFLAX .... *Linaria canadensis*
PURPLE FALSE PIMPERNEL .... *Lindernia crustacea*
ROUND-LEAVED FALSE PIMPERNEL .... *Lindernia grandiflora*
FIELD SPEEDWELL .... *Veronica agrestis*
PERSIAN SPEEDWELL .... *Veronica persica*
CHRISTMAS BERRY .... *Lycium carolinianum*
BRAZILIAN NIGHTSHADE .... *Solanum seaforthianum*
CORN SNAKEROOT .... *Eryngium aquaticum*
MATTED BUTTON SNAKEROOT .... *Eryngium baldwinii*
LANCE-LEAF BUTTON SNAKEROOT .... *Eryngium integrifolium*
GLORY-BOWER .... *Clerodendrum fragrans*
TALL MATCH-HEAD .... *Phyla lanceolata*
WIDE-MOUTH EUROPEAN VERVAIN .... *Verbena halei*
BLUE VERVAIN .... *Verbena hastata*
EUROPEAN VERVAIN .... *Verbena officinalis*
CHASTE TREE .... *Vitex agnus-castus*
FLORIDA VIOLET .... *Viola affinis*
SEVEN-LOBED VIOLET .... *Viola septemloba*
SOUTHERN DOWNY VIOLET .... *Viola villosa*

Species with flowers that may be GREEN or BROWN ...

RATTLESNAKE MASTER .... *Polianthes virginica*
MOTHER-IN-LAW'S TONGUE .... *Sansevieria hyacinthoides*
BEAR-GRASS .... *Yucca flaccida*
CARPET WEED .... *Mollugo verticillata*
GIANT AMARANTH .... *Amaranthus australis*
SMOOTH PIGWEED .... *Amaranthus hybridus*
COTTON WEED .... *Froelichia floridana*
BLOODLEAF .... *Iresine diffusa*
PERENNIAL BLOODLEAF .... *Iresine rhizomatosa*
WINGED SUMAC .... *Rhus copallina*
POISON IVY .... *Rhus radicans*
SMALL-FRUITED PAWPAW .... *Asimina parviflora*
GREEN DRAGON .... *Arisaema dracontium*
GREEN ARUM .... *Peltandra virginica*
CURLY MILKWEED .... *Asclepias amplexicaulis*
FRAGRANT MILKWEED .... *Asclepias connivens*
SCRUB MILKWEED .... *Asclepias curtissii*
FLORIDA MILKWEED .... *Asclepias longifolia*
PEDICILLATE MILKWEED .... *Asclepias pedicillata*
VELVET-LEAF MILKWEED .... *Asclepias tomentosa*
WHORL-LEAF MILKWEED .... *Asclepias verticillata*
SOUTHERN MILKWEED .... *Asclepias viridula*

BLODGETT'S CYNANCHUM .... *Cynanchum blodgettii*
LEAFLESS CYNANCHUM .... *Cynanchum scoparium*
SALTWORT .... *Batis maritima (only species)*
AMERICAN HORNBEAM or BLUE BEECH .... *Carpinus caroliniana*
SOUTHERN NEEDLE-LEAF .... *Tillandsia setacea*
SMOOTH-FRUITED WATER STARWORT .... *Callitriche heterophylla*
WIDE-FRUITED WATER STARWORT .... *Callitriche peploides*
HEARTS-A-BUSTIN' .... *Euonymus americanus*
SPINY HORNWORT .... *Ceratophyllum muricatum*
MUSK-GRASS .... *Chara species*
MEXICAN-TEA .... *Chenopodium ambrosioides*
ANNUAL GLASSWORT .... *Salicornia bigelowii*
BUTTON WOOD .... *Conocarpus erectus*
COMMON RAGWEED .... *Ambrosia artemisiifolia*
NARROW LEAF RAYLESS GOLDENROD .... *Bigelowia nuttallii*
NARROW-LEAVED ELDER .... *Iva microcephala*
PRETTY DODDER .... *Cuscuta indecora*
PERSIMMON .... *Diospyros virginiana*
DOTTED DANGLEBERRY .... *Gaylussacia frondosa*
BLACK-HEAD PIPEWORT .... *Eriocaulon ravenelii*
BROWN BOG-BUTTON .... *Lachnocaulon engleri*
THREE-SEEDED MERCURY .... *Acalypha gracilens*

SAND DUNE SPURGE	*Chamaesyce bombensis*
ROUND-LEAVED SPURGE	*Chamaesyce cordifolia*
ERECT SPURGE	*Chamaesyce hypericifolia*
SPOTTED SPURGE	*Chamaesyce maculata*
SILVER LEAF CROTON	*Croton argyranthemus*
ELLIOTT'S CROTON	*Croton elliottii*
PINELAND CROTON	*Croton linearis*
ABNORMAL PHYLLANTHUS	*Phyllanthus abnormis*
LONG-STALKED PHYLLANTHUS	*Phyllanthus tenellus*
CASTOR-BEAN	*Ricinus communis*
QUEEN'S DELIGHT	*Stillingia sylvatica*
SOUTH FLORIDA TRAGIA	*Tragia saxicola*
NETTLE-LEAVED TRAGIA	*Tragia urticifolia*
SAND LIVE OAK	*Quercus geminata*
DWARF LIVE OAK	*Quercus minima*
RUNNING OAK	*Quercus pumila*
WOODS GRASS	*Oplismenus setarius*
PARROT'S-FEATHER	*Myriophyllum aquaticum*
WATER MILFOIL	*Myriophyllum heterophyllum*
MARSH MERMAID WEED	*Proserpinaca palustris*
LOVE VINE	*Cassytha filiformis*
GROUND NUT	*Apios americana*
PIMPLED DUCKWEED	*Lemna perpusilla*
FEW-ROOT DUCKWEED	*Spirodela punctata*
COMMON ASPARAGUS	*Asparagus officinalis*
INDIAN CUCUMBER ROOT	*Medeola virginiana*
FEATHER-SHANK	*Schoenocaulon dubium*
MISTLETOE	*Phoradendron serotinum*
SOUTHERN CLUB-MOSS	*Lycopodium appressum*
NODDING CLUB-MOSS	*Lycopodium cernuum*
COMMON WAXWEED	*Cuphea carthagenensis*
BUSHY POND WEED	*Najas guadalupensis*
PRIVET	*Forestiera ligustrina*
WINGED WATER-PRIMROSE	*Ludwigia alata*
LITTLE SEEDBOX	*Ludwigia curtissii*
POND WATER-PRIMROSE	*Ludwigia spathulata*
CORAL ROCK ORCHID	*Basiphyllaea corallicola*
ROSE ORCHID	*Cleistes divaricata*
SPRING CORAL-ROOT	*Corallorhiza wisteriana*
BUTTERFLY ORCHID	*Encyclia tampensis*
UMBELLED EPIDENDRUM	*Epidendrum difforme*
RIGID EPIDENDRUM	*Epidendrum rigidum*
NORTHERN WILD COCO	*Eulophia ecristata*
LONG-STALKED HABENARIA	*Habenaria distans*
WATER-SPIDER ORCHID	*Habenaria repens*
SOUTHERN TWAYBLADE	*Listera australis*
GREEN ADDER'S MOUTH	*Malaxis unifolia*
SMALL GREEN WOOD ORCHID	*Platanthera clavellata*
GYPSY-SPIKES	*Platanthera flava*
PALE FLOWERED POLYSTACHYA	*Polystachya flavescens*
SHADOW WITCH	*Ponthieva racemosa*
GREEN LADIES'-TRESSES	*Spiranthes polyantha*
LEAFLESS BEAKED ORCHID	*Stenorrhynchos lanceolatus*
SHORT STALKED NODDING CAP	*Triphora gentianoides*
SQUAW ROOT	*Conopholis americana*
CORKY-STEMMED PASSION-FLOWER	*Passiflora suberosa*
GUINEA-HEN WEED	*Petiveria alliacea*
LARGE BRACTED PLANTAIN	*Plantago aristata*
ENGLISH PLANTAIN	*Plantago lanceolata*
NARROWLEAF PLANTAIN	*Plantago purshii*
WHITE BACHELOR'S BUTTON	*Polygala balduinii*
WHITE BACHELOR'S BUTTON	*Polygala brevifolia*
LARGE-FLOWERED POLYGALA	*Polygala grandiflora*
WILD BACHELOR'S BUTTON	*Polygala nana*
PIGEON PLUM	*Coccoloba diversifolia*
GIANT SMARTWEED	*Polygonum densiflorum*
MARSH PEPPER SMARTWEED	*Polygonum hydropiper*
STUBBLE SMARTWEED	*Polygonum setaceum*
CURLY DOCK	*Rumex crispus*
TROPICAL DOCK	*Rumex obovatus*
FIDDLE DOCK	*Rumex pulcher*
RATTAN VINE	*Berchemia scandens*
PURPLE GALIUM	*Galium hispidulum*
HAIRY FRUITED BEDSTRAW	*Galium pilosum*
DULL-LEAF WILD COFFEE	*Psychotria sulzneri*
WAFER ASH TREFOIL	*Ptelea trifoliata*
MOSQUITO FERN	*Azolla caroliniana*
INDIAN PAINT-BRUSH	*Castilleja coccinea*
STICKY GROUND CHERRY	*Physalis viscosa*
NARROW LEAF CAT-TAIL	*Typha angustifolia*
BLUE CAT-TAIL	*Typha x glauca*
COMMON CAT-TAIL	*Typha latifolia*
COINWORT	*Centella asiatica*
RATTLESNAKE MASTER	*Eryngium yuccifolium*
SNAKEROOT	*Sanicula canadensis*
SMALL'S SNAKEROOT	*Sanicula smallii*
BOG HEMP	*Boehmeria cylindrica*
FLORIDA PELLITORY	*Parietaria floridana*
EUROPEAN PELLITORY	*Parietaria officinalis*
WHITE PELLITORY	*Parietaria praetermissa*
PEPPER VINE	*Ampelopsis arborea*
POSSUM GRAPE	*Cissus sicyoides*
VIRGINIA CREEPER	*Parthenocissus quinquefolia*
SUMMER GRAPE	*Vitis aestivalis*
RED GRAPE	*Vitis palmata*
CALUSA GRAPE	*Vitis shuttleworthii*

Species with flowers that may be ORANGE ...

BLACK-EYED CLOCKVINE	*Thunbergia alata*
SCARLET MILKWEED	*Asclepias curassavica*
RED MILKWEED	*Asclepias lanceolata*
BUTTERFLY WEED	*Asclepias tuberosa*
TRUMPET VINE	*Campsis radicans*
FLAME VINE	*Pyrostegia venusta*
TRUMPET HONEYSUCKLE	*Lonicera sempervirens*
HAWKWEED	*Hieracium gronovii*
SERINIA	*Krigia cespitosa*
DWARF DANDELION	*Krigia virginica*
FALL CONEFLOWER	*Rudbeckia fulgida*
SOUTHERN RAGWORT	*Senecio anonymus*
GOLDEN RAGWORT	*Senecio aureus*
BALSAM GROUNDSEL	*Senecio pauperculus*
COMMON MARIGOLD	*Tagetes erecta*
RED MORNING-GLORY	*Ipomoea hederifolia*
CHANDELIER PLANT	*Kalanchoe tubiflora*
RED BASIL	*Calamintha coccinea*
LION'S EARS	*Leonotis nepetifolia*
SWEET ACACIA	*Acacia farnesiana*
ANIL INDIGO	*Indigofera suffruticosa*
DAUBENTONIA or SPANISH GOLD	*Sesbania punicea*
HORNED BLADDERWORT	*Utricularia cornuta*
CATESBY'S LILY	*Lilium catesbaei*
CARTER'S FLAX	*Linum carteri*
YELLOW INDIAN MALLOW	*Abutilon hirtum*
VELVET LEAF	*Abutilon theophrasti*
PRICKLY SIDA	*Sida spinosa*
YELLOW FRINGELESS ORCHID	*Platanthera integra*
CANDY WEED	*Polygala lutea*
PURSLANE	*Portulaca oleracea*
LANTANA	*Lantana camara*

Species with flowers that may be PINK ...

Species with flowers that may be PURPLE ...

Species with flowers that may be RED ...

Species with flowers that may be WHITE ...

Species with flowers that may be YELLOW ...

ACHENE – A small, hard, dry, one-seeded fruit.

ACUMINATE – Tapering at the end to a gradual point.

ACUTE – Terminating in a sharp or definite angle.

ADNATE – Having unlike parts united.

ALTERNATE – Arrangement of parts in which one occurs at each node.

ANNUAL – Of one year or less duration.

ANTHER – The pollen-bearing part of a stamen.

APETALOUS – Having no petals.

APPRESSED – Lying flat against.

AQUATIC – Living in water.

ASCENDING – Rising or curving upward.

AURICULATE – Having an ear-shaped appendage or lobe; in some leaves and petals.

AWN – A bristle-shaped appendage; in many grasses.

AXIL – The angle between any two plant parts.

AXIS – The lengthwise support around which the parts of a structure are arranged.

BEAKED – Ending in a firm slender tip; on certain carpels and fruits.

BEARDED – Having long or stiff hairs.

BERRY – A pulpy fruit with many seeds.

BIENNIAL – Of two years duration.

BIPINNATE – Twice pinnatified (divided).

BISEXUAL – Having both stamens and pistils.

BLADE – The expanded portion of a leaf.

BRACT – A modified leaf.

BRISTLE – A stiff hair.

BUD – Shoot that contains young leaves or flowers.

BULB – Modified underground fleshy stem that functions in food storage; found in many monocots.

CALYX – Sepals; the outer perianth of a flower.

CAMPANULATE – Bell-shaped.

CAPSULE – A dry dehiscent fruit developed from more than one carpel.

CARNIVOROUS PLANTS – Plants that trap insects in order to obtain nutrients.

CARPEL – Female reproductive part which consists of an ovary, style and stigma.

CATKIN – Unisexual spikelets of flowers.

CAULINE – Associated with the stem.

CHASMOGAMOUS – Fertilization of open flowers.

CILIA – Marginal hairs.

CLASPING – When lower portion of leaf encloses the stem.

CLEISTOGAMOUS – Self-fertilized in the bud; the flower not opening.

COMPOUND – Composed of two or more similar parts.

COROLLA – Petals; the inner part of the perianth.

CORONA – A crown.

CREEPING – Trailing and rooting on the ground.

CRENATE – Having shallow, scalloped, rounded teeth on leaf margin.

CYATHIUM – A cup-like structure bearing flowers; in *Euphorbia*.

DECIDUOUS – Having leaves or floral parts that are not persistent.

DEHISCENT – Splitting open along definite lines at maturity.

DENTATE – Leaf margin with sharp teeth pointing outward.

DIADELPHOUS – Stamen filaments united into two groups.

DICOT – A dicotyledonous plant.

DICOTYLEDONS – A group of flowering plants that are characterized as having two cotyledons, leaves with venation that is netted and flowers that usually have parts in four's and five's.

DIDYNAMOUS – Stamens or fruits that occur in two pairs of unequal length.

DIOECIOUS – Unisexual; with male and female reproductive parts on separate plants.

DISC – Fleshy extension of the receptacle; interior flowers of the Daisy Family.

DISSECTED – Divided into narrow segments.

DRUPE – A fleshy, pulpy fruit outside; inside a hard, stony pericarp.

DUNES – Coastal sandy ridges.

EMERSED – Having some parts under water and some parts above water.

ENDEMIC – Confined to a single area.

ENTIRE – Leaf margins without teeth or divisions.

EPIPHYTE – Plants growing on other plants for support only.

EQUITANT – Leaves that enfold each other in two ranks; as in *Iris*.

EXOTIC – Refers to plants that are not native to an area.

EXSERTED – Projecting beyond an enclosure.

FASCICLE – A bundle or cluster.

FOLLICLE – A 1-carpel fruit dehiscing on the side facing the axis.

GLAND – A depression or projection secreting or containing a fluid.

HABITAT – The kind of location where a plant grows.

HAMMOCK – Ecosystem with deep, fertile, moist soil dominated by mixed hardwood species such as oaks and hickories.

HEAD – A dense cluster of flowers or fruits on a receptacle or short axis.

HERBACEOUS – Having no persisting woody stem above ground.

HYBRID – A cross between two different (unlike) species.

HYPANTHIUM – A floral receptacle that has developed from the fusion of floral parts.

IMMERSED – Growing under water.

INDEHISCENT – Persistently closed.

INFERIOR – Generally referring to an ovary which lies below the other flower parts.

INFLORESCENCE – The flowering part of a plant.

INSECTIVOROUS – Plants which capture insects for nutrients.

INTERNODE – The part of a stem between two nodes.

INTRODUCED – Brought intentionally from another region.

INVOLUCRE – The small leaves or leaf-like bracts surrounding a flower cluster, head, or single flower.

KEEL – A central, dorsal ridge. The two anterior petals of some legumes.

LANCEOLATE – Broadest near base and narrowed to the apex.

LEAFLET – One division of a compound leaf; a small subordinate leaf.

LEGUME – Fruit of the Pea Family, splitting along both sutures.

LIGULE – The flattened ray flower of many composites.

LINEAR – Long, narrow, with parallel sides.

LIP – The upper or lower division of a two-parted calyx or corolla.

MARSH – Wetland in which emergent herbaceous vegetation dominates.

MIDRIB – The central or main rib (vein) of a leaf.

MONADELPHOUS – When stamen filaments are grown together.

MONOCOT – A monocotyledonous plant.

MONOCOTYLEDONS – A group of flowering plants that are characterized as having one cotyledon, leaves with parallel venation and flowers that have parts in three's.

MONOECIOUS – Plants that have separate male and female reproductive parts in separate flowers but on the same plant.

NATURALIZED – A non-native that has become established.

NODE – A joint or place on a stem that normally bears a leaf or leaves.

NUT – A hard, one-seeded, indehiscent fruit.

NUTLET – Small nut.

OBOVATE – Ovate with wider portion towards tip.

OCREA – A tube around stem at base of leaf; occurs in some members of the Buckwheat Family.

OVARY – The female reproductive structure that contains the ovules.

OVATE – Egg-shaped; the broad end at the base.

OVULE – After fertilization it becomes a seed.

PALMATE (LEAF) – Radially divided; lobes or veins spreading like fingers on a hand.

PAPPUS – A crown of hairs or scales that occur on top of the achenes of the Daisy Family.

PARASITE – Growing on and obtaining nourishment from a living plant.

PERENNIAL – Living few to many years.

PERFECT – Flowers having both male (stamens) and female (pistil or pistils) reproductive parts.

PERIANTH – The sepals (calyx) and petals (corolla) if present.

PETAL – A division of the corolla, usually brightly colored and showy.

PETIOLE – The support or stalk of a leaf.

PHYLLARY – A bract of the involucre of the Daisy Family.

PINE FLATWOODS – Ecosytem typically dominated by longleaf or slash pines with saw palmetto as the dominant understory plant. Soil ranges from dry to wet.

PINNATE (LEAF) – Compounded with leaflets on each side of a common axis.

PINNATIFID – Pinnately cut, lobed, or divided.

PISTIL – The ovary, style and stigma of a flower.

POME – The fleshy fruit type of apples and pears in the Rose Family.

PRICKLE – A sharp extension of the outermost cell layer.

RACHIS – The axis of an inflorescence or of a compound leaf.

RAY – A ligule of a Daisy Family flower.

ROSETTE – A circular cluster of leaves, etc.

SALT MARSH – Wetland in which salt tolerant herbaceous vegetation dominates.

SAMARA – An indehiscent winged fruit.

SANDHILL – Dry, open ecosystem in which dominant longleaf pines and turkey oaks are found with other scrub oaks.

SCALE – A small leaf or any small scarious body.

SCRUB – Dry coastal ecosystem in which several species of oaks dominate. Vegetation is typically dwarfed or thicket-forming.

SEPAL – A division of the calyx that is the outermost portion of the flower.

SERRATE – Margins that have sharp teeth pointing forward.

SESSILE – Without a stalk.

SHRUB – A woody perennial usually with several main stems.

SIMPLE LEAF – Leaves that are not divided.

SPADIX – A spike having a fleshy axis and surrounded by a spathe as is typified in Arum Family members.

SPATHE – Leaf-like bract that surrounds spadix; typical in the Arum Family.

SPIKE – Sessile or subsessile flowers on a common axis.

SPINE – A rigid, sharp outgrowth on the stem.

SPUR – A sac-like or tubular extension of a flower part.

STAMENS – The pollen-bearing parts of a flower.

STANDARD – The large upper petal of a legume.

STIGMA – The part of a pistil that receives the pollen.

STIPULE – A leaf-like appendage at the base of a leaf.

STYLE – Structure that connects the stigma and ovary.

SUBMERSED – Plants in which all parts are under water.

SUPERIOR – Generally referring to an ovary which lies above the other flower parts.

SWAMP – Forested wetland.

TAPROOT – A single main vertical root with several smaller lateral roots.

TEPALS – Sepals and petals that are quite similar.

THORN – Branch that is modified into a sharp point.

TUBER – A short, thickened underground branch with several buds or eyes.

UMBEL – The stalks of each flower cluster come from the same point.

UTRICLE – A fruit that is one-seeded, small and bladdery.

WHORL – Leaves arranged in a circle around the stem.

WING – Any thin, papery expansion of an organ. The side petal of a legume flower.

Baker, Mary Francis. 1926 and 1938. *Florida Wild Flowers.* The Macmillan Company, New York, NY. 245 pp.

Barrett, Mary F. 1956. *Common Exotic Trees of South Florida.* University of Florida Press, 15 NW 15th St., Gainesville, FL. (Out-of-print). 245 pp.

Bell, C. Ritchie and Bryan J. Taylor. 1982. *Florida Wild Flowers and Roadside Plants.* Laurel Hill Press, Chapel Hill, NC. Available from FWF Books, Box 2744, Winter Park, FL 32790. 308 pp.

Burgis, D. S. and J. R. Orsenigo. 1969. *Florida Weeds, Part One.* Circular 331. Florida Cooperative Extension Service, Institute of Food and Agricultural Sciences, University of Florida, Gainesville, FL 32611. 11 pp.

Clewell, Andre F. 1985. *Guide to the Vascular Plants of the Florida Panhandle.* University Presses of Florida, 15 NW 15th St., Gainesville, FL 32603. 616 pp.

Cronquist, Arthur. 1980. *Vascular Flora of the Southeastern United States. Vol. 1. Asteraceae.* The University of North Carolina Press, P.O. Box 2288, Chapel Hill, NC 27514. 261 pp.

Dressler, Robert L., David W. Hall, Kent D. Perkins and Norris H. Williams. 1987. *Identification Manual for Wetland Plant Species of Florida.* IFAS Special Publication-35. University of Florida, Publications, Bldg. 664, Gainesville, FL 32611. 297 pp.

Duncan, Wilbur H. 1967. Woody Vines of the Southeastern United States. *Sida* 3(1): 1-76. Reprinted 1975 by the University of Georgia Press, Athens, GA 30602. 76 pp.

Duncan, Wilbur H. and Marion B. Duncan. 1987. *The Smithsonian Guide to Seaside Plants of the Gulf and Atlantic Coasts from Louisiana to Massachusetts, Exclusive of the Lower Peninsular Florida.* Smithsonian Institution Press, Washington, DC. 409 pp.

Duncan, Wilbur H. and Marion B. Duncan. 1988. *Trees of the Southeastern United States.* University of Georgia Press, Athens, GA 30602. 322 pp.

Duncan, Wilbur H. and Leonard E. Foote. 1975. *Wildflowers of the Southeastern United States.* The University of Georgia Press, Athens, GA 30602. 296 pp.

Duncan, Wilbur H. and John T. Kartesz. 1981. *Vascular Flora of Georgia. An Annotated Checklist.* The University of Georgia Press, Terrell Hall, Athens, GA 30602. 143 pp.

Eleuterius, Lionel N. 1981. *An Illustrated Guide to Tidal Marsh Plants of Mississippi and Adjacent States.* Publication No. MASGP-77-039. Mississippi-Alabama Sea Grant Consortium, Gulf Coast Research Laboratory, Ocean Springs, MS 39564. 131 pp.

Elias, Thomas S. 1980. *The Complete Trees of North America. Field Guide and Natural History.* Outdoor Life/Nature Books, Van Nostrand Reinhold Co., 135 West 50th St., New York, NY 10020. 948 pp.

Elmore, C. D. et al. <years vary> *Weed Identification Guide.* Southern Weed Science Society, 309 West Clark Street, Champaign, IL 61820. <Published in unbound sets and maintained in 3-ring binder.>

Fleming, Glenn, Pierre Genelle and Robert W. Long. 1976. *Wild Flowers of Florida.* Banyan Books, Inc., Miami, FL. 96 pp.

Godfrey, Robert K. 1988. *Trees, Shrubs, and Woody Vines of Northern Florida and Adjacent Georgia and Alabama.* The University of Georgia Press, Athens, GA 30602. 734 pp.

Godfrey, Robert K. and Jean W. Wooten. 1979. *Aquatic and Wetland Plants of Southeastern United States. Monocotyledons.* The University of Georgia Press, Athens, GA 30602. 724 pp.

Godfrey, Robert K. and Jean W. Wooten. 1981. *Aquatic and Wetland Plants of Southeastern United States. Dicotyledons.* The University of Georgia Press, Athens, GA 30602. 864 pp.

Hall, David W. 1978. *The Grasses of Florida.* Ph.D. Dissertation, Department of Botany, University of Florida, Gainesville, FL 32611. Available as #79-13,279 from Dissertations Abstracts International (Vol. 39(12), 1979), P.O. Box 1764, Ann Arbor, MI 48106. 498 pp.

Hall, David W. and Vernon V. Vandiver. 1989. *Weeds in Florida.* University of Florida, Institute of Food and Agricultural Sciences, Special Publication-37. 57 pp.

Hardin, James W. and Jay M. Arena. 1974. *Human Poisoning from Native and Cultivated Plants.* 2nd ed. Duke University Press, Durham, NC. 194 pp.

Hitchcock, A. S., 2nd ed. revised by A. Chase. Issued May 1935, revised February 1951. *Manual of the Grasses of the United States.* 2nd ed. Miscellaneous Publication No. 200, U.S. Dept. of Agriculture, Washington, DC. 1051 pp.

Kingsbury, John M. 1964. *Poisonous Plants of the United States and Canada.* 3rd ed. Prentice-Hall, Inc., Box 500, Englewood Cliffs, NJ 07632. 626 pp.

Kurz, Herman and Robert K. Godfrey. 1962. *The Trees of Northern Florida.* University of Florida Press, 15 NW 15th St., Gainesville, FL 32603. 311 pp.

Lakela, Olga and Robert W. Long. 1976. *Ferns of Florida.* Banyan Books, Box 431160, Miami, FL 33143. 178 pp.

Lakela, Olga and Richard P. Wunderlin. 1980. *Trees of Central Florida.* Banyan Books, Box 431160, Miami, FL 33143. 208 pp.

Long, Robert W. and Olga Lakela. 1971. *A Flora of Tropical Florida.* University of Miami Press, Coral Gables, FL. Reprinted 1976 by Banyan Books, Box 431160, Miami, FL 33143. 962 pp.

Luer, Carlyle A. 1972. *The Native Orchids of Florida.* New York Botanical Garden, Bronx, NY 10458. 293 pp.

Luer, Carlyle A. 1976. *The Native Orchids of the United States and Canada.* New York Botanical Garden, Bronx, NY 10458. 361 pp.

Muenscher, Walter C. 1944. *Aquatic Plants of the United States.* Comstock Publishing Company, Inc., Cornell University, Ithaca, NY. 374 pp.

Orsenigo, J. R. et al. 1977. *Florida Weeds, Part II.* Circular 419. Florida Cooperative Extension Service, Institute of Food and Agricultural Sciences, University of Florida, Gainesville, FL 32611. 19 pp.

Perkins, Kent D. and Willard W. Payne. 1978 and reprint with addendum 1981. *Guide to the Poisonous and Irritant Plants of Florida.* Circular 441. Florida Cooperative Extension Service, Institute of Food and Agricultural Sciences, University of Florida, Gainesville, FL 32611.

Radford, Albert E., Harry E. Ahles and C. Ritchie Bell. 1968. *Manual of the Vascular Flora of the Carolinas.* University of North Carolina Press, Chapel Hill, NC. 1183 pp.

Rickett, Harold W. 1967. *Wild Flowers of the United States, Volume Two, The Southeastern States.* Parts 1 and 2. Printed by McGraw-Hill Book Company, New York, NY, for the New York Botanical Garden, Bronx, NY 10458. 688 pp.

Scurlock, J. Paul. 1987. *Native Trees and Shrubs of the Florida Keys: A Field Guide.* Laurel Press, Pittsburgh, PA. Distributed by The Florida Keys Land Trust, Inc., P.O. Box 1432, Key West, FL 33041-1432. 220 pp.

Small, John K. 1933. *Manual of the Southeastern Flora.* Published by the author, New York, NY. Also by University of North Carolina Press, Chapel Hill, NC. Reprinted in 1972 by Hafner Publishing Co., 866 Third Ave., New York, NY 10022. 1554 pp.

Stevenson, George B. 1974. *Palms of South Florida.* Published by the author. Distributed by Fairchild Tropical Garden, 10901 Old Cutler Road, Miami, FL 33156 and Horticultural Books, Inc., P.O. Box 107, Stuart, FL 33495. 251 pp.

Taylor, Walter K. 1992. *The Guide to Florida Wildflowers.* Taylor Publishing Company, 1550 West Mockingbird Lane, Dallas, TX 75235. 320 pp.

Tomlinson, P. Barry, with illustrations by Priscilla Fawcett. 1980. *The Biology of Trees Native to Tropical Florida.* Published by the author, Harvard Forest, Petersham, MA 01366. 480 pp.

Ward, D. B. (editor). 1979. *Rare and Endangered Biota of Florida. Volume Five. Plants.* University Presses of Florida, 15 NW 15th St., Gainesville, FL. 175 pp.

West, Erdman and Lillian Arnold. 1956. *The Native Trees of Florida.* University of Florida Press, 15 NW 15th St., Gainesville, FL 32603. 218 pp.

Wherry, Edgar T. 1964. *The Southern Fern Guide.* Doubleday and Company, Inc., Garden City, NY. Reprinted by the New York Chapter of the American Fern Society, New York Botanical Garden, Bronx, NY 10458. 349 pp.

Willis, J. C. 1973. *A Dictionary of the Flowering Plants and Ferns.* 8th ed. Rev. by H. K. Airy Shaw. Cambridge University Press, London, England. U.S. address: 32 East 57th Street, New York, NY 10022. 1245+ pp.

Wunderlin, Richard P. 1983. *Guide to the Vascular Plants of Central Florida.* University Presses of Florida, 15 NW 15th St., Gainesville, FL 32603. 480 pp.

Yarlett, Lewis L. (no date given, ? about 1967.) *Important Native Grasses for Range Conservation in Florida.* U.S. Dept. of Agriculture, Soil Conservation Service, Gainesville, FL. 163 pp. Also includes: "General Map of Ecological Communities of Florida."